RURAL INDIA

About the Editors

Madaswamy Moni is a senior technocrat of National Informatics Centre (Department of Information Technology, Government of India) currently working as Deputy Director General. With more than 24 years of professional experience he has held several prestigious assignments. He pioneered to establish the District Information System of National Informatics Centre in 27 sectors of importance.

His forte is ICT Diffusion and Infusion for achieving sustainable agricultural and rural development in India. He has been propagating an integrated approach to ICT and agricultural, and rural and backward area development.

He is the General Secretary of the Bhoovigyan Vikas Foundation and has written several papers and edited over six books. Moni has travelled extensively to a number of countries on various technical missions. His research interests span over a wide area of agricultural economics and informatics, micro-level planning, regional and sustainable development, information support systems and networking, decision support system and fusion of technologies for increased agricultural productivity.

Suresh Misra is currently working with Indian Institute of Public Administration as Chair Professor Consumer Affairs. He holds his Masters Degree in Political Science from Jawaharlal Nehru University, New Delhi, masters in Human Resource Management from Pondicherry University and Ph.D. from Allahabad Central University.

He had his advance training in Rural Management from Development Study Centre, Rehovot, Israel. Well versed in training techniques and methodology, he has been organising a variety of training programmes in rural development and decentralized planning. His areas of interest include rural development, local governance and decentralized planning.

Dr. Misra has written twelve books and contributed several papers in journals of repute. He has carried out a number of research studies and consultancy assignments. He is currently associated with the Capacity-Building of Afghanistan Civil Services in association with USAID. He is a Life Fellow of Bhoovigyan Vikas Foundation.

RURAL INDIA

Achieving Millennium Development Goals and Grassroots Development

Edited by

M. Moni
Suresh Misra

CONCEPT PUBLISHING COMPANY, NEW DELHI-110059

Cataloging in Publication Data—DK
Courtesy: D.K. Agencies (P) Ltd. <docinfo@dkagencies.com>

Rural India : achieving millennium development goals and grassroots development / edited by M. Moni, Suresh Misra.

p. cm.

Papers presented at the Third International Conference on Rural India : achieving Millennium Development Goals and Grassroots Development, held at New Delhi.

Includes bibliographical references.
Includes index.
ISBN 13: 9788180695285
ISBN 10: 818069528X

1. Rural development--India—Congresses. 2. Rural development—Government policy—India—Congresses. 3. Communication in rural development—India—Congresses. 4. Information technology—Social aspects—India—Congresses. I. Moni, M. (Madaswamy) II. Misra, Suresh, 1958-

DDC 307.141 209 54 22

ISBN-13: 978-81-8069-528-5
ISBN-10: 81-8069-528-X

First Published 2009

Published and Printed by
Ashok Kumar Mittal
Concept Publishing Company,
A/15-16, Commercial Block, Mohan Garden,
New Delhi-110059 (INDIA)

Phones : 25351460, 25351794
Fax : 091-11-25357103
Email : publishing@conceptpub.com

Preface

We are all aware that Rural India presents a baffling multiplicity of images and problems. During the last more than fifty years of our "Learning Curve" in development planning, we have initiated a large number of development programmes, with changes in focus, style and content. How much of this effort has yielded the intended benefits is a subject-matter of debate.

In a large country, like ours, with more than six lakhs of villages, caught in 'poverty trap', a mere linear expansion of plan outlays alone cannot ensure the required order of socio-economic change and impact on development. It has to go beyond outlays to 'outcomes'. For this, we need to adopt a holistic and integrated approach, involving the interplay of policy, strategy, planning and programme design, adapting the effective means and methods of implementation with people's participation in development.

This volume is an outcome of the Third International Conference on "Rural India: Achieving Millennium Development Goals and Grassroots Development". It has been edited by Moni and Suresh Misra and consists of eighteen thought-provoking papers on various themes arranged in three parts. It examines the current rural development policies, programmes and strategies in the context of the Millennium Development Goals.

I would like to thank the Editors who have, laboured to bring out this volume in its present form. I am grateful to the authors of the various papers, for their contributions. A large number of organizations and individuals have helped the Bhoovigyan Vikas Foundation to organize this thematic international conference. I would like to specially acknowledge the help and support received from our Conference Partner, Sri Ram Prasad of Sahanav, who made every effort to make the Conference a big success. I am particularly thankful to the Planning Commission, Department of Science and Technology, NABARD and SAIL for providing generous grants to organize this Conference.

I am thankful to Sri Ashok Kumar Mittal of Concept Publishing Company for bringing out this volume in a record time.

K.V. Sundaram
Chairman
Bhoovigyan Vikas Foundation, New Delhi

Contents

List of the Contributors

Amarnath Kshatriya, Founder Director, SERVIS, 2441 Nelson Avenue, West Vancouver, British Colombia, Canada.

Anand Sagar, K., Associate Manager, PSI Data Systems, Bangalore.

B. Yerram Raju, Managing Director, Sandilya Consultants Pvt. Ltd.

C.S.R. Prabhu, National Informatics Centre, Hyderabad.

Eswarappa Kasi, ICSSR Doctoral Fellow, Department of Anthropology, University of Hyderabad, Hyderabad.

G. Palanithurai, Rajiv Gandhi Chair for Panchayati Raj Studies, Gandhigram Rural Institute, Gandhigram (T.N.).

K.V. Raghavulu, CGM, NABARD, Mumbai.

K.V. Rao, Computer Programmer, MANAGE, Hyderabad.

K.V. Sundaram, Chairman, Bhoovigyan Vikas Foundation, Vasant Kunj, New Delhi.

Kamla Nath, A 5/56, Vasant Vihar, New Delhi.

Madaswamy Moni, Deputy Director General, National Informatics Centre, Department of Information Technology, Government of India, New Delhi.

Mahi Pal, Associate Professor, Haryana Institute of Rural Development, Nilokheri, Karnal, Haryana.

Mogali J. Nandan, Regional Research Laboratory, CSIR, Bhopal.

Mohini Saxena, Regional Research Laboratory, CSIR, Bhopal.

N. Ramakrishnan, Regional Research Laboratory, CSIR, Bhopal.

Nira Ramachandran, Consultant, UNICEF, New Delhi.

S. Mahendra Dev, Director, Centre for Economic and Social Studies, Hyderabad.

Satya Bhushan Rath, CEO, Renewable Energy and Agricultural Development, (READ) Foundation, Bhubaneswar.

Shiramdas Murali, Regional Research Laboratory, CSIR, Bhopal.

Suresh Misra, Chair Professor Consumer Affairs, Indian Institute of Public Administration, New Delhi.

V. Nath, Late Member of the Indian Administrative Services, Rajasthan Cadre.

V. Raghupathy, Rajiv Gandhi Chair for Panchayati Raj Studies, Gandhigram Rural Institute, Gandhigram (T.N.).

V. P. Sharma, Director, (Information Technology, Documentation and Publication), National Institute of Agricultural Extension Management (MANAGE), Rajendranagar, Hyderabad.

Vijayanand Kommaluri, Senior Lecturer, Department of Computer Science, Pondicherry University, Pondicherry.

Introduction

M. MONI and SURESH MISRA

Background

A livelihood comprises people, their capabilities and their means of living, including food, income and assets. A livelihood is environmentally sustainable when it maintains or enhances the local assets on which livelihood depends and has net beneficial effects on other livelihoods. There are many elements that are related to the development of sustainable livelihood. Creation of gainful employment, on-farm and off-farm, poverty reduction and enhancement of income are the important elements. A sustainable livelihood is simply a job. But it is a job that is meaningful and remunerative and does not destroy the resource base of the community. Sustainable livelihoods produce goods and services that are needed to better the lives of the people. At the same time they create purchasing power and with it generate economic and social equity.

It is now recognized that development planning in India has to concentrate on generation of more employment as it is the only effective way of poverty alleviation, reduction of inequality and meaningful growth. Poverty can be defined as a social phenomenon in which a section of the society is unable to fulfil even its basic necessities in life. The Planning Commission has defined the poverty line on the basis of recommended nutritional requirements of 2400 calories per person per day for rural areas and 2100 calories for urban areas.

Rural areas currently are home to 70 per cent of India's population and have historically accounted for more than half of Indian consumption. Even with increasing urbanization and migration, it is estimated that 63 per cent of India's population will continue to live in rural areas in 2025. In terms of economic output, rural India accounts for almost half (48%) of the country's economy, and the rural markets have the potential to reach $500 billion by 2020. Thus rural areas will continue to remain vitally important to the Indian economy.

Challenging the prosperity of rural India, is the unfortunate truth, that a large percentage of this population lives in high-risk and unpredictable environments. They are particularly vulnerable to floods, droughts and other natural disasters which threaten their livelihoods and often deplete whatever assets and resources they have. Providing rural India with better access to services, technology, education and health care will reduce inequality, alleviate poverty for hundreds of millions of India's citizens, and add an additional engine of growth to India's prospering economy. It is critical that rural initiatives be linked to empowerment and capacity-building to achieve a sustainable and inclusive rural development. Further appreciating India's diversity; these initiatives must be tailored to different sections of the society.

Declining Contribution of Agriculture: Cause of Concern

The share of rural activities in the Gross National Product of the country is, however, secularly declining. The problem is further compounded with the declining contribution of agriculture in India's GDP. Agriculture's share in the GDP in 1951 was 65 per cent; it declined to 45 per cent in 1971 further to 31 per cent in 1993 and is currently at 24 per cent. The agricultural growth rate has remained around 1.4 per cent in the last two decades. The agricultural production is declining. The growth rate in the production of the cereals declined from 4.13 per cent in the period 1984 to 1994-95. From here it dipped to 1.09 per cent between 1994-95 and 2004-05. Annual growth rate in the agriculture sector as a whole declined from 3.69 per cent in the period 1990-96 to 1.65 per cent during 1996-2005. Net sown area, gross cropped area, fertilizer use and electricity consumption all have declined. Naturally then, the per capita income of rural population is far lower than that of its counterpart living in urban areas. There are more poor in rural than in urban areas whether one looks at poverty in terms of absolute number or in terms of per cent of population below the poverty line. A substantial number of urban poor have deep roots in rural areas.

The condition of the growing number of small farmers in the country is appalling with lack of support in terms of investment, technology, friendly markets and rural infrastructure. Small and marginal farmer households account for 84 per cent of all farmer households. Fifty-seven per cent of India's total employment and 73 per cent of total rural employment come from this sector. Farming in our country still is the backbone of the livelihood security system for a majority of our population. A mere 5 per cent of the population controls 25 per cent of the agricultural land. The remaining land is broken up into very small family plots, some smaller than a single acre. The current fragmentation worsens under the land inheritance methods, dividing a man's land among all of his sons. The land reform measures following independence might have rectified some of the problems and inequities in the system, however, not surprisingly the large landowners still wield political power and have effectively blocked all attempts to reapportion agricultural land. The situation is not likely to change for decades to come if appropriate strategies are not adopted. If India wants to reduce poverty and achieve the millennium development goals their condition has to be reversed.

Rural Employment and Underemployment

On the basis of NSS data, the Special Group on Employment Generation set up by the Planning Commission has estimated the number of unemployed to be of the order of 26.58 million in 1999-2000, as against 20.13 million in 1993-94. The unemployment rate in India has increased significantly since 1993-94 and was 7.32 per cent in 1999-2000 compared to 5.99 per cent in 1993-94. According to the special group nearly 74 per cent of unemployed are in the rural sector. The Report mentions "The present rising unemployment is primarily an outcome of a declining job creating capacity of growth observed since 1993-94."

Among the sectors outside the organized ones, agriculture, which was the major employment generating sector in the past, has almost stopped absorbing any more labour in the later period. As a result, in spite of higher overall GDP and agriculture growth in the 1990s the overall growth of employment fell from 2.7 per cent per annum between 1993-94 and 1999-2000.

This trend of a slow down in the employment growth and increase in the incidence of unemployment is of greater concern, especially in the case of scheduled castes and scheduled tribes, other weaker sections of the population including women, youths and the uneducated. Therefore, to break away from the present poverty-population trap, India has to create some million sustainable livelihoods by creating jobs in the rural areas. The enactment of the National Rural Employment Guarantee Act, 2005 (NREGA) creating jobs for a period of hundred days in the rural areas is to be welcomed. The involvement of the Panchayati Raj Institutions in operationalisation of the scheme is a step in the right direction. Unless people's organizations and institution are strengthened the problems of rural India cannot be solved. People have to be the agents of rural transformation and not only a process in rural transformation.

A decade after economic reforms, in 2003, according to NSS Household Consumer Consumption and Expenditure and Employment-Unemployment situation in India, the all India average Monthly Per Capita Consumer Expenditure (MPCE) was Rs. 554 for rural India. Those below Rs. 470 constituted fifty per cent of the rural population. For urban India it was Rs. 1022, that is about 84 per cent higher than rural areas. The proportion of rural households living in *kutcha* structures was 21 per cent.

When we look at the question of poverty in terms of access to work opportunities and basic needs, the rural parts of India are in far worse situation. Self-reliance which made rural people meet their basic needs in the most adverse circumstances stands totally destroyed. Governments, which professed and promised to uplift them from the miserable living conditions, have failed to do much, if the records of the last five decades is any indicator. The less said about the implementation of the rural development schemes the better it is. Corruption has become the major cause of worry. The Panchayats in large number of States continue to remain dormant and the local level bureaucracy still controls the rural development process. Why rural India remains under-developed in spite of sixty years of Independence needs to be understood? Perhaps we have misunderstood the very notion of rural development. Poverty, unemployment, hunger, subsistence agriculture and poor public services are the hallmark of our rural development process. There is an urgent need to find solutions to the problems and empower the rural population.

Poor Quality of Rural Life

Rural underdevelopment has produced a vicious circle of poverty and unemployment; it does not allow people to get out of it. Today the small and marginal farmers are committing suicide as they are in severe debt. They have very little options left, a few migrate to over crowded cities but the cities offer nothing better and they return only to repay the debt they had taken to go to the cities. Only those who are educated and highly skilled make migration a success story and the rest are condemned to live in slums and *jhuggi jhopadis* and to eke out a living from what is known as informal sector of the urban economy. No less than 30 per cent of the rural people of India live below the poverty line. But the quality of life of the other 70 per cent is no better. The 30 per cent means 300 million people, which is too huge to be ignored. These are the people without food security; who cannot afford to send their children to school; who have no money for health care; who have just a roof over their head; and who are partially or completely dependent on Government's direct or indirect support.

Apart from these the public service delivery mechanism is also very poor in the rural areas. Sectors like education, health, environment, sanitation, and infrastructure continue to be largely neglected, even though crores of rupees are being spent in the name of the rural poor. In a recent report of the Human Resource Ministry, Government of India, it has been found that nearly 33,000 primary schools in India do not have students. Many of these schools are in the rural areas. In a study on Corruption in Education by UNESCO's International Institute of Educational Planning, it has been found that teacher's absenteeism in India is the highest in the world. While the global average of teacher absenteeism is about 20 per cent, in India it is 25 per cent. While in Bihar two of every five teachers were reported absent, the figure in UP was reported to be one in every three. These are the two States where the bulk of the population still lives in rural areas. The condition only marginally varies in many other States.

Gender discrimination continues to be a major area of concern particularly in the rural areas. Even in prosperous States like Punjab and Haryana the sex ratio is declining. According to the Mortality Statistics of 2006 recently released by the Central Bureau of Health Intelligence of the Health Ministry, GoI, nearly half of all female deaths in rural Rajasthan are of girls below the age of 20. The precise percentage is 49.4 per cent. Out of these, 42 per cent of the deaths are of girls who haven't yet celebrated their fifth birthday. What's worse is that virtually the situation is same for girls in most Hindi heartland including Uttar Pradesh, Bihar, Madhya Pradesh and Chhattisgarh. In UP, for instance, nearly half the female deaths in a year (47%) are of girls below 20 years. Bulk of these deaths are in the rural areas.

In spite of several programmes and schemes the state of rural healthcare system is no better. In Rajasthan over 41 per cent of all those who die in the State (male and female) are within the age of 19. In contrast, 37 per cent of deaths are of those in the 60 plus age category. More young people are dying in Rajasthan than the elderly. In rural Rajasthan, boys under 20 comprise 44 per cent of the male deaths, indicating the abysmal state of rural healthcare, especially for children. In UP, boys below 20 account for over 40 per cent of the rural male deaths and nearly a third of urban male deaths, showing that in this state healthcare for children is almost as bad in urban as in rural areas. In UP, too, more children die than old people. In contrast to these States in Kerala over 7 per cent of deaths are of girls below 20 years of age. Even in rural Kerala, only 8 per cent of female deaths are of girls less than 20 years. In Tamil Nadu and Maharashtra, about 13 per cent and 17 per cent of deaths are of people below 20 years of age. To an extent the vulnerability of the poor to diseases is not just because of financial poverty but also due to paucity of information, knowledge and access to facilities. Millions of children are malnourished and undernourished. In the light of the MDG's there is an urgent need for a rights-based approach in health policies, protection of women's rights at every stage, care for the girl child and gender sensitization.

The Government of India, in its Vision 2020, has proposed knowledge powered village complexes—PURA (Provision of urban amenities in rural areas) for effective planning and implementation of rural development programmes with people's participation which will lead to inclusive rural development. But to bring about inclusive rural development, with people's participation the growth impulses need to be structured and built up at the village clusters. The new concept of rural development can be translated into action only through social animators and change agents joining hands with the local bureaucracy/technology and local people and their institutions. The need is to create an economically viable agricultural and industrial base for a village by establishing locally integrated economic circuits to achieve the full development of the natural resources and human skills available at the local level itself. This can be achieved by guarding the basic physical well-being of the villagers by ensuring food and nutrition security, public health and sanitation: development of functional life skills, including upgrading of basic practical literacy skills at all age levels, providing practical training in marketable business and crafts and providing access to agricultural services including inputs and markets: breaking the isolation of the village and establishing external support relations with public and private sectors: leveraging ICT for transforming the village into a knowledge-based society and others.

We need to reorient our rural development programmes and policies as well as strategies in light of the millennium development goals. The need is to eradicate poverty, target hunger, promote gender equality and empower women and improve the quality of services being delivered in the rural areas. The Millennium Declaration Goals must be the guiding force if we are serious about improving the quality of rural life. The MDG's should help us achieve a sustainable rural development as this is what we have been aiming at, for the last six decades without much success.

Millennium Declaration Goals

The Millennium Declaration adopted by the General Assembly of the United Nations in September 2000 reaffirmed its commitment to the right to development, peace, security and gender equality, to the eradication of many dimensions of poverty and to overall sustainable development. These are intended for the Member Countries to take efforts in the fight against poverty, illiteracy, hunger, lack of education, gender inequality, infant and maternal mortality, disease and environmental degradation.

The Declaration includes eight development goals, eighteen time-bound targets and forty-eight indicators, each with quantified targets, to motivate the international community and provide an accountability mechanism for actions taken to enable millions of poor people to improve their livelihoods.

About 70 per cent of the MDGs' target group lives in rural areas, particularly in Asia and Africa, and for most of the rural poor agriculture is a critical component in the successful attainment of the MDGs. Thus, a necessary component in meeting the MDGs by 2015 in many parts of the world including India is a more productive and profitable agricultural sector. The MDG's aim at a holistic and integrated development, these goals are:

1. Eradicate extreme poverty and hunger,
2. Achieve universal primary education,
3. Promote gender equality and empower women,
4. Reduce child mortality,
5. Improve maternal health,
6. Combat HIV/AIDS, malaria and other diseases,
7. Ensure environmental sustainability, and
8. Develop a global partnership for development.

The first Millennium Development Goals—India Country Report for the year 2005 captures India's achievements, challenges and policies with reference to the goals and targets and reveals that there have been substantial improvements in the lives of people of the countries over the years. This has been possible due to the planned implementation of programmes despite the enormous and complex problems and diversities of our nation. The Central and State Governments have set up goals more ambitious than the MDGs. With the well thought out planning, comprehensive development strategies devised in the national policy, and matching implementation process, it is hoped that India will be able to meet the challenges and achieve all the MDG targets much earlier than the targeted dates.

India's position with reference to the various Goals is given below:

1. To achieve the Goal of eradicating extreme poverty and hunger, India must reduce by 2015 the proportion of people below poverty line from nearly 37.5 per cent in 1990 to about 18.75 per cent. As on 1999-2000, the poverty headcount ratio is 26.1 per cent with poverty gap ratio of 5.2 per cent, share of poorest quintile in national consumption is 10.1 per cent for rural sector and 7.9 per cent for urban sector and prevalence of underweight children is of the order of 47 per cent. National Rural Employment Act is a positive step to reduce the poverty ratio further.
2. To achieve universal primary education under Goal-2, India should increase the primary school enrolment rate to 100 per cent and wipe out the drop-outs by 2015 against 41.96 per cent in 1991-92. The drop-out rate for primary education during 2002-03 is 34.89 per cent. The gross enrolment ratio in primary education has tended to remain near 100 per cent for boys and recorded an increase of nearly 20 percentage points in the ten years period from 1992-93 to 2002-03 for girls (93%). The literacy rate (7 years and above) has also increased from 52.2 per cent in 1992-93 to 65.4 per cent in 2000-01.

   To ensure gender parity in education levels in Goal-3, India will have to promote female participation at all levels to reach a female male proportion of equal level by 2015. The female male proportion in respect of primary education was 71:100 in 1990-91 which has increased to 78:100 in 2000-01. During the same period, the proportion has increased from 49:100 to 63:100 in case of secondary education.
3. Goal-4 aims at reducing Under five Mortality Rate (U5MR) from 125 deaths per thousand live births in 1988-92 to 42 in 2015. The U5MR has decreased during the period 1998-2002 to 98 per thousand live births. The Infant Mortality Rate (IMR) has also come down from 80 per thousand live births in 1990 to 60

per thousand in 2003 and the proportion of 1 year old children immunised against measles has increased from 42.2 per cent in 1992-93 to 58.5 per cent in 2002-03.

4. To achieve Goal-5, India should reduce Maternal Mortality Rate (MMR) from 437 deaths per 100,000 live births in 1991 to 109 by 2015. The value of MMR for 1998 is 407. The proportion of births attended by skilled health personnel has been continuously increasing (from 25.5% in 1992-93 to 39.8% in 2002-03) thereby reducing the chances of occurrence of maternal deaths.
5. In so far as Goal-6 is concerned, though India has a low prevalence of HIV among pregnant women as compared to other developing countries, yet the prevalence rate has increased from 0.74 per thousand pregnant women in 2002 to 0.86 in 2003. This increasing trend needs to be reversed to achieve MDG 6. The prevalence and death rates associated with malaria are consistently coming down. The death rate associated with TB has come down from 67 deaths per 100,000 populations in 1990 to 33 per 100,000 populations in 2003. The proportion of TB patients successfully treated has also risen from 81 per cent in 1996 to 86 per cent in 2003.
6. Goal-7 aims at ensuring environmental sustainability. As per assessment made in 2003, total land area covered under different forests has been 20.64 per cent due to Government's persistent efforts to preserve the natural resources. The reserved and protected forests together account for 19 per cent of the total land area to maintain biological diversity. The energy use has declined consistently from about 36 kilogram oil equivalent in 1991-92 to about 32 kilogram oil equivalent in 2003-04 to produce GDP worth Rs. 1000. The proportion of population without sustainable access to safe drinking water and sanitation is to be halved by 2015 and India is on track to achieve this target.
7. Goal-8 is regarding the developing global partnership for development. It is basically meant for the developed countries to provide development assistance to developing countries. The Government of India holds the following views regarding the role of the developed countries in achieving this goal:
   - The financial support needed to achieve the targets under this Goal had been estimated for the least developed land locked and small countries by a high-level panel on

'Financing for Development, at an additional amount of US $ 50 billion which would be required for this purpose every year till 2015.

- However, a huge gap still exists for those countries between the development assistance required to meet the MDGs and what has been pledged by the developed countries so far.
- Recent months have seen new commitments towards reaching the internationally accepted 0.7 per cent of Gross National Income (GNI) target. We have reminded that these potential increases still leave development assistance donor countries as a group well short of 0.7 per cent.
- It is also a matter of satisfaction that actual disbursements of ODA, in recent years, have shown a welcome reversal of the declining trend that lasted for almost a decade since the early 1990s. In this regard, it is important to realize that unless aid commitments translate into actual delivery, securing MDGs will remain elusive goals. We do hope that all the developed countries would scale up the ODA to realize the goals reaffirmed at the Monterey Consensus.
- It has also been our consistent position that additional resources for implementing the development agenda should be channelized through the existing multilateral agencies. Moreover, allocations must be based on pre-defined and transparent criteria. Our own development experience clearly indicates that, ultimately, it is the availability of untied additional resources for use in accordance with national development strategies, which is most beneficial for recipient countries.
- To deal with the problems of debt, the Heavily Indebted Poor Countries (HIPC) Initiative was launched by the World Bank and IMF and endorsed by 180 Governments. In regard to the HIPC Initiative, India is of the view that the Initiative should be met by additional funding from the developed countries and the flow of concessional assistance to other countries should not be reduced. India also opposes the concept of "equitable burden sharing" since some of the non-Paris Club creditor countries are themselves poor countries.
- We have supported the G8 initiative on irrevocable debt cancellation for the HIPC countries, which has now been adopted by IMF and the World Bank as the Multilateral

Debt Relief Initiative (MDRI). We have always been supportive of all efforts being extended to the Low-Income Countries (LICs), including those in Africa, where debt burdens are serious threats to attainment of the MDGs.

8. With regard to one of the targets of the Goal-8, i.e. in cooperation with the private sector, make available the benefits of new technologies, especially information and communications, India has made substantial progress in recent years. The overall tele-density has remarkably increased from 0.67 per cent in 1991 to 9.4 per cent in June 2005. Use of Personal Computers has also increased from 5.4 million PCs in 2001 to 14.5 million in 2005 and there are 5.3 million Internet subscribers as on March 2005 (2.3 Internet users per 100 populations and 0.5 per 100 Internet subscribers).
9. The National Employment Guarantee Act, Sarva Siksha Abhiyan, Total Literacy Campaign of the National Literacy Mission, 73rd and 74th Constitutional Amendments providing reservation for women, commitment for women empowerment in the NCMP, National Health Mission, Total Sanitation Campaign and Bharat Nirman are some of the important steps taken by the Government which will help in achieving the Millennium Development Goals.

The basic objective of sustainable rural development is to develop a balanced and ecologically sound micro-regional economy and polity. In a country like India, where the majority of the population still lives in rural areas, improving the quality of rural life should be on top priority. The productivity and the service delivery mechanism need to be improved and targeted programmes to achieve the MDGs should be charted in full sincerity. We also need to address the issues and problems being faced by institutions of local governance. The service delivery mechanism in the rural areas needs to be restructured and local level grassroots institutions involved and empowered to deliver quality services. ICT has a major role in delivery of services, by bringing down both cost and reach. Integrating MDGs with the activities of the Panchayats may help to some extent.

Education, employment generation, empowerment of women and gender equality hold the key in achieving the millennium development goals. Education is fundamental to health care and awareness. This would help in reducing maternal mortality, infant mortality and equip people to take care of their health. For India to make impact on the world map the rural areas need to be taken care of. In this endeavour the grassroots organizations have a major role to play. The government and the bureaucracy need to transform their role from provider of services to a facilitator in which large number of stakeholders have a role to play.

This Volume

The present volume is a collection of theme papers of the *Third International Conference on "Rural India: Achieving Millennium Development Goals and Grassroots Development"* organized by Bhoovigyan Vikas Foundation in association with Sahanav and Government of Andhra Pradesh, from 10-12 November, 2005 at Hyderabad. The Conference touched upon several economic, social and technological issues that impact rural India through scores of presentations, lectures and debates. It was able to sensitise the governments, administrators, industry and the academia towards rural well-being and development and achieving millennium development goals. The book consists of nineteen papers including the present one and is divided into three parts and each part deals with a specific theme. Part I: deals with *Transformative Rural Development and MDG* and consists of nine papers, Part II: deals with *Participation and Empowerment* and consists of five papers and Part III: focuses on *Bridging the Digital Divide* and has four papers. The papers are written by well known academicians, scientist, technocrats, researchers and policy-makers. The papers review the current scenario, actions and draw lessons for developing a professional development strategy to strengthen the rural base of the country.

Nira Ramachandran, in her paper on *Rural Food Security: Unresolved Issues and Emerging Challenges,* attempts to identify some of the unresolved issues underlying the persistence of rural food insecurity in the country. The author has analysed the ineffectiveness of government intervention to address these issues and indicates directions for policy change. The paper has highlighted four issues which impact considerably on the rural food security status—gender, seasonality, common property resources, bio-diversity and self-help initiatives. Nira advocates that food security can only become sustainable if it is built on local self-help initiatives. Policy focus on initiating and nurturing such initiatives with the ultimate objective of making them self-sustaining could provide a long-term solution to this persistent problem and make for a food secure India.

Energy plays a critical role in underpinning efforts to achieve the MDGs and improving the lives of poor people across the world. World-wide 2.4 billion people rely on traditional biomass as their primary source of energy and 1.6 billion people do not have access to electricity. Satya Bhushan Rath in his paper *Renewable Energy and Sustainable Development,* asserts that if approached as an integrated part of MDG strategies, access to energy services can be an important instrument in helping promote economic growth, social equality, and environmental sustainability. According to him renewable sources of energy can be used to provide modern energy services to the poor, contribute to meeting their increasing global energy demand, reduce air pollution, mitigate climate change and delay the eventually fossil fuel depletion. Rural India has enough of renewable energy resource potential and traditional knowledge to utilize it. It is clear that renewable energy is the only way out, which can navigate to reach at energy independence, which means a way to sustainable development.

The number of people suffering from poverty has increased, although there is a decline in percentage of the poor. India still houses 33.33 per cent of the worlds poor. We are not in a position to curb poverty and reduce unemployment. The paper on *Facets of Rural Economy and Development of SMEs: A Paradigm Shift is Imperative* by B. Yerram Raju addresses the issues relating to poverty and rural employment problem in India on small and medium enterprise development platform. The author argues that rural India is beyond agriculture, although its inclusion predominates. The paper is divided into three section, each section deals with specific problems like climate for private investment, employment and entrepreneurship and issues related to sustainable development of rural areas—rural entrepreneur nucleus estates. The author opines that one of the thrust areas for policy intervention is to facilitate increase in value addition in agriculture with a view to stimulate rural employment, generate higher incomes for rural households, promote export, and at the same time, increase the knowledge and skill base of those living in rural areas. Issues of enterprise employment and efficiency are critical to the growth of the economy.

Utilisation of plant fibres for generating employment in rural sector is cost effective and ecologically sustainable. India has vast resource for different natural fibres. Mohini Saxena, Shriandas Murali, Mogali J. Nandan and N. Ramakrishnan discuss various aspects of *Sisal*, its application and potentialities for sustainable rural livelihood in their paper on *Sisal: Potential for Employment Generation and Rural Development.*

Among the plant fibre *Sisal* has many advantages like thriving in wastelands and yielding superior fibre continuously for 6-8 years with least management inputs. Its various applications make it a viable option for employment generation and rural development.

Drought is a major problem that needs to be addressed. The unchecked exploitation of groundwater borewells loom large in Andhra Pradesh. This problem has been highlighted by Anand Sagar K. and Vijayanand Kommaluri in their paper on *Rural Tsunami: a Famine in 21st Century*. The paper highlights issues relating to rural disaster, effective measures for drought preparedness and response. The role of GIS in providing cost effective information at various stages of the disaster life cycle is also discussed. The authors have also attempted with regionally differentiated strategy integrating Web Enabled GIS and the Decision Support System to provide a better solution for effective disaster management. They are of the view that various stakeholders should recognise the fundamental flaw that forces farmers to death. The idea of industrial farming should be kept away since the majority of the population in India makes their livelihood with small portions of land and gainful employment from agricultural sector.

Mahendra Dev in his paper *Can Andhra Pradesh Achieve Millennium Development Goals,* discusses the various steps taken by India in meeting the MDGs. According to him alleviation of poverty in all its forms has been a major goal of development policy in India since the beginning of planning. The study is focussed on Andhra Pradesh and examines the progress made by the State towards achieving the MDG targets. It is pointed out that in the case of Andhra Pradesh economic growth particularly pro-poor growth in terms of concentrating on agriculture and employment are important in order to reduce poverty but economic growth alone will not be sufficient to lift some above poverty. The paradigm shift in poverty eradication through women empowerment can provide a solution to this problem of poverty. Based on analysis of data it has been suggested that there is a need to increase public expenditure on health and education. The author concludes by observing that it is depressing to note that if Andhra Pradesh follows the current rate of progress, it cannot achieve MDGs in 10 out of the 14 indicators. Sector specific interventions and strengthening of Government, civil society institutions and PRIs are needed for better delivery system particularly in the rural areas.

Self-Sustaining Development in Rural India: A Study of Technological, Geographic and Social Changes in Three Villages of

Sawai Madhopur District of Eastern Rajasthan, is authored by V. Nath. The study has been carried out in two time intervals (1961 & 2001) to show the changes that have taken place in the district. The two studies show changes in production techniques, increase in output of cereals and commercial crops, major increase in incomes and consumptions of people of all castes, reduction in social disparities and increase in education, improvement in health and major changes in aspirations of the people of the area. All this has been achieved by the intervention of the government. Further the study shows that the technological and economic changes have not been at the expense of natural resources of soil, water, pasture and woodland.

Gender Equality and Women's Empowerment is authored by Kamla Nath in which the author stresses that equality between women and men should be promoted in ways that are appropriate to each particular context. The paper analyses the current status of women and men, their relations to each other and the interconnections between gender and other factors. According to the author gender analysis can provide an understanding of the social relations and how inequality can be addressed. Gender equality is not only a social/gender justice issue and a measure of equal participation; it is an important component for economic development.

In the milieu of the new banking culture fostered by reforms, lending to agriculture or priority sectors has generally, became unfashionable asserts K.V. Raghavulu in his paper on *Reaching the Unreached: Public Services*. According to him the neglect of agriculture and the rural sector generally was total during the 1990s, and there is a deterioration of the health of all institutions involved in rural credit. The financial sector reforms which were transplanted from the so-called Basle norms were ill-conceived and unrelated to the specific socio-economic milieu of the Indian soil and they succeeded in creating an adversarial environment for rural credit. He has favoured the formation of SHG's as SHG-Bank linkage programme increases outreach to poor, it is cost-effective for both banks and the poor.

G. Palanithurai and V. Ragupathy in their paper *Governance for Development at Grassroots* emphasise that rural India has undergone a metamorphosis after the introduction of the 73rd Amendment to the Constitution of India and the positioning of the Panchayats in place for governance, administration and management of development. A number of new initiatives have been introduced and the paper explores the way these new initiatives have been working. The 73rd Amendment provides an opportunity to change the paradigm of development from the supply driven to the demand driven mode. The outcome of these opportunities depends on how they are utilized by the people. They call upon various stakeholders to play a vital role in the transformation process.

Mahi Pal in his paper on *Rural Governance and Panchayati Raj Institutions,* reviews the functioning of the Panchayats from the point of view of ascertaining as to what extent these institutions have enabled the people to participate in delivering good governance at the local level. The author assesses the empowerment and entitlement of the PRIs, effectiveness of these institutions in delivering rural good governance with people's active and functional participation. The paper has been divided into seven parts and each part deals with a specific sub-theme of importance. The author has highlighted that some States like Andhra Pradesh and Haryana are implementing decentralized programmes mainly through the bureaucracy which is against the spirit of the 73rd Amendment. He has analysed reports and recommendations of various committees to highlight the institutional weaknesses of Panchayats. According to the author, the affirmative action for the marginalized groups in local governance has resulted in social identities and political awareness among them and created an urge to become part of the mainstream political, economic and social life. He concludes by saying that it is time that the local level representatives unite and assert for their right.

In his paper on *Transforming Village Clusters into Territorial Production Complexes,* K.V. Sundaram sets the PURA concept in the wider context of Holistic Village Cluster development planning to achieve "Inclusive Development" and advocates a nation-wide Village Renewal Movement, based on a wider understanding of the PURA concept. The programme PURA is essentially a Habitat Design to achieve the quality of life in rural places and lead to inclusive rural development. He has argued that in our present development context, the Indian village has ceased to be an economic entity for planning and therefore, there is a need to think in terms of 'viable unit area of development in rural space'. However, he cautions that successful implementation of the Holistic Cluster development strategy requires complex decentralized matrix structures, with permanent mechanism for vertical and lateral integration, a combination of specialists and generalists skill, institutional leadership and social intervention capability.

Eswarappa Kasi, in his paper, *Socio-Cultural Dimensions of Sericulture: A Village Study from Andhra Pradesh* examines the significance of socio-cultural factors in sustainable development. The paper seeks to explore the role of education among sericulturists and how it has become a sustaining activity in spite of certain difficulties. It also reveals as to how the practice of sericulture has transformed Kotha Indlu Village of Chittoor District of Andhra Pradesh. He advocates that there is more than mere economic development. In spite of improved economic status of farmers from different castes, the lower caste members are still denied entry into public places like temples and such social issues also need to be addressed. He opines that the lessons drawn from such experiments would be of great significance for promotion of sustainable development.

The paper on *Empowering Rural India with SERVIS* by Amarnath Kshatriya focuses on the issue of energizing the vast 'Human Capital' scattered in the villages of India. He has proposed the establishment of a voluntary organization SERVIS that seeks to activate the dormant spirit of Service to the Nation among dedicated professional in all fields of knowledge to energies and activate the dormant potential of the school age children in rural India. The focus will be on stimulating a child's curiosity and engage them in tasks and activities to develop their understanding or ways of thinking about the world and themselves. An organization of people dedicated to the cause of empowerment is need and provides a blueprint to achieve it.

E-governance is the latest buzzword as it provides for solutions to various problems. C.S.R. Prabhu in his paper *Cost Effective Solution for Effective e-Governance—e-Panchayat,* explores a cost effective solution at the village level. He has discussed the advantages of an electronic knowledge-based system e-Panchayat which has been successfully designed, developed and implemented with the active involvement of the stakeholders at lower administrative level in Andhra Pradesh. The project involves—e-governance right at the grassroots level and offers a very cost effective solution. According to the author the methodology adopted for implementation of the system is novel, cost effective and self-sustaining. Prabhu has highlighted that the Government has to act as a catalyst for creating a cooperative and friendly e-climate for achieving and sustaining the success, rather than a purchaser of hardware/outsourcing agency.

Madaswamy Moni's paper, *Mainstreaming ICT for Fostering Agricultural Growth, Poverty Reduction and Sustainable Resource Use to Usher in Rural Prosperity*, focuses on integrating the two approaches that the policy-makers and planners have adopted to remove inequality and poverty. In poor rural areas where agricultural productivity is low and unreliable and there is food insecurity, better information and knowledge-exchange can be important in lessening poverty. In this ICT has an important role to play. He suggests that science and technology offers tremendous opportunities in simultaneous achievements, towards the goal of sustainable agriculture and improving rural livelihoods. Digital opportunities have emerged as "powerful tool" for fostering agricultural growth, poverty reduction and sustainable resources use in developing countries. In his view the digital development in rural areas of India facilitates rural prosperity, rural empowerment and a warehousing of data for development. A action plan to bridge the digital divide has also been suggested.

V.P. Sharma and K.V. Rao, in their paper on *Bridging the Digital Divide: Information Kiosks in Rural India—Challenges and Opportunities,* point out that if rural India can be connected and the masses are empowered with information, the Indian economy will take a leap forward into the Digital Millennium with a great speed. The process has already started. The paper discusses in brief the basic objectives, funding pattern, outreach, content, community orientation of the Information Kiosks in three States. It also addresses the issues of capacity-building, linkages with other public and private institutions for information support and sustainability. They are in favour of meeting the information needs of the farmers and its regular updating to make it relevant. Good quality agricultural information should be available at the information kiosks in the rural areas. To enable rural India to take the benefits of ICT the authors have suggested that there is an urgent need to connect all the district and block level officers of agriculture and other line departments, to facilitate regular two-way flow of information between the research and development wings of the agriculture.

Water, as a resource, is one and indivisible: rainfall, river waters, surface ponds, lakes and ground water are all part of one system. Water is part of a larger ecological system. This important subject had been dealt by Madaswamy Moni in his paper on *Informatics Blueprint for Integrated Water Resource Planning and Management at Grassroots Level: A Quintessential Requirement for Adaptation to Climate Changes and Sustainable Agricultural Development in India.* The paper makes a powerful plea for an informatics Blueprint for Integrated Water Resource Planning and Management at grassroots level in relation to climate change and sustainable agricultural development. A comprehensive analysis has been attempted, which starts with a broad-brush picture of existing water resource development strategies and policies. The author pleads for a plan of action to protect, allocate and manage water resources judiciously with a water accounting and budgeting system at various levels. With water now a scarce commodity, the paper covers a wide spectrum of issues and problems relating to integrated water resource planning and management, providing a blue print for accelerated broad-based rural growth, leveraged through the new information and communication technology to achieve sustainable development.

REFERENCES

National Sample Survey Organisation; *Household Consumer Consumption and Expenditure and Employment—Unemployment situation in India*, January-December, 2003, New Delhi.

United Nations; *The Millennium Development Goals Report, 2007*, New York.

M.S. Swaminathan; Not a Grain to Spare, India's policy on wheat reserves is flawed. It's time it is corrected, *Hindustan Times*, Dated 6-9-2007, New Delhi.

Hindustan Times; Strengthening India's Backbone; The Key is to Link its Villages to the World Economy, Dated 25-9-2007, New Delhi.

Government of India, Ministry of Rural Development; *Performance Budget of Department of Rural Development, 2002-03*, New Delhi.

The Times of India; Spending Rs. 2500 a Month? You are in top 1 per cent, dated 10-9-2007, New Delhi.

The Times of India; In Rajasthan, Death Prefers Young Girls, dated 16-8-2007, New Delhi.

P.N. Sankaran; Rural and Urban Development: A Partnership Perspective, *Kurukshetra*, Vol. 53, No. 2, December 2004, New Delhi.

PART I

TRANSFORMATIVE RURAL DEVELOPMENT AND MDGs

1

Rural Food Security: Unresolved Issues and Emerging Challenges

NIRA RAMACHANDRAN

Intorduction

More than 24 million rural Indians still live below the hunger line as defined by the availability of two square meals a day throughout the year (NSS 55th Round, 1999-2000). While chronic food insecurity has been sharply curtailed, seasonal food shortages are common and tend to persist for several months at a time in most parts of the country. For the rural householder, the fear of hunger overshadows his very existence severely limiting effort, enterprise and achievement. Even more severe is the impact of actual food scarcity reflected in the country's dismal nutrition and health indicators. Reports of large numbers of children succumbing to malnutrition or malnutrition related ailments and isolated adult starvation deaths in remote locations surface every year, but little has been done to address them. While not perhaps, as momentous as large scale famine, the eradication of which has been one of our major achievements, these incidents strikingly highlight the fact that food insecurity in rural areas persists and remains a major concern of public policy.

This paper attempts to identify some of the unresolved issues underlying the persistence of rural food insecurity in the country, analyse the ineffectiveness of Government interventions to address these issues and indicate directions for policy change.

Food Security Status of the Rural Population

The rural population in India has consistently recorded poorer levels of food security in comparison with their urban counterparts. This may be attributable to lower income levels, fewer employment opportunities or even sheer lack of physical access to food in the form of home grown supplies or access to markets or the Public Distribution System. This is a feature common to rural populations in most developing countries, where the producers of food are the very ones unable to access sufficient food to meet their own needs.

Estimates of calorie requirements for rural and urban populations place rural calorie needs at a higher level than those of the urban population. These estimates are based on the premise that the larger amounts of energy expended by the rural population in agriculture and other manual labour based occupations necessitate higher calorie consumption. However, average urban calorie consumption at 2156 Kcalories exceeds rural calorie consumption, which is only 2149 Kcalories (1999-2000).

The official poverty line in India, which is itself, based on the purchasing power to acquire sufficient calories together with some basic non-food items, firmly places over one-quarter of the country's population among the under-nourished.[1] On the other hand, the National Sample Survey (NSS) which collects basic information on the availability of two square meals a day has reported a quantum improvement in reducing hunger. While in 1983, 19 per cent of the rural population reported lack of access to regular food supply, by 1993, this segment had reduced to only 7 per cent. Currently, 1 per cent of the rural and 0.5 per cent of the urban population report chronic food insecurity (NSS 56th Round 2000-2001). This amounts to about 8.6 million people in the country with lack of access to sufficient food to meet their daily requirements. Of these, 7.2 million live in rural areas. It must, however, be pointed out that these figures are considered vastly underestimated.

While the rural areas of the country are undoubtedly disadvantaged in their access to food, the level of food insecurity is not uniform. The most disadvantaged groups—the marginal farmer and the landless labourer, the scheduled castes and scheduled tribes and the especially vulnerable among these, i.e., the women and women-headed households, the old and infirm and those unable to work are the most under-nourished.

While government policies have so far focused on boosting food production at the national level and have achieved tremendous success by any standard, the focus must now shift to ensuring that no individual remains hungry or malnourished. Table 1.1 brings out the disparities in calorie consumption within rural areas. Only 33.4 per cent of the rural population consumes a diet with more calories than the minimum requirement of 2400. This segment also belongs to the middle and higher income groups with an average monthly expenditure of more than Rs. 525. Those in the expenditure range of less than Rs. 470, which accounts for two-thirds of the rural population, consume less than 2289 calories or about 95 per cent or less of the norm.

Table 1.1: Percentage of Persons in Rural India with Calorie Intake below the Nutrition Norm of 2400 Calories (1999-2000)

MPCE Class (in rupees)	*Average Monthly Per Capita Expenditure (in rupees)*	*Calorie Intake per diem*	*Percentage of Total Persons*
470-675 and below	496.70	2289 and below	66.6
525-615 and below	566.62	2403	10.3
615-775 and above	686.00	2581 and above	23.1

Source: NSS Report No. 471, Nutritional Intake in India, 1999-2000.

Monthly Per Capita Expenditure (MPCE)

The data discussed above seems to point to the fact that while a reduction in starvation and absolute hunger has taken place, under-nourishment in terms of actual calorie consumption remains a serious problem. What is even more disquieting is the fact that there appears to be a declining trend in rural calorie consumption from the seventies till date. Table 1.2 depicts the change in calorie consumption from the early seventies to the end of the nineties in fifteen major states. In the case of four states, i.e., Kerala, Maharashtra, Orissa and West Bengal there has been an increase in per capita calorie consumption by more than 100 calories per day, contrary to the declining trend in the remaining eleven states. In the case of Kerala, the increase is as much as 423 calories per capita per day, which represents a substantial increase. In the remaining three states, the trend has been a fluctuating one as against the steady increase in the case of Kerala.

The sharpest decline has taken place in two of the richest agricultural states, Punjab and Haryana by as much as 1112 calories and 760 calories respectively. However, this is the result of a gradual decline over three decades representing perhaps a shift in dietary patterns. Both states are now close to the rural consumption norm of 2400 calories as are Rajasthan and Uttar Pradesh.

Table 1.2: Change in Average Calorie Consumption by States (Rural Kcal Per Capita Per Day)

States	*1972-73*	*1983*	*1993-94*	*1999-2000*
Andhra Pradesh	2103	2204	2052	2021
Assam	2074	2056	1983	1915
Bihar	2225	2189	2115	2121
Gujarat	2142	2113	1994	1986
Haryana	3215	2554	2491	2455
Karnataka	2202	2260	2073	2028
Kerala	1559	1884	1965	1982
Madhya Pradesh	2423	2323	2164	2062
Maharashtra	1895	2144	1939	2012
Orissa	1995	2103	2199	2119
Punjab	3493	2677	2418	2381
Rajasthan	2730	2433	2470	2425
Tamil Nadu	1955	1861	1884	1826
Uttar Pradesh	2575	2399	2307	2327
West Bengal	1921	2027	2211	2095

Table 1.3 shows the proportion of the population consuming less than 2400 calories per day by state as also the change in size of this under-nourished group over two decades.

Table 1.3: Per cent Rural Population with Less than 90 per cent Recommended Calorie Intake

States	*1983*	*1993-94*	*1999-2000*
Andhra Pradesh	39.3	48.4	51
Assam	48.8	53.9	61.5
Bihar	39.6	41.8	45.9
Gujarat	46.7	53.7	53.6
Haryana	27.5	28.4	30.2
Karnataka	42.3	48.6	54
Kerala	61.2	55.1	51.6
Madhya Pradesh	33.7	41	50.7
Maharashtra	44.4	57.4	52.4
Orissa	46.2	35.4	42.1
Punjab	24.1	27.6	31
Rajasthan	28.8	23	24.3
Tamil Nadu	62.9	61.3	65.2
Uttar Pradesh	31	31	32.9
West Bengal	53.6	37.2	47.8

The table shows that in eight of the fifteen States, more than half the population is under-nourished. Only in one State, Rajasthan is under-nourishment restricted to one-quarter of the rural population, the other states recording between one-third and half the population as consuming less than 90 per cent of the required calories. Over time, the proportion of the rural population consuming less than the required calories has increased in all but four states, implying growing food insecurity. The four states where food security levels have improved include the three states of Kerala, Orissa and West Bengal, where the overall consumption of calories has also gone up and the State of Rajasthan, where the proportion of under-nourished rural population has decreased, despite an accompanying decrease in overall calorie consumption. This would, perhaps, suggest decreasing disparity in food consumption in this state.

What underlies the persistence of food insecurity and the poor nutritional indicators in most parts of the country? The following sections assess the probable factors underlying the issue and seek to identify pointers for policy formulation.

The Gender Factor[2]

The role of women as producers and providers of food is often overshadowed by their primary role as care-givers. However, in most of the developing world, including India, large number of women are engaged in agriculture, primarily the production and processing of food. With male-selective migration from rural areas on the increase, women are often left behind to take care of both family and farm on their own. With women-headed households being more prone to poverty, wages being unfavourable to women in general and access to financial, technical and other support services being denied to them, the poor nutritional status of the rural population should not evoke surprise.

According to current statistics (Census of India 2001), 27.5 per cent of cultivators in the rural areas are female, while in the case of agricultural labour, as much as 46.9 per cent are women.

Of the rural female workforce, an overwhelmingly large proportion, i.e., 80 per cent are employed in the agricultural sector. About 36.5 per cent (40.6 million) work as cultivators on their own/family landholdings, while about 43.4 per cent (48.4 million) are engaged as hired agricultural labour. It is, therefore, obvious that women play no small role in food production. This is further corroborated by the fact that agriculture in India is subsistence based and 93.8 per cent of the country's cultivated area is under food crops.

When women workers, in such large numbers, form the core of the food producing workforce, it becomes obvious that "the starting point of food security must be the security of the food producing communities. Women in agriculture have to be among the beneficiaries if the promise of food for all is to be achieved (Rengem, undated). Unfortunately in India, women farmers face considerable discrimination with regard to agricultural as well as other assets and resources. "They do not have the rights of ownership of land, draught and milch animals. Men own productive assets such as tube-wells, winnowers, tractors, bullock carts, generators, etc. Except in matrilineal societies, land ownership patterns favour men. Land reform measures have ignored women's interests as co-owners or cultivators of land" (WFP-MSSRF, 2001). Consequently, women also face discrimination with regard to availability of loans and access to technology and training.

In many parts of the country, where male selective migration is a dominant feature, it is the women who are left behind to maintain the family land and perform all agricultural tasks independently. This holds true in such diverse regions as the mountain villages of the Himalayas, on the one hand and the tribal villages of Chhattisgarh, on the other (Ramachandran, 2000; Barme and Ramachandran, 2002). In Uttaranchal, men restrict their farming activities to ploughing the fields alone, with all other tasks being left to women. Yet access to credit, training and other facilities remains out of reach for them.

'Given women's role in food production and provision, any set of strategies for sustainable food security must address their limited access to productive resources. Women's limited access to resources and their insufficient purchasing power are products of a series of inter-related social, economic and cultural factors that force them into a subordinate role to the detriment of their own development and that of society as a whole' (FAO, 1996).

A serious constraint for women farmers is their lack of access to security of tenure or ownership of land. As Agarwal (2002) notes 'In largely agrarian economies, arable land is the most valued form of property and productive resource. It is a wealth creating and livelihood-sustaining asset. For a significant majority of rural households, it is the single most important source of security against poverty'. Traditionally, women have been denied equal inheritance rights to property both under the Hindu and the Islamic systems of law. With the recent Amendment of the Hindu Succession Act (August, 2005), women have been given equal rights of inheritance in all property including joint family property, agricultural land and tenancy rights throughout the country. However, the law regarding the inheritance rights of Muslim women remains unchanged.

Also, where agricultural land is concerned, agriculture being a state subject, the law will come into effect only if the concerned state takes necessary action. The positive impact of this amendment thus remains to be seen.

Till recently, under the Hindu system, for example, a woman could inherit property only in the absence of four generations of men in the male line of descent. Even then, her rights were limited to a lifetime interest with no right to mortgage or dispose of the property, except in exceptional circumstances. The Hindu Succession Act of 1956 sought to resolve this issue by giving equal shares to sons, daughters and widows in a man's own property and the same in his share of joint family property. However, agricultural land was exempted from the purview of the Act. Women's inheritance in agricultural tenancy land depends on state-level tenurial laws, which strongly favour succession in the male line.

Under Muslim law, daughters are allowed only half the share of sons in the property bequeathed by their father. The Muslim Personal Law Shariat (Application) Act of 1937 also sought to enhance the property rights of Muslim women, but excluded all agricultural land, both tenanted and owned from its purview, except in some of the states of South India.

Regional variations in women's right to land are, in part, due to pockets of matrilineal systems prevailing in some states such as the Southern State of Kerala and the North-Eastern State of Meghalaya where land is inherited in the female line and in part due to the underlying social constructs in these regions. Social norms in the Southern part of India permit marriage within kinship and in the same village, whereas in the Northern part of India, marriage within the family is strictly prohibited with intra-village marriage relationships being frowned upon. The impact of these social strictures can be clearly seen in the strictly patrilineal inheritance patterns in the North as compared to the more flexible rules in the South. In fact, in the Southern States of Andhra Pradesh, Tamil Nadu and Karnataka, as well as in Maharashtra in Central India, the Hindu Succession Act has been amended to make daughters coparceners in joint family property, while in the Southern State of Kerala, the concept of joint family property has been abolished altogether.

Gender equality in inheritance, Agarwal (*Ibid*) writes, must be promoted as most agricultural land is privately held. In India, for example, 86 per cent of the arable land is in private hands. While gender disaggregated records are not maintained, sample surveys have brought to light the sharply skewed pattern of inheritance in rural areas of India. Chen's 1991 (Chen, 2000) survey of seven states in India, quoted in

Agarwal, found that of 470 women with land owning fathers, only 13 per cent inherited land as daughters. This ranged from 18 per cent in South India to only 8 per cent in the North, indicating that 87 per cent of women did not receive their due as daughters. Of the 280 widows surveyed, only 51 per cent received land and most often their shares were not recorded in official land records. Other studies reveal that even when women's land rights are recorded, it is usually in joint ownership with their sons.

Land reform programmes and resettlement schemes are also overtly male biased. Under 'Operation Barga', a scheme implemented by the West Bengal State Government to secure the rights of tenants by registration in the late seventies and early eighties, few women received land. A study conducted in a village in Midnapur district (Gupta, 1993 quoted in Agarwal *Ibid*) showed that 98 per cent of the 107 holdings distributed, went to men and in 9 out of 10 female-headed households, it went to the sons. Married women did not even receive joint titles. Under the Eighth Five Year Plan (1992-97), State Governments were directed to allot 40 per cent of ceiling surplus land to women alone and the rest as joint titles to spouses. However, ceiling surplus land accounted for just about one-half per cent of India's arable land in the early nineties and today is reduced to merely 0.2 per cent. Also, the implementation of the scheme rests with State Governments, which may or may not follow through.

Why is ownership of land so essential for women farmers? The rapid feminisation of agriculture in the region has thrown into prominence the issue of land rights for women. Increasing migration by males from rural to urban areas in search of livelihoods has followed the fragmentation of land holdings, lack of wage opportunities in rural areas and deepening poverty. What is often overlooked in policy formulation is the increasing number of *de facto* women heads of households struggling to eke out a livelihood and ensure the food security of their families, without access to credit, technology or extension services. Denied security of tenure, they lack the collateral required for credit or the social status to deal with extension workers on an equal basis. Their needs tend to be ignored, even in agricultural research and technological innovations. FAO statistics show that worldwide, only 5 per cent of extension services have been addressed to rural women (FAO, 1996).

In order for women farmers to use land more efficiently and thereby make a greater contribution to food security, they need access to land, management control of land based resources and the economic incentives that security of tenure provides (FAO *Ibid*). "Land rights can serve multiple functions in rural women's lives, which are not easy to replicate through other means". Apart from the direct benefits in terms of crop output, trees, fodder, fuel and garden produce and indirect advantages in terms of collateral for credit or an asset, which can be sold or mortgaged when needed, title to land also enhances the probability of finding supplementary wage employment and serves as an asset base for rural non-farm enterprises. Chadha (1992) in a study of the rural non-farm sector, found that those with land generated much higher rural non-farm earnings from self-employment, than the totally landless. Agarwal concludes that, women's access to even a small plot can be a critical element in a diversified livelihood system and can significantly improve women's and the family's welfare, even if the plot is not large enough to provide full family subsistence.

Denied ownership and tenancy rights to land, the women farmers' lot is further exacerbated by the gender disparity in wages, both on and off-farm. Micro-studies undertaken in various states of India reveal various levels of wage disparity. Surveys undertaken in West Bengal (cited in IDRC *Ibid*) show that gender wage disparities have begun to decline along with the general rise in wages. Female wages in the state rose from 75 per cent of male wages to about 86 per cent between the mid-sixties and the early seventies. The study states that on an average, across India, rural women's wages rose from 52 per cent of their male counterparts in 1972 to as much as 69 per cent by 1983. The gap between male and female wages was highest and fluctuated most in the least agriculturally developed areas. However, a more recent micro-study of 54 casual labour households in the Karimnagar district of Andhra Pradesh (Lingamurthy, 2002) found that while both male and female wages have increased over time, the relative wage differentials by gender have not reduced. Females are paid less than half the wages given to men across all agricultural and non-agricultural operations. At best, women receive one-half of the male wages for the same operation.

In the context of food security, the need for gender parity in wages assumes greater significance. Households, wherein women have access to their own incomes and can exercise decision-making powers, tend to have an expenditure pattern different to the one existing in male dominated households. Research in several developing countries of Asia, Africa and Latin America has found that improvements in household food security and nutrition are associated with women's access to income and their role in household decisions on expenditure. This is because women tend to spend a significantly higher proportion of their income than men on food for the family (IDRC, 2004). Dyer and Bruce cited in Agarwal (2002a) in a 1988 study found that women in poor households of India and other parts of the world, spend most of the earnings under their control on basic household needs, while men tend to spend a significant part of theirs on personal good such as alcohol, tobacco, etc.

Studies undertaken in India and neighbouring countries reveal that women invest larger amounts of money in nutrition and health. Among marginal farmer households in the Indian State of Kerala, the mother's cultivation of a home garden (the output of which she controlled) was found to have a consistently high positive effect on child nutrition (Kumar, 1978).

Women's access to adequate food security, both for themselves and their families is dependent, not only on their economic status, but on their own health, education and social status within the family and in society. Women with low status tend to have weaker control over household resources, tighter time constraints, less access to information and health services, poorer mental health and lower self-esteem. These factors are, thought to be, closely tied to women's own nutritional status and the quality of care they receive and in turn to children's birth weights and the quality of care they receive (Smith, Lisa, *et al.* 2003). Smith concludes that 'making a decision at the policy level to improve women's own nutritional status produces significant benefits. Not only does a woman's nutritional status improve, but so does the nutritional status of her young children. Raising women's status today is a powerful force for improving the health, longevity, mental and physical capacity and productivity of the next generation of adults'.

Seasonal Food Insecurity

Seasonality is now accepted as a central concern in the poverty and under development problems in low income developing countries (LDCs). There have been various attempts to analyse and explain the relationship between seasonality and employment, under-nutrition, poverty, health, working capacity, stress and gender bias. But little effort has been made to translate these explanations and conclusions into tangible counter seasonal development strategies (Chambers *et al.*, 1981; Chambers, 1995; Longhurst, 1986). The all-pervasive influence of seasonality on food security, income and all other facets of life in rural areas and the need for interventions to smooth the pattern of income and consumption emerges strongly from research in the area. Jodha even explains an increase in well-being in rural areas as a function of decline in the seasonality of existence. The overwhelming influence of the seasons in rural areas extends beyond farmers and agricultural labourers or those directly dependent on the land. It also affects not only food security and livelihoods, but much broader areas of rural life. As Spitz writes "The whole social fabric is adjusted to this qualitatively predictable seasonal pattern of food availability and employment, indebtedness and migration with its likelihood of year to year fluctuations aggravating or partly alleviating food insecurity" (Spitz, 2002).

Existing policies for temporary employment creation often fail to take into account the seasonality factor in rural life. Seasonality coupled with low mean incomes and high seasonal variance in incomes, is one of the primary causes of distress in rural livelihoods. Seasonality of rural livelihoods is not confined to the economics of income. Seasonal changes also aggravate the gamut of existing economic, social, political and cultural handicaps that directly or indirectly influence the well-being of the rural population. Seasonality is, and will continue to be, a major stumbling block in rural development in every LDC where agriculture and rural livelihoods depend on seasonal fluctuations in access to food, employment, income and seasonal changes in status of health/morbidity (Raina, 2000).

NSS statistics show that the proportion of the population suffering from chronic food insecurity has considerably declined and currently stands at a mere one per cent of the rural population (2000-01). However, there is still a considerably large segment which does not have access to adequate food seasonally. While the NSS classification is limited to estimates of population facing food insecurity seasonally, the actual duration of this lean period is best assessed from micro-studies. A participatory study of twelve tribal villages in four States, i.e., Rajasthan, Madhya Pradesh, Chhattisgarh and Orissa (Barme and Ramachandran, 2002) revealed that the hunger period extended from a minimum of two months to as much as eight months of the year. The level of food insecurity varied from shortage to actual starvation. Similar results have been obtained from more detailed surveys carried out in the rural areas of Jharkhand (Ramachandran and Karan, 2000).

Seasonality also has a significant impact on women's nutritional status. The seasonal dimension reflects in fluctuations of workloads during the agricultural cycle as well as varying availability of food. The period of greatest nutritional stress for rural women is the 'lean months' of the pre-harvest period when household stocks and energy intake are low, but the energy demands of agricultural work tend to be highest. Heavy work during pregnancy can lead to premature labour and without increased calorific intake, to low birth weight babies, perpetuating the vicious cycle of weak infants, malnourished children, adults both physically and mentally deprived and the inability to break out of the vicious cycle of poverty and ill-health. The participatory study referred to above reiterates these findings and concludes that the impact of several months of reduced food intake on the health and nutritional status of women, particularly pregnant and lactating mothers and on their new born infants is reflected in the persistently high levels of low birth weight infants, malnourished and anaemic mothers and maternal mortality in the country (Ramachandran, Nira, 2003).

While seasonal food insecurity affects almost all the rural population, the intensity of impact varies with economic and social factors. Usually, only the large land holders escape seasonal hunger. Caste-wise and occupation-based differences also exist. An in-depth survey of twelve villages in Jharkhand (Ramachandran and Karan, 2002) reveals the food security scenario in relation to class and caste factors.

The threat of food insecurity overshadows the tribal villages of Jharkhand, particularly in the case of the landless, engaged as casual labour in agricultural or non-agricultural occupations. Seasonal food shortages affect almost every household cutting across caste and class barriers, with the sole exception of large land-holders. As food insecurity is an intrinsic part of rural life, traditional coping mechanisms have evolved which include the collection of edible forest products as well as gleanings from fields and other commons. The level of food distress varies from a few months of food shortages when harvest supplies run out, to near starvation levels for small landholders and the landless during the monsoons, when both food and employment are in short supply. A large proportion of the households face food shortages for at least part of the year. As land holdings are small and irrigation facilities inadequate, the area is single-cropped, and productivity is low. Food production even in a good agricultural year meets food requirements for barely six months. For the rest of the year, food intake is reduced to minimal levels, and in extreme situations, the diet consists of berries, leaves and other forest gleanings.

A seasonal food calendar prepared on the basis of participatory techniques carried out in twelve sample villages of Jharkhand (Table 1.4), reveals that food distress is an integral part of village life. While there are variations based on location, essentially, an assured food supply exists for only about three to four months of the year, i.e., in winter following the harvest in late October-early November. Food supplies tend to run short by the end of winter, and the starvation period begins by mid-summer (June) and in many cases, continues till the end of October.

Table 1.4: Seasonal Food Calendar (Jharkhand)

Village	*Availability*	*Shortage*	*Starvation*
Andhradih	Nov–Feb (4)	Sep-Oct, Mar-Apr (4)	May–Aug (4)
Baradih	Dec–Mar (4)	Sep-Nov, Apr-May (5)	Jun–Aug (3)
Sarapmunda	Oct–May (8)	Aug-Sep (2)	Jun–Jul (2)
Gujharia	Nov–Feb (4)	Mar-Jun (4)	Jul–Oct (4)
Udaipura	Dec–May (6)	Sep-Nov (3)	Jun–Aug (3)
Ghutwa	Apr–Jul & Dec–Jan (6)	Feb-Mar, Nov (3)	Aug–Oct (3)
Baherakalan	Nov–Mar (5)	Oct, Apr-May (3)	Jun–Sept (4)
Karma	Dec–Mar (4)	Apr-May, Sep-Nov (5)	Jun–Aug (3)
Rampur	Dec–May (6)	Jun, Oct-Nov (3)	Jul–Sep (3)
Maheshpur	Dec–Apr (5)	May-Jun, Nov (3)	Jul–Oct (4)
Karmatand	Nov–Feb & May–Jul (7)	Mar-Apr (2)	Aug–Oct (3)
Indravani	Dec–Mar (4)	Apr-Jun, Oct-Nov (5)	Jul–Sep (3)

Note: Figures in brackets indicate number of months.

Source: Ramachandran, Nira and Anup Karan (2001).

During periods of starvation, people consume small amounts of coarse grains and home-grown vegetables and turn to the forest for supplies of berries, herbs, edible leaves and wild vegetables. As Mukerjee (2000) states 'the commons provide a safety net for the poor villagers to tide over the hunger period'. Even among the sample districts, there are variations in the level of food distress. While in Gumla, people turn to coarse cereals like maize during periods of food shortages, Dumka and Palamau do not have even maize stocks available. The villagers in these districts thus resort to consuming inferior cereals such as goda, gondli and mahua. The response to food shortages is in the form of cutting down amount of intake and number of meals, as well as turning to forest gleanings.

Table 1.5 brings out the level of food insecurity among the sample households. Only 70 per cent of the households felt that they had access to sufficient food throughout the year. Inter-district variations are evident with Dumka being worst off with almost 41 per cent of the sample households claiming to be food insecure for at least some part of the year. Across the caste groups, not even the upper caste households are totally food secure. Almost 8 per cent felt that food supplies were not sufficient throughout the year. Among the other groups, the scheduled caste households were the worst off with 48 per cent being food insecure, followed by the Christian

Scheduled Tribes, the Muslims and the Scheduled Tribes (37, 31 and 31 per cent respectively). Across occupational groups, the proportion of food insecure households ranged from 7 per cent (Other Occupations) to as much as 59 per cent of the households engaged in agricultural labour. Based on land ownership, the landless, as expected were the worst off with 48 per cent being food insecure. The proportion reduced progressively with the increasing size of land owned till households owning 10 acres and above recorded no food insecurity.

Table 1.5: Food Sufficiency by Caste, Occupation of HH, Land Size and Districts (Jharkhand)

		Per cent Households with Sufficient Food throughout the Year
Caste Group	Upper caste	92.3
	OBC I	80.0
	OBC II	76.9
	SC	52.2
	ST	69.1
	ST (Ch)	63.2
	PTG	76.5
	Muslims	69.0
Occupation of HH	Self-employed in agriculture	71.2
	Agricultural labour	41.2
	Non-agricultural Labour	66.2
	Salaried worker	88.9
	Petty job	72.7
	Petty business	70.0
	Other occupation"	93.3
	Non-worker	69.2
Land Size (in acres)	Landless	51.7
	Up to 1	62.4
	1 to 2.5	78.0
	2.5 to 5	80.0
	5 to 10	84.6
	10 and more	100.0
District	Gumla	76.7
	Palamau	78.0
	Dumka	59.5
All		70.3

While 30 per cent of all households reported some degree of food insecurity, the extent to which the households are food insecure is depicted in Table 1.6. The table reveals that two-thirds of the households experience at least two months of food distress every year, while about 8 per cent of the overall households report as much as four months of food scarcity.

This seems to corroborate the findings from the food calendar, where seasonal food shortages with starvation periods from mid-summer to autumn have emerged in the exercises conducted in all twelve villages. Similar experiences have been recorded in other areas of the country. In the hill districts of Uttaranchal, the monsoon months are the critical period with not only food stocks running out, but also the drying up of the main source of income from agriculture or construction work which also stops during this period (Ramachandran, 2000).

Table 1.6: Extent of Seasonal Food Insecurity in Jharkhand

	Per cent Households by Number of food deficient months in a year			
	1	*2*	*3*	*4*
Caste Group				
Upper Caste				
OBC I	0	100.0	0	0
OBC II	0	75.0	25.0	0
SC	0	50.0	16.7	33.3
ST	0	63.6	27.3	9.1
ST(Ch)	10.9	67.4	17.4	4.3
PTG	0	57.1	28.6	14.3
Muslims	0	50.0	25.0	25.0
Occupation of HH				
Self-employed	22.2	77.8	0	0
in agriculture	6.3	75.0	12.5	6.3
Agricultural	20.0	60.0	20.0	0
Labour	12.5	58.3	16.7	12.5
Non-agricultural	0	50.0	50.0	0
Labour	0	66.7	0	33.3
Salaried Worker	0	75.0	25.0	0
Petty job	0	.0	0	100.0
Petty Business	0	50.0	50.0	0
Other Occupation	0	71.4	21.4	7.1
Non-workers	7.9	55.3	21.1	15.8
Land Size (in acres)				
Landless				
Upto 1	0	70.0	20.0	10.0
1 to 2.5	6.9	69.0	13.8	10.3
2.5 to 5	10.2	63.3	20.4	6.1
5 to 10				
District				
Gumla	8.0	65.9	18.2	8.0
Palamau				
Dumka				
All				

Common Property Resources as a Natural Food 'Safety Net'

Common Property Resources (CPRs) often play a significant, though unacknowledged role in ensuring the food security of households living on the verge of hunger. This is particularly so in the case of remote villages which lack access to markets and Government safety net provisions. Tribal populations with their traditional dependence on forest-based resources and access to knowledge of food and medicinal plants and herbs handed down through the generations have a particularly close link with CPRs.

CPRs may be defined as 'private property for a group' with organisational rules circumscribing the nature of rights and responsibilities existing within the group with respect to them. Multiple uses and inter-related rights are common in the rural areas with sets of resources being characterised by complementarity of use. For example, grazing rights on private land are often given to nomadic communities after the harvesting of the monsoon crop in recognition of the contribution of animal herds to the rejuvenation of soil fertility (Chopra and Gulati, 2001). Agricultural labour is likewise free to access the gleanings from fields after harvest in a mutually beneficial relationship.

Table 1.7: Identification of Common Property Resources

Classification	*Included in CPR*	*Source of Sanction for Land Access*
Net Sown Area	No	On uncultivated owned land: limited user rights
Current Fallow	No	On uncultivated owned land: limited user rights
Other Fallow	Yes	User rights by convention
Cultivable Waste	Yes, partial	User rights by convention
Pastures and Other Grazing Land	Yes	User rights by law
Barren and Uncultivable Land	May be included	No access
Area put to Non-Agricultural Use	No	No access
Forest Area		
1. Reserved	No	No access
2. Protected	Partial	Partial user rights
3. Unclassed	Yes	User rights by law

Source: Chopra and Gulati (2001).

A useful classification of land use categories and their inclusion in common property resources as well as the type of access permitted to them has been developed by Chopra and Gulati (Table 1.7).

The extent of common property resources available by state is given in Table 1.8. Total CPR land in 16 States is about 70 million hectares. Of this 64 per cent is non-forest land. If estimates of CPR land in the North Eastern states, for which land records are not reliable, is included, the total availability of CPR land would go up to about 75 million hectares. The distribution pattern clearly indicates that the proportion of CPR land is least in the agriculturally developed States like Punjab and Haryana and highest in the hill state of Himachal Pradesh. In the other states, CPR land ranges between 10 and 34 per cent of the total land area (Chopra and Gulati, *Ibid*). Proper management of these resources in conjunction with the community could contribute significantly to inproving livelihoods and food security.

Table 1.8: State-wise Extent of CPRs and Wasteland

State	*Total CPR (000 ha)*	*Non-Forest CPR (000 ha)*	*Total Wasteland (000 ha)*
Andhra Pradesh	5989	624	5932
Bihar	5267	2850	2474
Gujarat	3269	2707	4189
Haryana	190	44	357
Himachal Pradesh	5188	1619	1069
Jammu & Kashmir	278	278	3714
Karnataka	3207	2203	2680
Kerala	331	207	163
Madhya Pradesh	13890	6446	8872
Maharashtra	8039	5926	6209
Orissa	4882	1537	2045
Punjab	359	73	370
Rajasthan	11977	11697	9605
Tamil Nadu	2773	2387	2272
Uttar Pradesh	3756	2221	5007
West Bengal	647	164	435
Total (16 States)	70042	44983	57438

Source: Chopra and Gulati (2001).

Lack of access to forests and shrinking CPRs have particularly impacted on tribal people. Tribal societies are defined by their close relationship with nature. The very existence of remote tribal groups depends on the sustenance they derive from forests and other common property resources, whether directly in terms of food or indirectly as a livelihood option. Even today, while there is much greater interaction between tribal and non-tribal populations, which have become a part of the monetised economy, the strong links of the former with the forest remain. A recent survey, undertaken by the FAO-WFP (November 2001) in tribal villages of the States of Chhattisgarh, Madhya Pradesh, Orissa and Rajasthan revealed the extent of tribal dependence on the forest as a source of food. The survey found that in tribal villages located in the vicinity of forests, food supplies may run short because of lack of access to the PDS or markets, or lack of purchasing power as off-farm employment oppportunities are not readily available, but actual starvation periods are much shorter than in other tribal villages which do not have access to forest resources. Also, the food basket in these villages is much more varied and balanced than in those dependent on markets.[3]

The commons have always provided a natural food safety net for the most vulnerable among the rural poor, i.e., those without land of their own as well as those with landholdings too small to support themselves. The poor have traditionally held the right to gleanings from both common and privately owned land. Mukherjee (2000) writes that degradation of surrounding forests, trees and CPRs and rigid forest laws have adversely affected food accessibility, livelihood options and quality of life of local communities. Such degradation has often caused periodic phases of hunger and malnutrition in the absence of supportive means of livelihood for acquiring the basic needs of life.

Two developments have resulted in the large scale destruction of this additional source of food. One, the introduction of intensive chemical fertilizer-based high yielding varieties, have boosted yields for the large farmer, but destroyed the diverse eco-system flourishing in traditional multi-cropping systems, which provided food in the form of weeds, ants, termites, mudfish, crabs, shrimps, frogs. Two, the conversion of grazing lands to short-cycle firewood plantations with state support under the social forestry programme created vast stretches planted with eucalyptus and Australian Accacia. Many forests were also converted to mono-species plantations by private corporations and the Forest Department. While this has undoubtedly help restore forest cover, the impact on the food security of the poor has been negative[4]. Mukerjee *et al.* cite the case of 15 villages in West Midnapur district of West Bengal, where not only direct food security, but also livelihoods dependent on a variety of NTFPs, varying over the seasons, have sharply declined with the introduction of mono-forestry. More than 80 per cent of the population of these tribal villages were directly dependent on the forest for both food and livelihoods.

While dwindling common property resources have had sharply negative impacts on rural food security, equally dramatic improvement in the situation can result from promoting a revival of these resources. A success story from Birbhum district of West Bengal resulted from the handing over of 3-5 m wide strips of public land along canals to groups of landless people on a 25-30 user rights basis by the Panchayats. Within 5-6 years, a variety of fodder, fuelwood and food trees and crops were cultivated improving food supply as well as the output of animal products. Members of the community were provided with skills training and guidance on the choice of species for cultivation.[5]

On a larger scale, the Bhartiya Agro Industries Development Corporation (BAIF) set out to rejuvenate pasture land in the village of Kavlas in Bhilwara district of Rajasthan in 1979 (Infochange News and Features, 2003a). Grazing lands have always been an important lifeline for Rajasthan's livestock-based farmers. In the Princely States, these were well maintained by *Thikandedars*, but handing over of management to the *Gram Panchayats* resulted in sharp deterioration. Shoıtage of fodder and fuel, drinking water scarcity and falling water tables were impacting negatively on the livelihoods and food security of the area. After initial hesitation, the villagers were persuaded to help develop pasture on 180 ha of temple land, with the initiative of allowing free grazing in a limited area. Grasses and trees suited to the area were introduced and within two rainy seasons, the pasture was established and began yielding profits. The villagers were allowed to cut grass, provided they deposited 50 per cent with the community, which was later sold to other villagers. After a fifteen year period, the average income from the pastures rose to Rs. 2500-4000 per ha. In addition, the members were allowed to forage for fodder and fuel on payment of a small fee. Proper pasture management techniques resulted in improved soil and water levels and the once arid, single cropped area began yielding two crops a year. The project was later extended to cover 320 ha across 9 villages in the district. The Governmennt has also decided to develop pastures on the same pattern on 2500 ha of land under the *Swarnajayanti Gram Rozgar Yojana* (SGRY).

A study of the tribal belt of the southern Aravalis in Rajasthan (Vyas, 1995) reveals how the deterioration of the natural resource base in the region has led to a change in the livelihood patterns of the tribal population accompanied by impoverishment and increasing inequities. With the introduction of modern legislation in the area, the common property resources of forests were converted into state property. The rise in economic value of timber and forest products led to large scale felling of trees and a major change in the conservationist role of the forest department to a revenue earning one. The tribal population had to resort to either of two options: to migrate to the nearby plains or to extend cultivation on fragile lands. Those who resorted to the former found low-paying menial jobs in farms, factories and mines, while the latter group brought progressively larger areas of infertile and marginal lands under the plough with disastrous consequences. This was compounded by the growing dependence on livestock to supplement the meagre income from agriculture. While the numbers of livestock have remained steady, the grazing area has significantly declined, thus over burdening the carrying capacity. Thus both groups of tribal population have faced a considerable drop in living standards.

Maintaining Bio-diversity for Food Security

Bio-diversity is not simply a tool to measure sustainable development; it also affects communities in the poorest regions of the world whose well-being relies on the diversity of nature — plants, animals, forests, water. Poor people are extremely dependent on these diverse resources, and are increasingly vulnerable to the loss of bio-diversity (FAO, 2003). Examples abound from different agro-climatic regions of the country, where the loss of crop diversity in agriculture has led to food insecurity, nutritional loss, loss of livelihoods and enforced migration.

It is a well-known fact that the narrow focus on wheat and rice during the Green Revolution has virtually eliminated local varieties of foodgrains. The emphasis on chemical fertilizers, irrigation and high yielding varieties, while boosting the overall production of foodgrain in the country, has impacted adversely on the small and marginal farmer, who no longer has access to traditional varieties of seeds and cannot afford the cost of the high priced inputs required to obtain good results from the new ones. The resulting monoculture, has not only reduced the rich diversity of locally adapted species, but has also adversely affected the nutritive content of local diets. The higher nutritive value of coarse grains has been forsaken in the quest for fine varieties of grain sponsored by the Public Distribution System (PDS).

A case in point is the change in cropping pattern in the mountains of the Hindu-Kush Himalayas. Traditional agriculture focused on mixed cropping, a large variety of crops being grown and consumed. All this changed when the Government introduced its 'Grow More Food' campaign and the public distribution system which limited the number of crops to three or four which could be sold through the PDS. Government help came only if the farmer grew high yielding varieties that required intensive cultivation. Mixed cropping decreased, fertility reduced and the farmers became dependent upon chemical fertilizers supplied by the Government. The impact on diet and nutrition have been equally dramatic. Food varieties that contributed to the biological diversity of the region have completely disappeared. The farmer grows only what he can sell or what he gets external support to grow. As a result, households that used to consume over twenty varieties of food items, now consume only five. The food items which have disappeared are those which were environment friendly, resilient, grown to increase soil fertility or grown at a time when the soil was being rested (Nagpal, 1999).

However, growing awareness of the negative impacts of monoculture on the food security and nutritional levels of the rural population have led to attempts at restoring traditional agriculture. Success stories from different parts of the country raise the possibility of improving rural food security through this option. The *Beej Bachao Andolan* (BBA) began in Jardhargaon Village of Tehri Garhwal in Uttaranchal in an attempt to reverse the problems creaed by monoculture (Infochange News and Features 2003b). Varieties of seeds were sought for in untouched remote locations, where the old methods of agriculture were still practiced. These practices included 'Baranaja' or twelve grains, an intercropping method where as many as twelve different varieties of cereals and legumes are planted on the same plot of land. The method uses only natural pesticides and the wide range of crops replenishes nitrogen and maintains soil fertility. With a growing collection of traditional seeds (as much as 150 varieties of paddy, for example), the method was reintroduced in the area with visible impacts. Not only is there now, year round supply of food, the risk of crop failure is reduced with the large varieties of crops being cultivated, livestock health has improved with the elimination of chemical use and the human population is reaping the benefits of a varied and balanced diet.

Self-Help Initiatives for Rural Food Security

While poverty and seasonal constraints serve to keep a large proportion of the rural population in a perpetual state of hunger, even those families who manage to climb out of the hunger trap remain vulnerable to natural hazards and disasters. For families teetering on the edge of survival, the loss of a single harvest or other assets in tems of home or livestock can serve to push them back below the hunger line. The regular ocurrence of floods, drought, cyclonic storms and other calamities in one part of the country or another needs immediate redressal, if rural food security is to be ensured. While Government relief measures exist and are put into action at the earliest, the time lapse between need and supply is often inordinately long, and many times, too late. This is particularly so in remote or relatively inaccessible areas or those rendered inaccessible by the monsoons or flood waters, etc. In such cases, the revival of the community grain gola or grain bank could protect against immediate food shortages and long-term lapse into penury and persistant hunger.

NGO attempts at reviving grain banks in the hunger-prone district of Rayagada in Orissa have met with great success and enthusiatic community participation (Infochange News and Features, 2003c). The NGO Agragamee observed that certain villages in the district never suffered starvation deaths, despite persistant droughts. On enquiry, it was found that these villages had community grain banks, which insured them against hunger. Agragamee persuaded a group of 40 families in the small hunger prone village of Ranjaguda to contribute 9 kilos of ragi to set up a grain bank. A remote village surrounded by forested hills, the villagers were illiterate except for three and lived off a single crop grown during the monsoons. For the remainder of the year, they survived on forest produce, the hunger season leading to distress migration or survival on mango kernels and tamrind seeds. With the setting up of the grain bank and the first collection of 6 quintals of grain, the villagers gained confidence and turned to other activities including the construction of a weir which has helped to irrigate 20 acres and allowed double cropping for the first time. The bank has grown from strength to strength and now has a cash savings of Rs. 5000. There is work available for longer periods of the year and no one dies from lack of food. Several such successful grain banks exist in Madhya Pradesh and Orissa.

Summing Up

The foregoing discussion has highlighted four issues, which are often overlooked in policy formulation, yet may impact considerably on the rural food security status—gender, seasonality, common property resources, bio-diversity and self-help initiatives.

That women's empowerment and equality of status is essential for economic devlopment is conceded, yet the fact that the key to family and individual food and nutrition security lies in the woman's hand is yet to be acknowledged. Enabling women through ownership rights, equal wages and education is as essential as providing nutrition supplements and health care in the effort to achieve universal food security.

The issue of seasonal scarcities, which in most semi-arid regions of the country extends for the greater part of the year, could more aptly be termed the issue of perpetual food scarcity. It also serves to prevent expectant and nursing mothers and their infants from accessing adequate nutrition when most needed thus perpetuating the vicious cycle of low birth weights, stunting and inadeqate mental and physical capacities as adults.

Common property resources, which served as insurance against seasonal hunger for the rural poor are dwindling and with them is also dwindling the already meagre amounts of animal protein, green vegetables and fresh fruit in the diet of the poor. Government-sponsored food safety nets may replace the loss of foodgrain from traditional structures, but cannot replenish the diversity of the food basket. Exacerbating this loss of diversity is the reducing crop diversification in agriculture with the practice of high-yielding monocultures firmly entrenched. The time has now come to question the necessity of imposing fertilizer and irrigation based crops in regions unsuited to these practices and on small and marginal farmers, who get caught in the debt trap dealing with small outputs versus mounting water, electricity and seed input costs. The subsistance farmer may have more to gain from the limited, but diverse outputs of multiple crops, than from high yielding varieties.

In the final analysis, food security can only become sustainable if it is built on local self-help initiatives, as can be seen from the case studies cited above. It is also evident that once the process has been initiated, it does not stop short at ensuring food security alone, but expands in many directions ultimately changing the entire quality of life in the area. Policy focus on initiating and nurturing such initiatives with the ultimate objective of making them self sustaining could provide a long-term solution to this persistant problem and make for a food secure India.

NOTES

1. An average per capita expenditure of Rs. 328 for rural areas and Rs.454 for urban areas per month corresponding to a total household expenditure sufficient to provide 2400 calories per day in rural areas and 2100 calories per day in urban areas and some basic non-food items.
2. This Section draws heavily on Ramachandran, Nira (2005) 'Women and Food Security in South Asia: Current Issues and Emerging Concerns,' Paper presented at a Workshop on Food Security, UNU-WIDER Helsinki, 12-15 October 2005.
3. Other occupations includes the gathering of firewood, cowdung for fuel,

animal grazing, *jajmani* system, etc. The *jajmani* system is a traditional form of employment in villages, where a professional household provides its services to the villagers in return for foodgrain, a piece of land or other payment. The range of services provided is varied including the services of priests, carpenters, masons, barbers, etc.

4. Barme, Catherine and Nira Ramachandran, Report of the PRA on Food Security conducted in selected states for the FAO-WFP Grain Banks Project, January 2002.
5. This section is based on a communication from Chatterjee, Ardhendu S., DRCSC, West Bengal on the Food Solution Exchange.

REFERENCES

Agarwal, Bina (2002). Are We Not Peasants Too? Land Rights and Women's Claims in India, *Seeds* No. 21, 29. www.popcouncil.org/publications/seeds/seds21.pdf

Barme, Catherine and Nira Ramachandran (January 2002). *Food Grain Bank Project: Report of the PRA Conducted in Selected States and Final Recommendations*, submitted to the WFP-FAO.

Chadha, G.K. (1992). *Non-Farm Sector in India's Rural Economy: Policy, Performance and Growth Prospects*, mimeo, CRD, JNU, Delhi.

Chambers, Robert (1995). Poverty and Livelihoods: Whose Reality Counts! IDS Discussion Paper, Insitute of Development Studies, University of Sussex, Sussex.

Chambers, Robert, Richard Longhurst and Arnold Pacey (Eds.) (1981). *Seasonal Dimensions to Rural Poverty*, Frances Pinter, London.

Chen, Martha (2000). *Perpetual Mourning: Widowhood in Rural India*, University of Pennsylvania Press, Philadelphia.

Chopra, Kanchan and S.C. Gulati (2001). *Migration, Common Propert Resources and Environmental Degradation: Interlinkages in India's Arid and Semi-arid Regions*, Sage Publications, New Delhi.

Dyer, David J. and J. Bruce (Ed.) (1998). *A Home Divided: Women and Income in the Third World*, Stanford University Press, Stanford.

FAO (1996). *Focus: Women and Food Security*, Women and Population Division. www.fao.org.

Gupta (1993). Land, Dowry, Labour: Women in the Changing Economy of Midnapur, *Social Scientist* 21 (9-11):74-90.

IDRC (2004). *Gender Inequality and Poverty Eradication: Promoting Household Livelihoods*, http://web.idrc.ca/fr/er-42966-201-1-DO_topic.html

Infochange News & Features, (July 2003 a). *BAIF Ropes In Farmers To Regenerate Community Pastureland* <http://infochangeindia.org/Agriculture>.

Infochange News & Features, (July 2003 b). *SOS: Save our seeds!* <http://infochangeindia.org/Agriculture>.

Infochange News & Features, (July 2003 c). *No Starvation Deaths in Villages with Grain Banks* <http://infochangeindia.org/Agriculture>.

Kumar, S.K. (1978). *Role of the Household Economy in Child Nutrition at Low Income*, Occasional Paper No. 95, Department of Agricultural Economics, Cornell University.

Lingamurthy, Sita V.N. (2002). Gender Inequalities in the Wages of Agricultural Labourers—A Micro Study, *The Asian Economic Review*, Vol. 444, No. 2, August.

Mukerjee, A. (2000). Perspectives of Rural Women on Food Security from a Tribal Village (in West Bengal) 1993 to 1998 (A longitudinal study through PRA), in S. Mahendra Dev, K.P. Kannan and Nira Ramachandran, *Towards a Food Secure India*, Institute for Human Development, Centre for Economic and Social Studies.

Mukerjee, Neela, Meera Jayaswal, Sumita Roy and Madhumita Parihari, *Forest as Safety Net: Listening to the Voices of the Poor from 15 Forest Villages of India*, <http://www.eldis.org/fulltext/ForestryAsSafety%20NetDevelopment%20Tracks.pdf>

Nagpal, S. (1999) Food Security in the Hindukush Himalayas, *Economic and Political Weekly*, September 18-24, pp. 2717-2720.

Raina, R.S. (2000). Countering Seasonal Rural Problems: Integrated Participatory Information System, *Economic and Political Weekly*, September 9, pp. 3335-3342.

Ramachandran, Nira (2005). Women and Food Security in South Asia: Current Issues and Emerging Concerns, Paper presented at the Project Meeting on Food Security: New Challenges and New Opportunities, UNU-WIDER, Helsinki, 12-15 October 2005.

Ramachandran, Nira (2004). Seasonal Hunger: Implications for Food and Nutritional Security in Swaminathan, M.S. and Pedro Medrano (Ed.) *Towards Hunger Free India: From Vision to Action*, East West Books (Madras) Pvt. Ltd.

Ramachandran, Nira (2003 a). Learning from Micro-Level Experiences in Food Security: The Case of Mountain Villages in Uttaranchal in S. Mahendra Dev, K.P. Kannan and Nira Ramachandran, *Towards a Food Secure India*, Institute for Human Development, Centre for Economic and Social Studies.

Ramachandran, Nira (2003 b). *The Paradox of Gender Food Security*, Paper presented at the Gender Poverty Summit, Women's Political Watch and National Council for Applied Economic Research, New Delhi, 9-11 November.

Ramachandran, Nira (2003 c). Why Women Must Be a Central Concern in Food and Nutrition Policy, *Indian Social Science Review* (submitted for publication).

Ramachandran, Nira (2002). Hunger, Illiteracy and Child Deprivation in a Tribal Region: A Study of Jharkhand in Ramachandran, Nira and Lionel Massun (ed.) *Coming to Grips with Rural Child Work: A Food Security Approach*, Institute of Human Development, New Delhi.

Ramachandran, Nira (2000). Review and Field-Testing of Guidelines for Formulating and Implementing National Plans of Action for Nutrition, Prepared for the Nutrition Planning Assessment and Evaluation Service, Food and Nutrition Division, FAO, Rome.

Rengem, Sarojini (undated). Asia Pacific Division of the Pesticide Action Network.

Smith Lisa, C. *et al.* (2003). *The Importance of Women's Status for Child Nutrition in Developing Countries*, IFPRI Research Report 13.

Spitz, Pierre (2002). Bringing Rural Child Workers to School: Seasonal and Cultural Constraints in Ramachandran, Nira and Lionel Massun (Ed.), *Coming to Grips with Rural Child Work: A Food Security Approach*, Institute of Human Development, New Delhi.

Vyas, V.S. (1995). Agrarian Structure and Sustainable Livelihoods of Tribal People in Indian Forestry in Ahmed, I. and J.A. Doeleman (1995) *Beyond Rio: the Environmental Crisis and Sustainable Livelihoods in the Third World*, ILO Study Series, ILO.

WFP-MSSRF, (2001). *Food Insecurity Atlas of Rural India*, World Food Programme and M.S. Swaminathan Research Foundation.

2

Renewable Energy and Sustainable Development

SATYA BHUSAN RATH

Introduction

At the United Nations Millennium Summit in September 2000, World leaders agreed to a set of time-bound and measurable goals and targets for combating poverty, hunger, illiteracy, gender inequality, disease, and environmental degradation. The eight goals that were eventually agreed upon are what the development community now refers to as the Millennium Development Goals (MDGs). While there is no MDG specifically on conventional energy or on renewable energy, it has been taken for granted that access to any kind of energy services is a prerequisite to the achievement of all eight MDGs. This was recognized at the World Summit for Sustainable Development (WSSD) in Johannesburg in 2002. Unfortunately, much greater quantities and much greater quality of energy services will be required to meet these goals than are presently available in developing countries. Worldwide, 2.4 billion people rely on traditional biomass as their primary source of energy and 1.6 billion people do not have access to electricity.

To ensure that a lack of adequate energy (both sources) does not become a bottleneck to achieving the MDGs, urgent action is needed to move beyond the 'business as usual' approach to energy. It is no longer sufficient to think of energy usage as being driven by economic development. The time has come to begin assessing the role of energy *versus* other inputs as a means of stimulating human development. If approached as an integrated part of MDG strategies, access to energy services can be an important instrument in helping promote economic growth, social equality, and environmental sustainability.

Underpinning the Millennium Development Goals

Energy plays a critical role in underpinning efforts to achieve the MDGs and improving the lives of poor people across the world. Energy encompasses light, heat, mechanical power and electricity from a combination of fuels—fossil fuels such as coal, gas and kerosene, plus renewable energy sources such as solar power, wind power, hydropower and biomass—and fuel technologies—ranging from traditional three stone fires to efficient, clean electricity systems. Demand for energy is a 'derived demand': no one wants energy in itself but rather for the services it can provide. The wide range of 'energy services'—cooking, water heating, lighting, refrigeration, water pumping, transport and communications, etc.—made possible by fuels and fuel technology—can have a major impact in facilitating sustainable livelihoods, improving health and education and significantly reducing poverty. Conversely, the absence of adequate, affordable, reliable, safe and environmentally benign energy services can be a severe constraint on sustainable economic and human development.

Poor people often have a limited choice of technologies that convert energy to useful services. The technologies most readily available to them are typically inefficient or low quality, so they end up paying much more per unit of useful energy service than the rich. For example, light generated from a kerosene lamp costs 70 times more than the equivalent light from mains electricity, while light from a candle costs 150 times more; power from a battery costs 10 to 30 times more than from mains electricity.

No country has substantially reduced poverty in modern times without massively increasing the use of energy from either sources, replacing human and animal labour with more convenient and efficient sources of energy and technology. Different income groups have different requirements for which they use energy and, as income rises, they meet their needs with different energy sources and different conversion technologies. Some energy sources are better suited for a particular use than others: electricity, for example, is much sought after as the most effective source of lighting and for powering motors and communication technologies; but is rarely used by poorer people for cooking. Renewable sources are better suited for rural and agricultural set-up and can provide reliable and affordable energy supplies to millions of people in developing countries.

In the recent past, there has been an excessive focus on the provision of electricity to poor people. Electricity is not always the most appropriate form of energy, nor is it the quickest or most cost-effective way of providing energy services to the poor. For example, extending an electricity grid to households in rural areas can cost seven times more than for grid electricity in urban areas. In most of Sub-Saharan Africa, less than 10 per cent of the populations are connected to electricity. It is estimated that it would take over 250 years to supply all households in Uganda with electricity at current rates of electrification. In case of South Asian Countries, particularly in India, considering the current pace it may take a minimum of another 50 years to provide electricity in the remote inaccessible villages. The very poorest people, who cannot afford to pay for energy services and rely on energy sources that they can get free of charge, can gain indirect benefits from energy services provided to communities. For example, where it is cost-effective to provide electricity (through the grid or via decentralized systems or any other energy services) to remote communities such as schools, hospitals, trade and community centres. This can benefit a wider cross-section of the community, even if they are unable to afford household energy services themselves.

Energy's important role in underpinning the MDGs is yet to be fully recognized by the international development community. Developing a new approach to energy, where energy services are acknowledged not just as a result of development, but also as an actual driver of development, will be crucial if energy is to play a more prominent role in strategies aimed at achieving the MDGs. In 2006/2007 the Commission for Sustainable Development (CSD) will look at energy and sustainable development, presenting another opportunity to deepen the focus on energy and the MDGs. As such, awareness raised in 2005 on energy's linkages to the MDGs could be further elaborated into priority capacity building, financing, technical assistance, and technology issues to be adopted through the CSD process in 2006 and 2007 as means to support the achievement of the MDGs. Breakthroughs made in linking access to energy services to multiple development objectives would need to be further elaborated in terms of technical options available, the minimum energy inputs needed to support each MDG, the associated costs, and the related policy and financial options of these alternatives.

Renewable Energy and Sustainable Development

The Potential Role of Renewable Energy in Meeting the Millennium Development Goals, in achieving sustainable development was highlighted the 2005 World Summit at the United Nations. The report, produced and published by the "WorldWatch" Institute, brings together the expertise of the participants of REN 21, which provides a forum for international leadership on renewable energy and connects the wide variety of stakeholders that came together at the Bonn International Conference for Renewable Energies in 2004. Klaus Toepfer, Executive Director of the United Nations Environment Programme (UNEP), joined the "World Watch" Institute in UNEP's Green Room to discuss the report's findings with attendants of the UN Summit. "Delivering clean, efficient, reliable and renewable energy to developing countries is absolutely critical for poverty reduction and for meeting the internationally agreed development goals," he said. "Every time oil surges over $50 a barrel, the overseas aid of many African countries—money intended for health care, schools and other vital services—is gobbled up in paying the extra fuel costs. It is also vital because, over the next few decades, the world is likely to invest some 16 trillion dollars in new energy infrastructure. We need to ensure that this is low carbon technology that gives us a better chance to fight climate change," said Dr. Toepfer.

The news sources from PTI (August 20th 2005) said that the United Nations has made a strong plea for intensified efforts to promote use of renewable energy sources, stressing that some 1.6 billion people still lack access to power and 2.4 billion depend on fuel wood, crops and animal residues for cooking and heating.

Persistent "energy poverty," as the world body said, is seriously impending socio-economic development, particularly in Sub-Saharan African and South Asia.

In a report made public on August 19th 2005, UN Secretary-General Kofi Annan observed that accelerated development and increased use of new and renewable energy sources can offer multiple benefits for sustainable development.

"Renewable sources of energy can be used to provide modern energy services to the poor, contribute to meeting the increasing global energy demand, reduce air pollution, mitigate climate change and delay the eventual fossil-fuel depletion," he said. The report concludes the UN's follow-up to the World Solar Programme 1996-2005, the outcome of the

World Solar Summit organized by the UNESCO in Harare, Zimbabwe. Annan, remarked that while the initial plan of the Programme, which included assessment studies of renewable energy potentials and initiatives to boost investment in the area, was quite ambitious, the number of projects actually funded did not meet early expectations.

A snapshot of energy's linkages to the MDG

MDG	*Energy Linkage*
Eradicate Extreme Poverty and Hunger	Energy inputs such as electricity and fuels are essential to generate jobs, industrial activities, transportation, commerce, micro-enterprises, and agriculture outputs. Most staple foods must be processed, conserved, and cooked, requiring energy from various fuels.
Achieve Universal Primary Education	To attract teachers to rural areas electricity is needed for homes and schools. After dusk study requires illumination. Many children, especially girls, do not attend primary schools in order to carry wood and water to meet family subsistence needs.
Promote Gender equality and Empower Women	Lack of access to modern fuels and electricity contributes to gender inequality. Women are responsible for most household cooking and water boiling activities. This takes time away from other productive activities as well as from educational and social participation. Access to modern fuels eases women's domestic burden and allows them to pursue educational, economic, and other opportunities.
Reduce Child Mortality	Diseases caused by un-boiled water, and respiratory illness caused by the effects of indoor air pollution from traditional fuels and stoves, directly contribute to infant and child disease and mortality.
Improve Maternal Health	Burden of fuel collection and transport all contribute to poor maternal health. Women are disproportionately affected by indoor air pollution and water—and food-borne illnesses. Lack of electricity in health clinics, lack of illumination for nighttime deliveries, and the daily drudgery and physical conditions, especially in rural areas.
Combat HIV/AIDS, Malaria and other diseases	Electricity for communication such as radio and television can spread important public health information to combat deadly diseases. Health care facilities, doctors, and nurses, all require electricity and the services that it provides (illumination, refrigeration, sterilization, etc.) to deliver effective health services.

(Contd.)

(*Contd.*)

MDG	*Energy Linkage*
Ensure Environmental Sustainability	Energy production, distribution, and consumption has many adverse effects on the local, regional, and global environment; these effects include indoor, local, and regional air pollution; local particulates; land degradation; acidification of land and water; and climate change. Cleaner energy systems are needed to address all of these effects and to contribute to environmental sustainability.
Develop a Global Partnership for Development	The World Summit for Sustainable Development (WSSD) called for partnerships between public entities, development agencies, civil society, and the private sector to support sustainable development, including the delivery of affordable, reliable, and environmentally sustainable energy services.

The Renewable Energy Policy Network for the 21st Century (REN 21) released its report, "Energy for Development: The Potential Role of Renewable Energy in Meeting the Millennium Development Goals," on 15th September 2005, identifies renewable energy options that are currently in wide-use in some regions and that are now ready for large-scale introduction in many areas of the developing world. Through 26 case studies, the report cites biogas, small hydro, solar, wind, ethanol, and bio-diesel, among other technologies, as viable options for poverty alleviation in developing countries.

At local level, renewable energy services help improve the quality of life and facilitate sustainable livelihoods. At national level, they help to facilitate stable economic development, and have an impact on the national and global environment and can affect national budget allocations. At the same time, lack of access to energy can cause conflict. Co-operation between regions over energy supply and access can help establish links between countries that can help reduce tension and conflict. Renewable Energy services are essential ingredients of all three pillars of sustainable development—economic, social and environmental. In the past, there has been greater emphasis on technical and economic issues and, more recently, on the environmental and community aspects of renewable energy, with social concerns receiving attention. To redress the balance, it is important to take a people-centred approach, looking at how renewable energy affects peoples' lives directly and ensuring that all three pillars are considered in parallel.

Renewable Energy facilitates economic development at the local level by improving productivity and enabling local income generation through improved agricultural development (irrigation, crop processing, storage and transport to market) and through non-farm employment, including micro-enterprise development. An emphasis on productive uses of Renewable Energy services is important in helping people out of poverty.

A number of statistics show the very strong association between increasing commercial energy (in both forms) consumption and human welfare, as measured by indicators like the Human Development Index (which measures life expectancy and educational achievement as well as income). Renewable Energy services have the potential to help and to facilitate basic survival activities, for example, approximately 95 per cent of staple foods (such as rice, grains and green bananas) need cooking before they can be eaten. Renewable is the appropriate sources to provide such energy. Indeed, equity of access to basic energy services for cooking, space heating and lighting, like access to water, could be considered a human right. The rights-based agenda highlights inclusion of poor people, their participation in decision-making about their development, and the responsibility of Government, as well as the poor, to fulfil obligations. A recent study on "Energy and Poverty in China" found that access to electricity made the communities feel included in the modernizing process of the national economy.

Renewable Energy can contribute towards social capital, by powering transport (bio-diesel) and communications (Solar powered telephone) so that poor people can maintain contact with their extended family and friendship groups. Renewable Energy is also used to prepare meals (Community Solar Cookers) to celebrate special events, which is important for maintaining social capital. Poor households draw on their social capital in order to cope better at times of shock or stress.

Renewable Energy is strongly linked to the environment. Many renewable energy sources are drawn directly from the environment, requiring sound management for these sources to be sustainable. Renewable Energy use also affects the environment. Emissions from fossil fuels, excess use wood biomass for example, reach beyond the local and national levels to affect the global environment and contribute to climate change. The poorest people often live in the most ecologically sensitive and vulnerable physical locations. These areas may be the most affected by the predicted effects of climate change such as increased frequency of extreme seasonal effects. The risks facing poor people are often increased by the unsustainable use of bio-resources by themselves or others. The poor may also face difficult trade-offs in their search for livelihoods: in the words of a resident of Ha Tinh, Vietnam, 'We know that cutting down trees will cause water shortages and that making charcoal can cause forest fires, but we have no choice. Because we lack food, we have to exploit the forest'.

Many dilemmas of this kind are posed by the search for environmentally responsible renewable energy services. The World Commission on Dams highlights the fact that hydropower from large dams has made a significant contribution to economic development, but at unacceptably high prices with respect to environmental and social outcomes. Learning from the past and negotiating outcomes where the interests of the most powerful stakeholders are balanced with the needs of less powerful stakeholders, and are both more equitable and transparent can address such imbalances. Use of indigenous renewable resources combined with efficient supply and use of fossil fuels, with cleaner technologies, can help reduce the environmental effects of energy use and help developing countries make a 'technology leap' that avoids 'lock-in' to more polluting fossil fuel technologies. At the same time, careful management of Renewable Energy resources, such as forests for fuel wood and water for hydropower, is important to promote economic growth, protect eco-systems and provide sustainable natural resources.

A flexible approach that recognizes that 'one solution does not fit all', should have greater success in providing renewable energy services to meet people's needs. A full menu of renewable energy options should be considered since in some cases, efficient, clean wood for gasifier will be the best option, while in others, multi-utility bio-diesel or a hybrid system may be preferable.

Access to Energy and the Impact on People's Lives

'...In South Asia, rural women use the word "darkness" to express their drudgery...anxiety, also, of knowing that this is all you have to pass on to your daughter...'

Renewable Energy or energy in general interacts with people's lives in many different ways, from the basic survival activities to increasing productivity. Extending the working hours during the evening can increase productivity. With the help of lighting and by mechanization, several activities like for example, processing and packaging of food and spices, enterprises activities like weaving and stitching can be accomplished. When communities gain access to energy services, it can have a marked effect on their lives, particularly with respect to freeing up their time, improving their health and well-being, and opening up opportunities.

Most poor people currently meet the bulk of their inanimate energy needs by collecting fuel wood and other bio-mass. This costs very little in cash terms, but is hugely expensive in terms of the time it takes. Patterns of time use typical to Indian villages illustrate the impact of the absence of energy services. Typical families spend 2-6 hours each day collecting 10-30 kilograms of wood over distances of 4-8 km. In the Drass Region of Leh in the Himalayas, women sometimes have to camp overnight when collecting fuel-wood, as the distances they need to go are getting longer.

Energy for the Poor

Access to alternative or renewable forms of energy may also affect people's health. Use of bio-mass fuels for cooking and space heating creates indoor air pollution, which has been linked with increased rates of acute respiratory tract infection (ARI) in children and women. Energy is also important in facilitating access to health care by powering transport to health clinics. Fast access to medical services is important for the health and well-being of both mother and child when there are complications during childbirth. Access to good healthcare is also reliant on an efficient, functioning healthcare system. This requires a number of elements including appropriately trained and supervised personnel, a reliable supply of drugs and essential equipment, and an appropriate physical infrastructure. Renewable Energy services can contribute in a number of ways to the efficient performance of this system in remote public health centres, for example, through ensuring reliable heating, lighting, sterilisation and refrigeration, as well as safe disposal of medical waste.

Renewable Energy for lighting allows study at night and facilitates access to learning materials through radio, TV and other ICTs. There are at least 1.2 billion people in the world that cannot read and write, with the number of women far exceeding the number of men. Literacy can improve people's employment prospects, enabling them to increase household income. Access to energy services also opens up opportunities for income-generating activities like access to markets through transport and communications, and thus a way out of poverty. An example of the importance of energy comes from West Africa, where fish processing and trading at the artisan level provide diversified employment opportunities, especially for women in fishing communities.

The Scale of the Problem

In 1997, the OECD countries consumed 54 per cent of commercial primary energy compared to 12 per cent in transition countries, 11 per cent in China and 23 per cent in other developing countries. This means that slightly more than one billion people in industrialized countries consume about 54 per cent of total commercial energy supply and around five billion people in transition and developing countries consume the remaining 46 per cent. In practice, per capita energy use in Africa has barely increased since 1970 and is less than 10 per cent of that in North America.

The number of people without access to energy services to meet their basic needs is staggering: two billion people lack clean, safe cooking fuels and must depend on traditional bio-mass sources; and 1.7 billion are without electricity. Most of these people live in Asia and Sub-Saharan Africa. To meet the basic cooking needs of the two billion people not served by modern fuels would correspond to no more than 1 per cent of global commercial energy consumption. This figure should be achievable, especially if lessons are drawn from the past.

Given the importance of energy in facilitating poverty eradication, why are there still so many poor people without access to appropriate energy services?

In June 2001, a workshop organized by FAO was held to enable artisan, fish processors and traders to identify common problems in relation to fish processing and trade at the regional level. In a group discussion, participants reflected on barriers to fish processing and trade. Overall, a significant number of problems raised by the participants related to energy, particularly in terms of processing. These problems highlight the interconnected nature of livelihood needs related to energy. The workshop identified the need to have a reliable power supply at processing sites, to promote appropriate technology for greater fuel efficiency in ways that reduce the health hazards due to smoke. Women processors identified these and to facilitate access to land to be managed by women processors as woodlots for fuel supplies, poverty and improve their quality of life. All energy services cost money, and are bought and sold in various markets. At present, energy markets are not very effective in providing energy services in developing countries. Reform of these markets and improvement of their governance arrangements are central to the strategies currently adopted by the international community to increase the supply, improve the efficiency and extend the access of modern energy services.

Four issues are particularly important for poverty reduction:

1. How to improve the effectiveness of energy sector management.
2. How to attract more capital in order to increase access to energy.
3. How to make subsidies and other government policies work for poverty reduction.
4. How to make energy services affordable.

The challenge is to create a policy framework that allows Renewable energy to be introduced where it makes long-term economic sense, and to allow poor countries to benefit from the declining cost curves that characterize renewable energy markets in many industrial countries. Among the policies that have proven their ability to advance renewable.

New sources of capital are necessary to meet the growing demand for energy project financing in developing countries. It is currently difficult for RET project and companies to attract investment because they tend to be small transactions, require more time and effort to bring to investment quality than large conventional projects, and often involve less familiar technologies.

Innovative financing mechanisms, particularly when integrated with targeted confessional funding, can help to make renewable energy technologies more attractive for commercial funding and more accessible to customers. Several such mechanisms have been developed to complement cash sales. They include micro-credit from the company, potentially through dealers, or from a third party, such as a micro-finance institution or community-based organization.

Micro-credit approach can involve specially designed credit lines through larger financial institutions, potentially backed by the government. The new financing mechanisms also include micro-leasing or fee-services approach. For example, a utility-style concession in which concessionaire owns the system, charges a monthly fee, and is responsible for the services.

In order for Renewable-Energy services to advance the MDGs, they must be part of multi-sector oral approach, coupled with existing development activities for water, health, education, and enterprenurship. In particular, project planning should address poor women's needs for reducing labour, improving health, and providing security and income. In many developing countries, reducing indoor air pollution deserves particular attention.

In many situations, the largest consumers of energy services are big industrial unit and government. As Governments recognize with their own purchasing, the values of RETs, costs can be reduced through economies of scale, initial investment costs can be shared with private citizens and companies, awareness of renewable energy's potential can be more widely disseminated. By integrating renewable energy into rural electrification, rural development, poverty alleviation, and social welfare programmes, the effectiveness of these programmes can in many cases be advanced.

Renewable Energy and India

India's search for alternative source of energy that would ensure sustainable development on the one hand and energy security on the other began in 1970 itself.

The importance of the increased use of Renewable Energy to meet the increasing energy demand in a sustainable and environmentally sound manner was recognized in the early seventies. In fact, the renewable energy technologies (RETs), such as, bio-gas plants and improved cook stoves were available in India even in the late forties, though the renewable energy programme started in earnest only after the creation of CASE (Commission on Additional Sources of Energy) in 1980, and then the DNES (Department of Non-conventional Energy Sources) in September, 1982. Over the last twenty-five years or so, the initiative on increased use of Renewable Energy in various sectors and applications strengthened significantly. A continued thrust towards wider use of renewable energy devices at domestic, community, commercial and industrial levels not only resulted in greater awareness but also significant installed capacities. As shown in table above, a variety of renewable energy technologies are being used.

During the eighties, the DNES programmes were driven by direct Government subsidies and focused on the development, dissemination, and demonstration of various RETs. However, in comparison with the conventional sector, these programmes were low in investments in RET promotion. Between 1980 and 1992, the cumulative government expenditure for the renewable energy sector totalled only Rs. 11.55 billion, as compared to Rs. 812 billion for the power sector, Rs. 335 billion for the petroleum sector, and Rs. 158.5 billion for the coal sector (GoI, 1996). Likewise, in the Eighth Plan (1992-97), allocations for renewable energy were about 0.8 per cent of the total funds allocated for the energy sector (GoI, 1996). In the late eighties and early nineties realization grew that somehow private sector had to be involved to achieve greater penetration of renewable in the energy sector. Instead of pushing renewable in a limited way through a subsidy driven programme, one felt there was a need for appropriate policy framework and fiscal benefits that created an environment encouraging private sector players to invest in renewable energy projects. It was envisaged that the role of the DNES had to change from that of an implementing agency to a facilitator. As a result, the DNES was converted into a full-fledged Ministry (Ministry of Non-conventional Energy Sources, or MNES) in July 1992. Since 1992, the Ministry's initiative has been largely in the direction of creating an environment that facilitated greater penetration of renewable through a market led approach.

While the frequency of rise in oil prices is increasing each decade, the possibility of a downward trend in global oil prices seems bleak. Energy security of India, as an issue of national strategic importance, came to take the centre stage of the planning process against the backdrop of frequent rise in global crude oil prices. Our strong dependence on fossil fuels will continue for most part of the 21st century. National energy security has an important bearing on achieving national economic development goals and improving the quality of life of the people. The level of per capita energy consumption has for long been considered as one of the key indicators of economic growth. The continued dependence of the nation on fossil fuels is loaded against it with inherent price volatility linked to finite global reserves. In addition, global warming, caused largely by greenhouse gas emissions from fossil-fuel generating systems, is also a major concern. India needs to develop alternate fuels considering the aforesaid two concerns. In addition, indiscriminate and inefficient burning of fuel wood in traditional *chulhas* and in industries for thermal process heat has resulted in environmental pollution and health hazards.

To free the villager from the energy insecurity she faces, the Village Energy Security Programme is being conceived to exploit the opportunity offered by the technologies developed indigenously involving coordinated participation from several Ministries such as Power, Environment and Forests, Rural Development, Planning Commission and State Governments. Several meetings, workshops and seminars have taken place to discuss the modalities of such a programme. The Ministry plans to take up pilot projects in different States to demonstrate the concept, provide operational experience and test different institutional structures.

The village energy security plan based on bio-mass conversion has been mooted by the Union Ministry of Non-conventional Energy Sources in an attempt to equip villages to satisfy their energy needs. As many as

1.73 lakh villages located in fringe areas of forests and another 25,000 remote and un-electrified villages will be initially covered in the programme. For a village unit of 100 households, around 10,000 units of electricity are needed every year. An investment of Rs. 20 lakh @ Rs.20,000 per house hold would be required to meet all the energy needs of the village with the help of four dung-based plants, four leafy mass-based plants and two bio-mass gasifier.

Energy experts feel that to give a boost to the plan, the Government should revive the Wasteland Development Corporation, which earlier gave financial assistance of Rs. 4,000 per hectare for wasteland development. The entire wasteland in the country can thus be brought under bio-mass/bio-diesel plantation, which in turn will sustain the energy security plan. It is essential to note that no irrigation will be required for cultivating bio-mass/bio-diesel plantations. All the tree varieties are rain-fed (*Times of India*, 24th March 2005, Pune).

Now it is high time to give utmost importance to the renewable energy. Our country is heading towards an era of blackouts and burnouts and renewable energy source is the only solution. Currently the wind energy sector was growing at 30 per cent and felt that India needed to tap its solar energy potential since solar power from one sq. km. area can generate 20 MW power. At the same time it is also important to define the role of renewable energy in context of rural India, because if we are seriously thinking of energy security, renewable has a grater role to play in the rural sector. The Indian rural scenario is rich with natural resources. Looking at the previous experiences of Ministry and State Nodal Agencies, it is important to design an effective management system at the grassroots level giving priority to people or end users participation at every stage and hybridizing of technology on the basis of availability of resources and financial capabilities of users. Therefore, it is also very essential in built the resources mobilization, social capital and livelihood aspects in the energy security domain.

Rural India and Renewable Energy Need for Sustainable Development

Today India, a country with more than a billion people, faces the challenge of achieving growth and development in a sustained manner.

Economic growth and development call for huge capacity additions in the energy infrastructure of the country. The challenge is in achieving the developmental objectives without adversely impacting the environment, natural resources, wildlife and climatic conditions. Presently, fossil fuels, mainly coal and oils are being used in the commercial sector while bio-mass is being used in an inefficient manner in domestic and rural sectors. However, limitations of fossil fuels are all too obvious. Coal reserves will last a little over 200 years and the oil reserves would have dried up much before that. There is a need for a shift in our national energy strategy. More compelling than the diminishing reserves are the environmental reasons. Today, development based on commercial fuels with current rates of pollution and deterioration in natural resource base is not sustainable. Threat from GHG (Green House Gas) has caused world-wide concern. Kyoto Protocol, agreed at the Conference of Parties to the Framework Convention to Climate Change, in December 1997, is an indicator of global resolve to address this concern. In India electric power generation is the largest source of GHG emissions and accounts for 48 per cent of the carbon emitted. These concerns point towards more rational energy use strategies.

Perhaps, renewable energy based technologies, functioning in a sustainable manner is the way forward. The country having diverse climatic regions is well endowed with renewable energy resources. From the woody tropical forests of Andaman's to Sunny mountains in Ladakh, renewable energy can play a crucial role in meeting end use energy needs in a decentralized manner. In a country as vast as India, Renewable Energy is seen as an effective option for ensuring access to modern energy services. Dwindling fossil fuels, the impact of oil imports on foreign exchange reserves and the national energy security concerns are additional stimulants for greater thrust on renewable energy.

Rural Development in India and Energy Security: Role of Renewable Energy

The energy needs of the village, be it for cooking, motive power (operating pumps) or electrification, can be met through bio-mass conversion. Oil-producing trees or leafy mass can be used to run gasifier to produce energy which will help over 77 million people, who depend on kerosene for lighting, he said. For example, since the past few years over 2,000 engines were being run on bio-gas/bio-diesel in villages of Karnataka. The State Road Transport Corporation has been running three buses on Pongamia Oil (bio-diesel).

Plant species like *acada auriculiformis*, *acacia nilotica* and *bamboo* species can provide the bio-mass. Charcoal briquettes, vegetable oil, ethanol and organic waste can be used to create energy for the village. Wasteland in villages can be used for cultivating bio-mass plantations of non-edible oil bearing plants, which can supply oilseeds and bio-mass for the energy security plan of the village, this will also create employment opportunities for the local youth and women. On the economics of the energy security plan, the investment can be raised with the help of government grants and soft loans. The household contribution for maintaining the energy plant of the village would range between Rs. 65 and Rs. 120 per month.

In the rural, demand for energy (electrification) is a 'derived demand': mostly as a secondary because the rural people do not want energy in itself but the 'energy services' it provides—cooking, lighting, heating, water pumping or irrigation, transport, etc. A full menu of options should be considered for providing energy services to the poor, renewable energy sources, traditional bio-mass, and fossil fuels combined with cleaner, more efficient energy technology. The benefits of cleaner technology need to be balanced against the benefits that other technologies can provide. Analyzing the above facts Participatory Renewable Energy Management (PREM) should be the approach to overcome the barriers, which in turns ensure Sustainable Adoption of Renewable Energy Technology Application.

Participation of Primary Stakeholders: Evidence shows that if primary stakeholders are involved in the design and implementation of development initiatives they are much more likely to prolonged benefits. Local community possessed invaluable local expertise that should be taken into account in defining and implementing any energy project. As the end users of the technology, their involvement at early stage of planning will generate ownership that helps achieve long-term success. Project characterized by high levels of community engagement will typically generated a grater sense of community empowerment, ensures that improvements are tailored to a community specific needs, and create a much higher chance that the improvements will be well maintained by the community after installation.

PREM ensures to enhance livelihood options, such as forest and large plantation based enterprises, edible and non-edible oil exploration, products processing in the extra illumination hours. Plantation management and bio-mass production for energy itself can be assured as viable means of income for women. A viable means of livelihood ensures sustainability to bio-mass based rural energy programme.

Renewable Energy (RE) based Rural Electrification in a development context is not about technology provision, first it is about understanding the role that RE services play within people's lives, and responding to constraints in improving livelihoods. RE must be considered as one of the facets of service provision affecting the livelihood opportunities of the poor, and understood in terms of how the poor value and use it.

Today, when we are discussing on energy security of our country we must share the views of our scientist, His Excellency President of India, Dr. A.P.J. Abdul Kalam. On his address to the Nation during 59th

Independence Day he stressed upon that the energy security and its transition to energy independence is the call of the hour.

While dedicating his speech on 14th August evening, he said:

> "Energy Security rests on two principles. The first is to use the least amount of energy to provide services and cut down energy losses. The second is to secure access to all sources of energy including coal, oil and gas supplies world-wide, till the end of the fossil fuel era, which is fast approaching. Simultaneously we should access technologies to provide a diverse supply of reliable, affordable and environmentally sustainable energy."
>
> "Energy Security, which means ensuring that our country can supply lifeline energy to all its citizens, at affordable costs at all times, is thus a very important and significant need and is an essential step forward. But it must be considered as a transition strategy, to enable us to achieve our real goal that is—Energy Independence or an economy, which will function well with total freedom from oil, gas or coal imports. Is it possible?
>
> Hence, Energy Independence has to be our nation's first and highest priority. We must be determined to achieve this within the next 25 years, i.e. by the year 2030. This one major, 25-year national mission must be formulated, funds guaranteed, and the leadership entrusted without delay as public-private partnerships to our younger generation, now in their 30s, as their lifetime mission in a renewed drive for nation-building."
>
> "The strategic goals for Energy Independence by 2030 would thus call for a shift in the structure of energy sources. Firstly, fossil fuel imports need to be minimized and secure access to be ensured. Maximum hydro and nuclear power potential should be tapped. The most significant aspect, however would be that the power generated through renewable energy technologies may target 20 to 25 per cent against the present 5 per cent. It would be evident that for true Energy Independence, a major shift in the structure of energy sources from fossil to renewable energy sources is mandated."

It is evident that, India does not have any alternative unless it brings the renewable energy resources to center stage of national energy policy, planning and implementing process to secure its energy resources in a sustained manner.

The argument is that, if rural India is capable of generating alternative resources to address their current need, why we are imposing them to use fossil fuel, this will end up in resulting a huge foreign exchange burden from imported petroleum products. According to India Development Report 2004: Rural households in India consume about 314 million tonnes of bio- fuels in a year only for cooking. These bio-fuels are mostly procured through gathering. In a year about 352 hours are spent per household on gathering of bio-fuels. An estimated 85 million households spent 30 billion hours on gathering 314 million tonnes of bio-fuels per annum. While this provides livelihood to some of the poor who sell these in local markets, most household gather these for self-consumption.

Besides the drudgery of collection and transportation of these fuels, about 174 million females are exposed to an average 6.8 mg/cum of respirable Suspended Particulate Matter (RSPM) during cooking. Those who are in close proximity to the stove are some time exposed to even higher level of pollution (9.2mg/cum) than the person doing the cooking.

The bellow tables give statistical views of household energy scenario of rural India.

Household Energy Scenario in Rural India

- 96 per cent household use bio-fuels, 11 per cent use kerosene and 5 per cent use LPG for cooking. Most of them however use multiple fuels.
- Forest contributes 39 per cent of the fuel wood needs.
- 314 million tonnes of biofuels are gathered annually.
- 85 million households spend 30 million hours annually in fuel wood gathering.
- 15 million households use kerosene for cooking and light.
- 34 per cent out of 15 million kerosene users procure from open market per cent procure from fair price shop.
- 49 per cent households are willing to pay more than the market price to purchase kerosene for cooking.
- 0.2 per cent households use bio-gas for cooking.
- 63 per cent households are electrified.
- 33 per cent household are willing to contribute to reduce kitchen smoke through installing better technology.

Source: India Development Report, 2004.

Estimated Energy cansumption Pattern and Time and Efforts for Gathering

Items	*Estimate*
Use of any bio-fuels (household in Mn)	125.9
Fuel wood using household (Mn.)	103.9
Dung cakes using household	83.1
Croup residue using household (Mn)	53.6
Quantity of bio-fuels used per year. (Mn.Tonnes)	
Time and effort in bio-fuels gathering	
Bio-fuels gathering Household	85.1
Time spend on biofuels collection (Man hrs. per year)	29,998.0
House hold using petroleum product for cooking	
Household using kerosene for cooking (Mn)	14.9
LPG/Biogas using household (Mn)	7.4
Quantity of kerosene used for cooking exclusively (firing) and lighting (Mn. ltrs p.a.)	211.0
Quantity of kerosene used for cooking as mixed fuel, (firing) lighting per year (Mn. Ltrs.p.a.)	1233.0
Quantity of kerosene used only for lighting and the fire per year (Mn. ltrs p.a.)	6561.0
Cooking Involvement	
Involvement as chief cooks (Mn. Females)	173.8
Exposure for cook (No. of hrs. cooking)	1000 hrs. p.a.

The crucial need at this point is to bring technology, knowledge and experience together to provide right solution to the rural energy scenario. Undoubtedly renewable energy technology is the right option to provide a security to national energy resources. Not only that, sustainability can also be achieved.

The facts and figures given in this paper suggested that, currently Indian rural energy scenario is selling without a navigator. At the same time rural India has enough of renewable energy resource potential and traditional knowledge to utilize it. It is clear that renewable energy is the only way out, which can navigate to reach at energy independence, which means a way to sustainable development.

The policy-maker, technologist, development advocates and corporate must join hand together to make Energy Independence as national movement to achieve energy security. Secured energy being the prerequisites to achieve the MDG which sets at the down of the millennium, renewable energy is the only source which will ensure sustainable and secured energy for development.

REFERENCES

Achieving the Millennium Development Goals: The Role of Energy Services Case Studies from Brazil, Mali and the Philippines. UNDP 2005.
Energy for the Poor Underpinning Millennium Development Goal: DFID 2005.
MNES Annual Report—2004-05
www.renewingindia.org
www.presidentofindia.org
http://www.ptinews.com/pti/ptisite.nsf
http://timesofindia.indiatimes.com/
www.ren21.net

3

Facets of Rural Economy and Development of SMEs

A Paradigm Shift is Imperative

B. YERRAM RAJU

Introduction

Macroeconomic fundamentals of Indian Economy exuberate confidence like never before. GDP growth is on the upward trend since 1994-95 when it reached 5.4 per cent. Despite fluctuating monsoon trends interspersed with natural calamities in Maharashtra, in July 2005, Orissa and Andhra Pradesh, in September 2005, RBI estimated the current year's growth rate at around 6.9 per cent. The Quick Estimates of Index of Industrial Production (IIP) with base 1993-94 for the month of July 2005. The General Index stands at 212.1, which is 6.7 per cent higher as compared to the level in the month of July 2004. The Indices of Industrial Production for the Mining, Manufacturing and Electricity sectors for the month of July 2005 stand at 146.9, 224.3, and 184.1 respectively, with the corresponding growths of (-)0.4 per cent, 8.3 per cent, and (-)1.2 per cent as compared to July 2004. The cumulative growth during April-July, 2005-06 over the corresponding period of 2004-05 in the three sectors have been 3.1 per cent, 10.4 per cent and 5.3 per cent respectively, with the overall growth in the General Index being 9.3 per cent.

Indian Rupee appreciated against US Dollar and also unfazed by China moving to floating currency regime. Forex reserves have steadily been on the rise to touch around $144bn. FDIs and FIIs are also on gradual increase; all this, and much more. The Stock Markets have been on rising indices crossing 8000 mark. May be the budget season February-March

2006 could see it touching 10000 mark with steady P/E ratios. This exuberance would at once be short-lived when we watch the poverty scenario. Percentage of poor has, no doubt, been on the decline (from about 35 per cent in 1990s to 26 per cent in 2003-04) providing lever to the growth economists that the decline occurred mainly due to the improvements in general health of the economy, while the actual numbers have increased to around 357 million people (Population of the country at the time of Independence was just 250 million). We house 33.33 per cent of the world's poor and 70 per cent of such poor remain in Rural India. The rural-urban divide would seem to be widening during the last five decades despite ten times growth in urban population compared to three and half times growth in rural population. "It is often said that India lives in two different worlds separated, it seems some times by centuries. The reality of a globalised, industrialized India is as true as the reality of the bullock cart and the uneducated farmer. We are today both a developed and a developing nation at the same time. Extremely advanced forms of agriculture and industry coexist with other forms from a bygone era. Bridging this gap is the ultimate development challenge of the 21st century" (Manmohan Singh, 2005). Attention on the rural areas seem imminent not merely from the angle of vote bank but also from the points of view of stability and growth accompanied by peace. This is where rural full employment becomes an issue of great concern.

Still, the growth ended in increasing the jobless numbers in this country. Employment scenarios presented even by those who campaigned "India Shining" before the present Government came to power shook their seat of power. The new Government seized the opportunity to enact Rural Employment Guarantee Bill as "War on Poverty", that got President's assent on the 19th September 2005 although this is unlikely to impact efficiency-led cuts leading to deceleration in employment.

This is the *raison d'etre* for choosing the theme of this conference as "Rural India".

The 'Poor' comprising of around five billion people in the world always fascinate the rich, the researchers, and most to the politicians. Being at 'bottom of the pyramid' (Prahlad, 2005), they offer opportunities to corporates to discharge their 'social responsibility'; they are emerging markets for the flourishing private sector; they constitute the stable vote bank for the politicians; and they open windows of imagination and innovation to interact on a sustained basis. The National Sample Survey Organisation (NSSO) estimated that 26.1 per cent of the population was living below the poverty line in 1999-2000, down from 51.3 per cent in

1977-78. The criterion used was monthly consumption of goods below Rs. 211.30 per person for rural areas and Rs. 454.11 for urban areas. About 72 per cent of the poor are in rural areas (27.1% of the total rural population) with most of them comprising daily wagers, self-employed households and landless labourers. The major causes for poverty are unemployment or under-employment, low ownership of assets (especially productive assets like land and farm implements) and illiteracy.

Since the early 1950s, successive governments have implemented various schemes, under planning, to alleviate poverty that have met with partial success. All programmes have improved upon the strategies of the *Food for Work* programme and *National Rural Employment Programme* of the 1980s, which attempted to use the unemployed to generate productive assets and build rural infrastructure. S.S. Bhalla *et al.*, argue that the poverty estimates of both the Planning Commission (26%) and the World Bank (38%) are far above the actual level of 13 per cent.

The UN Report on the Millennium Development Goals (*See Note 1*) acknowledges "It is about unleashing the power of local entrepreneurs to reduce poverty in their communities and nations." This single sentence needs to be read carefully. What reduces the poverty? "The power of local entrepreneurs". Where it gets reduced? "In their communities and nations". How it is reduced? "By unleashing their power". The examples in the document did not, however, come from the poor graduating to the next tier or even a few tiers up above. But the goals set before the millennium and the means to reach the goals have attracted the attention of developing nations significantly. The renowned management guru Prahlad bothered rightly about 'the poor that became the most acceptable wards of the State'. "The problem of poverty must force us to innovate and not claim rights to impose our solutions." New and creative approaches are needed to convert poverty into an opportunity for all 'concerned'. It is this challenge that made me look at the issues, interventions and their effectiveness in pursuit of the goal of 'eradicating extreme poverty', from the Indian perspective. Extreme poverty leading to starvation, deprivation and death can be eliminated while the relative poverty would continue. Since both these are largely prevalent in rural India, attention of this paper is set on rural India. Solutions have to be sought where the problem exists and not else where. When we strategize into the future it is extremely important to envision it properly, review the past with a view to draw lessons and draw up an action plan not as a one time effort but as a dynamic intervention.

Employment Scenario presents not an encouraging picture at the moment. Agricultural and allied sectors accounted for about 57 per cent of the total work force in 1999-2000, down from 60 per cent in 1993-94. While agriculture has faced stagnation in growth, services have seen a steady growth. Of the total workforce, 8 per cent is in the organised sector, two-thirds of which are in the public sector. The NSSO survey estimated that in 1999-2000, 106 million, nearly 10 per cent of the population were unemployed and the overall unemployment rate was 7.32 per cent, with rural areas doing marginally better (7.21%) than urban areas (7.65 %).

Unemployment in India, like most other developing countries, is characterised by chronic under-employment or disguised unemployment. Government schemes that target eradication of both poverty and unemployment, attempt to solve the problem, by providing financial assistance for setting up businesses, skill honing, setting up public sector enterprises, reservations in governments, etc. The decreased role of the public sector after liberalisation has further undermined the need for focusing on better education and has also put political pressure on further reforms.

The National Common Minimum Programme (NCMP) of the Government attaches high priority to the development of social sectors to enable people to participate in and benefit fully from the development process. Major programmes initiated in this regard in 2004-05 include launching of the National Food for Work Programme in 150 most backward districts; provision of additional Rs.12,000 crore of Gross Budgetary Support (GBS) for Plan programmes like Food for Work, *Sarva Shiksha Abhiyan*, Mid-day Cooked-meal Scheme, Basic Healthcare, Accelerated Irrigation Benefit Programme, Drinking Water, and Roads; increase in coverage of *Antyodaya Anna Yojana* from 1.5 crore to 2 crore families; improving facilities for universal education by imposition of a cess of 2 per cent on Union taxes and duties; upgrading Industrial Training Institutes (ITIs) over the next 5 years; and introduction of a new Universal Health Insurance Scheme for the poor and a special group insurance scheme. The Outcomes of the Budget and the outputs are subject to a systemic review. This still falls short of addressing the low productivity and large-scale corruption plaguing the Government and public sector employment. Parliament would require a stronger resolve to solve such core issues to ensure that the development initiatives bear fruit to the desired levels.

About 18.7 million people work in Government, about 3 million in the organized private sector and another 353 million in the unorganized sector making a total of 375 million people working in India. There are a section of employees pampered by the Constitution under Article 211. All of them work in the Government and Public Sector. The productivity-pull of this section is a drag on the economy. In order to address the issue of rural unemployment, the Indian Parliament passed the *Rural Employment Guarantee Bill* in August 2005—the largest programme of this type, in terms of cost and coverage, which promises 100 days of minimum wage employment to every rural household in 2000 of India's 600-odd districts. The Bill despite raising debate over its curative effect left a hope among the rural unemployed youth and has been lauded by one of the noblest sons of India, Nobel Laureate Amartya Sen as a novel and bold intervention to resolve the poverty issue.

The Economic Survey 2004-05 of Government of India rightly raised five core issues that need to be addressed to step up investments. The first one relates to Public Investment in Agriculture and Allied Sectors that produce nearly 21 per cent of GDP supporting nearly 57 per cent of population. The second issue relates to Simplification of Procedures and Relaxing Entry-Exit Barriers. Even as recent as September 2005, the World Investment Report noted that while it takes 89 days for setting up an enterprise in India, it is just two weeks in Singapore and four weeks in China. The third issue relates to easy and ready Access to Finance by farmers, small and medium size entrepreneurs. While some policy announcements were made during the year, BASEL II compulsions put the bankers in jitters. The fourth issue relates to Infrastructure and its Inadequacies, more particularly in the area of roads, communication networks, transport networks and post-harvest technology infrastructure (*See Notes*). The fifth issue relates to the Foreign Investments—FDI and FII. During the current fiscal several steps have been taken to encourage FDIs and FIIs even in the wake of objections from the left wing in the UPA. The Right to Information Bill enacted in June 2005 hopefully responds to the information appetite once the rules framed under the Bill fulfil the objectives of the Bill. Entrepreneurship, employment and efficiency together should provide answer to endemic problems of Rural India. This paper would like to analyse these issues in sufficient detail to design project interventions leading to sustainable development. ***Section-I*** deals with Rural India and what it constitutes; vision and constraints and the progress thus far. ***Section-II*** deals with the issues of employment and entrepreneurship embedded in the SME sector along with the policy interventions so far and the way forward. ***Section-III*** deals with projectizing PURA—the sustainable interventions to ensure rural growth with the setting up of Rural Entrepreneurship Nucleus Estates (RENE).

SECTION-I

Rural India is beyond Agriculture

Omkar Goswami and Rama Bijapurkar have rightly made out that Rural India is much more than agricultural Bharat. Even in 1993-94, the Central Statistical Organisation's rural-urban classification of India's net domestic product (NDP, which is GDP minus depreciation) showed that rural India accounted for 54 per cent of the country's NDP, while agriculture contributed to 31 per cent of India's GDP. The share of rural economy is more or less the same seven years later at 52 per cent according to their analysis in 2000-01. "This is how rural NDP played out in 2000-01: agriculture accounted for 46 per cent; industry took up another 21 per cent; and services 33 per cent. Between 1993-94 and 2000-01, rural NDP at constant prices grew at an average of 6.2 per cent per year."

According to the Annual Survey of Industries, even in 1993-94, rural India accounted for 29 per cent of the country's organised manufacturing units; 30 per cent of its employees; 32 per cent of its output; and 30 per cent of its net value added. In 2000-01, over a rising base, rural India could speak of 36 per cent of organised manufacturing establishments, 38 per cent of its employees, 43 per cent of output; and 41 per cent of net value added.

In 2001, almost a fifth of rural households owned TV sets, and the rural regions of 10 States were significantly above the national average. Between 1991 and 2001, the percentage of rural households with electricity connections rose from 30 per cent to 43.5 per cent and 11 States ranked well over the national rural average. "In 2001, over 30 per cent of rural households kept their funds in banks or postal savings accounts."

(i) Rural India's Future: Vision and Constraints

Parties in power might have changed. VISION 2020 prepared by the Planning Commission being a dreaming exercise should not give rise to much difference between its architects and the present rulers. As a citizen, I would like to dream of an India where peace and tranquillity prevails, where transparency in governance exists in all fields, where there would be 100 per cent food security and 100 per cent self-sufficiency for food, where market forces do not devour the poor and the weak in society, where better water and farm management would lead to better employment and least migration to urban and metropolitan areas from the rural areas, where population growth would not stand as an impediment for further growth of the economy, where all the employable get fully employed and less employable would be endowed with appropriate skills and knowledge for full employment, where there is free entry and exit for firms in the economy with no parasites, where values of life fall in tune with the culture and ethos of the nation and where the digital divide between the rural and urban vanish, where information asymmetry and moral hazard do not exist and where all the sectors of the economy realize their mutual dependence to their mutual benefit and the growth rate of the economy would move to a double digit figure as a matter of practice. I recall what Swami Ranganadhananda said once: "I look forward to the day when rural people stop easing in public and start eating in public". The statement is highly profound and carries with it an agenda for action: provision of good sanitation, safe drinking water, crossing the caste and other societal barriers and food within the reach of all. This would also call for a synthesis between social and economic budgeting.

The barriers to realizing such vision would be:

- Lack of political will;
- High population growth;
- Poverty and low level of literacy;
- Institutionalization and harmonization of legal aspects to set up monitoring systems;
- Inadequate resources;
- Improper structural plans;
- Deficiencies in implementation.

Some of the strengths recognized world-wide are:

- A middle class estimated at 300 million out of a total population of over 1 billion providing a stable market;
- The second largest English-speaking scientific, technical and executive manpower in the world;
- An abundant supply of raw materials;
- An extensive rail and road network;
- A stable political system based on parliamentary democracy;
- A common law legal system with English as a court language;
- India is emerging as a major market and investment destination;
- The economic reforms initiated in 1991 have brought dramatic changes in international investment in India;

- The U.S. Department of Commerce has identified India as one of the world's top ten "Big Emerging Markets".

Another Major Strength is India's Ability to Respond to Crises

When there was crisis in meeting the food requirements in the backdrop of colonial misrule, severe famine and large patches of drought, we fought it out valiantly through Green Revolution and made India self-reliant in food; we are now on the threshold of food exports; when we had a crisis in Foreign Exchange and we ably steered through; most of the natural calamities—recurring floods in several States or hard-hitting recurring cyclones in Andhara Prdesh, Tamil Nadu; earthquakes of Latur in Maharashtra or Bhuj in Gujarat; the Tsunami of 2004 in Tamil Nadu—have been ably handled with domestic resources etc.

Gross inadequacies are noticed in terms of value addition due to inadequate attention to crop specific infrastructure and post-harvest technologies like pre-cooling, cold storages with assured power at uniform voltage, price hedging operations, market reforms in farm sector etc. Some States have initiated special studies in this regard to prioritize their investments in these areas and deploy the needed resources. The impacts of these initiatives would be felt in due course. However, there is a regulatory overhang in India with more than twelve Union Ministries, corresponding State Ministries, laws framed by the Union Government with rules framed by the State Governments for implementing them, still seventy-odd food control orders governing the production and trade of those commodities and crops into which the farmers would like to diversify, the farmer, rural industry and farm trade are virtually strangulated. While there is an awakening in respect of these areas, the speed of reforms and actions in these areas deserve an urgent attention (Raju, 2005).

Farmers benefit from more accurate weighing, faster processing time, and prompt payment, and from access to a wide range of information, including accurate market price knowledge, and market trends, which help them decide when, where, and at what price to sell. Farmers selling directly to ITC through an e-Choupal typically receive a higher price for their crops than they would receive through the *mandi* system, on average about 2.5 per cent higher (about US$ 6 per ton). The total benefit to farmers includes lower prices for inputs and other goods, higher yields, and a sense of empowerment. The e-Choupal system has had a measurable impact on what farmers chose to do: in areas covered by e-Choupals, the percentage of farmers planting so has increased dramatically, from 50 to 90 per cent in some regions, while the volume of soya marketed through *mandis* has dropped as much as half. At the same time, ITC benefits from net procurement costs that are about 2.5 per cent lower (it saves the commission fee and part of the transport costs it would otherwise pay to traders who serve as its buying agents at the *mandi*) and it has more direct control over the quality of what it buys. The system also provides direct access to the farmer and to information about conditions on the ground, improving planning and building relationships that increase its security of supply. The company reports that it recovered its equipment costs from an e-Choupal in the first year of operation and that the venture as a whole is profitable. It received the Development Gateway Award of $1lakh—ITC is the first company in India to have received this Award.[3]

Nearly 700 million Indians live in 600,000 villages across rural India. Connectivity of village complexes providing economic opportunities to all segments of people is an urgent need. We need to innovate to increase connectivities to the villages, making clusters out of them even while retaining their individuality. The integrated method that will bring prosperity to rural India is called PURA or Providing Urban Amenities to Rural Areas. This envisages four connectivities: physical connectivity through quality roads and transport; electronic connectivity through telecom with high bandwidth fibre optic cables; knowledge connectivity through education, skill training for farmers, artisans and craftsmen and entrepreneurship programmes.[4]

Varghese Kurien, Chairman of IRMA said, "Our experience so far suggests that the key to real development lies in enabling people to get the instruments of development in their own hands."[5]

The World Bank's first India Development Policy Review (DPR) pointed out that uneven development had resulted in poverty becoming concentrated in the slower growing states, with more than half of India's poor now living in four States—Uttar Pradesh, Bihar, Madhya Pradesh and Orissa.[6]

"India has undoubtedly done well in the last two decades, but with one-third of the world's poor and a billion people, it needs rapid growth and job creation to reduce poverty further and sustain the recent income increases. In the absence of comprehensive reform, growth will at best be moderate," said Michael Carter, the World Bank's Country Director for India.

The other challenges India faces are of improving the investment climate, and raising productivity in industry, agriculture, services and rural development. High levels of support for agriculture in developed countries also make Indian policy-makers cautious in considering either unilateral or multilateral reforms to agriculture. The Prime Minister called for stepping up public investments in the farm sector and irrigation significantly; rain-fed area authority had been announced to ensure development of 60 per cent of the area treated as dry and rain-fed; convergence of resources of the farm sector with those of the Bharat Nirman (Infrastructure Development Scheme) and the National Rural Employment Guarantee Scheme and ensuring quick reforms to agricultural marketing through appropriate monitoring mechanisms. (Manmohan Singh, September 27, 2005).

(ii) The Climate for Private Investment in Indian Agri-business

Despite faster growth and increased diversification in consumer food demand, available data suggest only sluggish growth in Indian farm output during the 1990s. Despite some gains in private domestic and foreign investment in agriculture in the 1990s, there is growing concern with the relatively low levels of both public and private investment in agricultural production, processing, and marketing. A number of factors have been identified that may be constraining private investment, including regulations governing domestic movement, storage, and marketing of major commodities, plant-scale restrictions in food processing, and restrictions on contract farming and land leasing.

SECTION-II

Issues of Employment and Entrepreneurship

'SMEs tend to be engines of job creation—seedbeds for innovation and entrepreneurship.'[7] Economies of scale no longer remain decisive in sustaining competitiveness, as they are being replaced by economies of scale, speed and networking. In most South Asian economies, if the first generation entrepreneurs suffered the wayward incoherent industrial policies, the current generation is suffering for want of a proper and sound policy standing on principles of transparency, responsibility and accountability.

The future of SMEs is of major policy concern at the present time, given their strategic importance in terms of resource-efficiency, capacity for employment-generation, reducing regional imbalances, raising exports, and fostering technological innovation and entrepreneurial skills.

(i) Re-shaping Policy Frameworks

The emerging view is that the State should intervene less and less in certain areas and more in others.[8] Changes in the framework of international trade through new international agreements, the increased pace of technology, capital and information flows in an integrating global economy and concerns over issues of intellectual property rights and the environment, demanded intervention by the Government. The interface between the Governments and industry is changing in countries across the globe as a result of the State re-inventing itself on one side and intense international competition on the other.

In India 'medium' enterprise has been defined in the SME Development Bill, 2005 currently under the consideration of the Parliament, which specifies them as those enterprises whose investment in plant and machinery valued at Rs.10 crores. There are only Small-Scale Industries (SSIs) and industry related Small-Scale Service and Business Enterprises (SSSBEs) and these are defined on the basis of investment in plant and machinery in the small-scale sector till the sector gets formally enlarged into the SME sector. This generally used definition can be said to include micro-enterprises that are engaged in some form of manufacturing such as food processing and handicrafts including textiles and other cottage based industries and Rural Non-Farm Enterprises (RNFS). "The RNFS comprises all non-agricultural activities that includes all manufacturing activities and services in villages and rural towns of up to 50,000 population."[9]

This inclusive approach to redefining the sector makes lot of sense for several future initiatives like the development of clusters.

SMEs vertically integrating with large enterprises at one end and micro-enterprises at the other end could definitely address the issue of job-oriented growth in the region on a more sustainable basis.

SMEs are pivotal to the current development of our economy on several other counts. First, since SMEs are liable to kick-start business with relatively small amount of capital, they have the potential to become a powerhouse for private investment pump-priming the growth of domestic demand. Second, with high degree of flexibility and resilience, SMEs are most suited to attuning their products and operational modes to the ever-changing market needs and diversification taking place in the international marketplace. Export-oriented SMEs prove an important driving force for sustainable growth. Third, in order to make the best use of limited operating resources, SMEs are more inclined to fit into team work, while pursuing specialization individually for vertical integration, as demonstrated in the UNIDO-assisted Cluster development programmes in India and Pakistan during the last 3-5 years. Because of this, they often act as a catalyst for the deepening of industrial division of labour, as well as cornerstones for the synergetic network that support and nurture large-scale enterprises. Fourth, SMEs can also become a cradle for innovations, in that knowledge-based economy is much predicated on individual researches and entrepreneurship.

However, it should be admitted that SMEs are generally less than solid, relatively slack in management, and lacking the experience and ability to guard against external shocks.[10] Among other things, financing, R&D, marketing, human resources, and information management are the foremost areas that often pose difficulties for them. In order to assist SMEs to overcome such inherent weaknesses, many governments have taken concrete measures to offer some relief.

India's SME profile broadly falls into two categories: Registered and Unregistered units under the manufacturing SSIs and SSSBEs.[11] 65.55 per cent (0.9 mn) of the registered units and 38.75 per cent (3.54 mn) of the unregistered units constitute SSIs and six million enterprises are SSSBEs. Nearly a million of the enterprises (9.46%) are run by women. While 88.9 per cent under registered and 96.9 per cent under unregistered categories are proprietary, 7.21 per cent and 1.13 per cent correspondingly are partnership entities. While 3 per cent of the enterprises are private limited companies, 2.61 per cent are public limited companies and others under both categories. Cooperatives are just 0.45 per cent of the total number of enterprises. 65.43 per cent of the SSI units are run with electric power while 26.23 per cent are run without using any power. The other forms of energy including non-conventional energy are used by 7.8 per cent of the units.

Employment per a lakh of Rupees (0.10 mn) of fixed investment has declined over the last Census (1988 data released in 1990-91) from 3.94 persons to 0.67 persons in 2001 and per unit of employment declined similarly from around 6.29 persons to 4.48 persons in 2001 during the same period. This is bound to happen when technology substitutes labour in factor productivity. They contribute to nearly 40 per cent of the exports from the manufacturing sector. Number of SSIs contributing to exports is just 7,344 in registered category and 43,262 in unregistered category. While the number of economic activities pursued by the units in the manufacturing sector is just 672, items reserved for the sector are 877 (this has been brought down to 597 in 2004). This sector produces 5,983 products, a decline from 7,449 in the second Census, with per unit of gross output of Rs.1.48 million an increase of almost 100 per cent during last 12 years. This reflects some efforts at consolidation of the product ranges. About 80 per cent of the Units are located in the States of Andhra Pradesh, Tamil Nadu, Kerala, Maharashtra, Gujarat, Madhya Pradesh, Uttar Pradesh, Haryana and Punjab.

(ii) Employment Effect of SSI Sector Growth

It is worthwhile doing a comparative study of the employment growth in public sector, organized private sector and SSI sector. We have taken the employment data of these three sectors for the period of 1973-79 to 2002-2003 from "Hand Book of Statistics on Indian Economy".

Table 3.1 shows the yearly total employment and its growth over previous year. Employment in public sector was much higher than employment is organized private and SSI sectors till 1994-95. Total employment in SSI crossed total employment in public sector in 1995-96 and after that in each year, total employment in SSI sector was higher than total employment in public sector. The growth rate of employment in SSI sector is always positive, whereas growth rate is negative for some years in case of public sector and organized private sector. This shows that SSI sector always generated employment opportunities throughout the period, but other two sectors sometimes showed reduction. The employment in SSI sector suddenly went up from 11.96 million in 1989-90 to 15.83 million in 1990-91 at a growth rate of 32.36 per cent reflecting highest growth in the period chosen for review.

Table 3.1: Employment and its Growth in Public, Organized Private and SSI Sectors

(In millions)

Year	*Public Sector*	*Organized Private Sector*	*SSI Sector*	*Growth Rate in Public Sector*	*Growth Rate in Organized Private Sector*	*Growth Rate in SSI Sector*
1973-74	12.73	6.75	3.94	—	—	—
1974-75	13.13	6.79	4.04	3.14	0.59	2.54
1975-76	13.63	6.79	4.59	3.81	0	13.6
1976-77	14.18	6.95	4.98	4.04	2.36	8.5
1977-78	14.73	7.11	5.4	3.88	2.3	8.43
1978-79	15.58	7.23	6.38	5.77	1.69	18.1
1979-80	15.12	7.24	6.7	-2.95	0.14	5.02
1980-81	15.48	7.4	7.1	2.38	2.21	5.97
1981-82	16.2	7.53	7.5	4.65	1.76	5.63
1982-83	16.75	7.39	7.9	3.4	-1.9	5.33
1983-84	17.22	7.36	8.42	2.81	-0.4	6.58
1984-85	17.58	7.43	9	2.09	0.95	6.89
1985-86	17.68	7.37	9.6	0.57	-0.8	6.67
1986-87	18.24	7.39	10.14	3.17	0.27	5.63

(Contd.)

(Contd.)

Year	Public Sector	Organized Private Sector	SSI Sector	Growth Rate in Public Sector	Growth Rate in Organized Private Sector	Growth Rate in SSI Sector
1987-88	18.32	7.39	10.7	0.44	0	5.52
1988-89	18.51	7.45	11.3	1.04	0.81	5.61
1989-90	18.77	7.58	11.96	1.4	1.74	5.84
1990-91	19.06	7.68	15.83	1.55	1.32	32.4
1991-92	19.21	7.85	16.6	0.79	2.21	4.86
1992-93	19.33	7.85	17.48	0.62	0	5.3
1993-94	19.45	7.93	18.26	0.62	1.02	4.46
1994-95	19.47	8.06	19.14	0.1	1.64	4.82
1995-96	19.43	8.51	19.79	-0.21	5.58	3.4
1996-97	19.56	8.69	20.59	0.67	2.12	4.04
1997-98	19.42	8.75	21.32	-0.72	0.69	3.55
1998-99	19.41	8.7	22.06	-0.05	-0.6	3.47
1999-00	19.31	8.65	22.91	-0.52	-0.6	3.85
2000-01	19.14	8.65	23.91	-0.88	0	4.36
2001-02	18.77	8.43	24.91	-1.93	-2.5	4.18
2002-03	18.58	8.42	26.01	-1.01	-0.1	4.42

Source: Hand Book of Statistics on Indian Economy, RBI, 2004-05.

The quinquennial average employment in both public, organized private and SSI sectors is shown in Table 3.2. The average employment in

Table 3.2: Five Yearly Average of Employment in Public, Organized Private and SSI Sectors

Year	Average Employment in Public Sector	Average Employment in Organized Private Sector	Average Employment in SSI Sector
1973-74 to 1977-78	13.68	6.878	4.59
1978-79 to 1982-83	15.826 (15.687)	7.358 (6.978)	7.116 (55.032)
1983-84 to 1987-88	17.808 (12.523)	7.388 (0.407)	9.572 (34.513)
1988-89 to 1992-93	18.976 (6.558)	7.682 (3.979)	14.634 (52.883)
1993-94 to 1997-98	19.466 (2.582)	8.388 (9.190)	19.82 (35.438)
1998-99 to 2002-03	19.042 (-2.178)	8.57 (2.169)	23.96 (20.888)

Note: Values in the parentheses give percentage growth over previous period.

public sector is always higher than the other two sectors till 1992-93. But during the period 1993-94 to1997-98, average employment in SSI sector (19.82 mn) was higher than the public sector (19.46 mn). During the period 1998-99 to 2002-03, average employment in SSI sector (23.96 mn) was still higher than the other two sectors.

Growth rate in SSI during Second Five-Year (Plan) period was 55 per cent whereas it was 15.7 per cent in case of public sector and 6.98 per cent in case of organized private sector. During Fourth Five-Year Plan period, the growth rate in SSI was 52.88 per cent, whereas it was below 10 per cent in other two sectors. During the last five-year plan period, the growth rate in SSI was 20.89 per cent whereas it was 2.18 per cent in public and 2.17 per cent in organized private sectors. Hence it is clear that the growth rates of average employment in SSI sector were much higher than the growth rate in other two sectors. Therefore, the analysis shows that the SSI sector has been creating employment opportunities at a higher rate than employment opportunities created by public sector and private sector.

Table 3.3: Average Annual Growth in GDP

	1981-2 to 1990-1	*1992-3 to 2001-2*	*1992-3 to 2002-3*	*1997-8 to 2001-2*	*1997-8 to 2002-3*
GDP	5.6	6.1	5.9	5.5	5.3
Agriculture	3.8	3.3	2.7	2.0	1.2
Industry	7.0	6.3	6.4	4.6	5.0
Service	6.7	7.8	7.8	8.1	8.0

Source: India Development Report, 2004-05.

Table 3.4: Growth of Employment (Usual Status)

(Per cent per annum)

	Rural		*Urban*	
	1983–93	*1993–2000*	*1983–93*	*1993–2000*
Agriculture and allied	1.38	0.18	1.54	-3.40
Mining and quarrying	3.84	-2.28	4.15	-3.71
Manufacturing	2.14	1.78	2.21	1.83
Electricity, gas and water supply	4.70	-5.65	4.46	-4.19
Construction	5.18	6.43	6.20	6.26
Trade, hotels and restaurants	3.72	1.18	3.94	5.54
Transport, storage and communication	4.58	7.29	2.90	3.91
Finance, insurance real estate	5.99	2.51	5.63	7.05
Public administration, community and personal services	3.13	0.32	4.16	0.13
Total non-agriculture	3.23	2.31	3.54	2.96
All sectors	1.75	0.66	3.27	2.27

To analyse whether establishment of small-scale industries (SSI) and increase in industrial production generate employment, we have carried out an empirical analysis through correlation and regression. To examine whether small scale industries have potential to increase employment, we have considered some of the employment and production related variables, such as, total employment of country, employment in the SSI sector, index of industrial production and number of units of SSI. We have taken yearly data on these variables starting from 1981-82 to 2002-03.

(iii) Correlation Results

The bivariate correlation among the growth of the above variables is carried out. The growth rate of a variable is calculated as the difference of the Natural Log of the variable. The correlation results in the following table show that there exists significantly high positive correlation between growth of total employment and growth of SSI (0.954). But there does not exist any significant correlation between growth of industrial production and growth of employment. As total employment constitutes a part of employment in the SSI sector, the correlations between growth of total employment and growth of employment in small-scale industries is higher and significant.

Corrclations

		DLTOEMP	*DLSSIEMP*	*DLIIP*	*DLSSI*
DLTOEMP	Pearson Correlation	1.000	.968**	.234	.954**
	Sig. (2-tailed)	—	.000	.307	.000
	N	21	21	21	21
DLSSIEMP	Pearson Correlation	.968**	1.000	.165	.990**
	Sig. (2-tailed)	.000	—	.474	.000
	N	21	21	21	21
DLIIP	Pearson Correlation	.234	.165	1.000	.154
	Sig. (2-tailed)	.307	.474	—	.504
	N	21	21	21	21
DLSSI	Pearson Correlation	.954**	.990**	.154	1.000
	Sig. (2-tailed)	.000	.000	.504	—
	N	21	21	21	21

** Correlation is significant at the 0.01 level (2 tabsle)

Notes: DLTOEMP: Growth of total employment,
DLSSI: Growth of SSI,
DLIIP: Growth of industrial production,
DLSSIEMP: Growth of employment in SSI sector.

Table 3.5: Population, Urban Population, Rural Population and Degree of Urbanization, India 1951-2001

Census Year	Population			Degree of Urbanization	
	Total	Urban	Rural	Per cent urban	Urban-Rural Ratio
1951	361,088,090	62,443,934	298,644,156	17.3	21
1961	439,234,771	78,936,603	360,298,168	18	22
1971	548,159,652	109,113,977	439,045,675	19.9	25
1981	683,329,097	159,462,547	523,866,550	23.3	30
1991	846,387,888	217,551,812	628,836,076	25.7	35
2001	1,027,015,247	285,354,954	741,660,293	27.8	39

Notes: 1. The total population and urban population of India for the year 2001 includes estimated population of those areas of Gujarat and Himachal Pradesh where census could not be conducted due to natural calamities during the appointed period.

2. The total population and urban population of India for the year 1991 includes interpolated population of Jammu and Kashmir where 1991 Census could not be conducted.

3. The total population and urban population of India for the year 1981 includes interpolated population of Assam where 1981 Census could not be conducted.

Source: http://www.censusindia.net/results/eci14_page2.html

(iv) Regression Results

To know whether the increase in SSI and increase in industrial production will lead to employment generation, we have estimated the following model, where growth of total employment is regressed on the growth of SSI and growth of industrial production (IIP).

$$\text{DLTOEMP} = \underset{(4.08)}{0.013} + \underset{(13.68)}{0.066\ \text{DLSSI}} + \underset{(1.29)}{0.063\ \text{DLIIP}} \quad \text{...(1)}$$

The above model shows that the coefficient of the growth of SSI is positive and statistically significant (*t*-value is given in the brackets). But the coefficient of the growth of industrial production is not statistically significant. Hence, the growth of employment is affected by growth of SSI, whereas growth of industrial production does not affect growth of employment. From model (1) it is seen that 1 unit increase in SSI growth rate leads to 0.06 unit increase in total employment growth rate. It means if the SSI sector growth rate increases from say, 5 per cent to 6 per cent, the corresponding growth rate of total employment will increase by 0.06. Hence we can say that establishment of more and more small-scale industries have potential to create considerable employment opportunities.

To know whether the growth of industrial production will have any impact on the growth of employment in SSI, we have estimated the following model.

$$DLSSIEMP = \underset{(1.332)}{0.039} + \underset{(0.731)}{0.316\,DLIIP} \qquad ...(2)$$

The result shows that growth of industrial production does not have any impact on the growth of employment in small-scale industries, as the DLIIP coefficient is not statistically significant. Again to examine whether the growth of employment will affect growth of industrial production, we have regressed DLIIP on DLTOEMP, and the regression result is:

$$DLIIP = \underset{(5.584)}{0.055} + \underset{(1.049)}{0.328\,DLTOEMP} \qquad ...(3)$$

Here the employment growth coefficient is not statistically significant. Hence growth of total employment does not significantly affect growth of industrial production. Therefore, we can say that increase in employment may not affect industrial production. Growth of industrial production can be obtained by adopting capital-intensive technology, adopting new production techniques or innovations etc. The above analysis clearly reinforces the general impression that the small-scale industries have considerable potential to generate employment.

Manufacturing would need to grow annually at 12 per cent compared to the trend growth rate of 7 per cent achieved during the last decade in order to create 1.61 million employments with present employment elasticity. If the employment elasticity were to improve to the earlier higher level of 0.59, the 12 per cent growth in manufacturing would result in 2.9 million new direct jobs per year in manufacturing sector alone. Both direct and indirect job creation would increase and the interdependence among the large and SME sectors would ensure sustainability. Stability of policy is a necessary but not a sufficient condition for achieving and sustaining high levels of economic development. However, integration of policies and regulations relating to interconnected sectors of the economy coupled with effective implementation would hold the key to eradication of poverty.

It is often argued that SME promotion is justified on grounds of the job-creating prowess of SMEs or of their greater efficiency and growth. Attempts are often made to draw a causal link between SMEs and poverty alleviation so as to justify policies and subsidies in favour of SMEs. But the empirical evidence supporting many of these claims is very mixed, making it difficult to justify SME promotion on the basis of inherent economic benefits of smallness. A survey of small- and medium-sized enterprises (SMEs) commissioned by the World Bank in 2000 cited the main constraints on their development: (i) lack of access to institutional credit and high interest rates for loans; (ii) poor infrastructure, including power, gas, and port facilities; (iii) government policies that allow too many imports, frequently devalue the currency, and keep tariffs high on intermediate inputs; (iv) government bureaucracy, corruption, and red tape; and (v) underdeveloped business management skills.

Belabouring on a historical review of policies at this juncture when new paradigm shifts are occurring and new policies are emerging as part of global integration could be a drag on this paper and therefore avoided. (See the Boxes below for synoptic view of rationale of regulation and policy framework).

Box 3.1: Rationale of Regulation in India

Law	*Rationale*
Incorporation	To allow enterprises to derive benefits as legally recognized entities, and to protect owners, investors and debtors.
Registration	To gather information for governing economic activities and allocating resources and services.
Licensing for production	To control modes of production, and to protect weaker economic actors (especially traditional and small producers), as well as workers, consumers and the environment.
Licences and quotas for raw materials	To control inter- and intra-sectoral resource allocations, and to allocate public or quasi-public goods.
Taxation (income and sales tax, excise, octroi)	To gather revenue for the various tier of government, from local to central; sometimes to change relative prices.
Labour laws	To ensure fair wages, welfare and safety for workers.
Environmental laws and consumer laws	To protect the environment and to protect consumers by ensuring quality and standards for products.
Laws applying to specific sub-sectors	All the above rationales may apply, depending on the specific laws introduced.

Source: Mahajan and Raju, 2004, RNFS.

Regulatory excesses have become endemic in India's body politic. These need an urgent correction. The fact that a regulation raises the cost of doing business doesn't necessarily mean that the regulation should be softened or eliminated. Environmental regulations, for example, impose a cost on the business sector, but these costs may be outweighed by the social benefits of improved environmental quality. Nevertheless, when doing the cost-benefit analysis of such regulations, one factor to consider is how the burden is distributed across different types of firms. In some cases, adding flexibility in the implementation of regulations can be an important way to ease the burden on small enterprises.

Government of India got an exclusive Bill-for Small Enterprise Development ready for enactment in the current financial year to address most of the issues relating to entry barriers, harassment, needless regulation, competitiveness of the sector and appropriate facilitation.

There are a number of States like Andhra Pradesh, Tamil Nadu, Karnataka, Maharashtra, Gujarat, Punjab, Haryana, Rajasthan, and Madhya Pradesh that have taken initiatives proactively to (i) simplify procedures, (ii) reduce the impact of inspector-raj, (iii) provide infrastructure at affordable cost to the entrepreneurs, (iv) provide easy exit rules within their realm, (v) rehabilitate potentially viable enterprises by dialoguing with the financial institutions and banks, and (vi) enter into partnership with credible Industry Associations to provide BDS. It will take time to gauge the benefits but the moves are worthy to note. States' experiences are illustrated in the following Box.

Box 3.2: Infrastructure Promotion and Development: States' Initiatives in Public Private Participation

Maharashtra State pioneered in providing the right type of infrastructure for the SMEs during 1970s by setting up Maharashtra Industrial Development Corporation (MIDC)-sponsored industrial estates wherein it provided comprehensive services and utilities to the SSIs in these estates at affordable costs that made them more competitive than their counterparts elsewhere.

Government of Andhra Pradesh formed 83 Industrial Area Local Authorities out of the 270 Industrial estates where the Industry Associations have to work in partnership with the Andhra Pradesh Industrial Infrastructure Corporation, to maintain the infrastructure in Industrial estates. The Government has empowered them to collect and share local taxes and user service charges for the utilities, with the local municipalities in 1996. A study commissioned by the Government in 2003 suggested a few amendments to ensure their functional effectiveness. These suggestions are being considered for review.

(*Contd.*)

Box 3.2 (*Contd.*)

Similarly, Government of Andhra Pradesh partnered with CII in the establishment of a Technological Development Centre and sub-contract exchange. It has also set up APSSI Centre in Association with Federation of Andhra Pradesh Small Industries Associations to provide the much-needed Business Development Services to the SSIs in November 2002.

Government of Karnataka has set up a sub-contract exchange in association with Karnataka State Small Scale Industries Association (KASSIA).

Coimbatore District Small Industries Association has set up a Large Exhibition Centre for the benefit of SSIs and has been part funded by the ZDH and Government of Tamil Nadu. Ambattur (Chennai) Industrial Manufacturers' Association had set up R&D centre in machinery manufacturing to benefit over 1000 manufacturing units in SME sector with the assistance of GoI and GoTN. Similarly, Fredrick Neuman Foundation helped the Tamil Nadu State Small Scale and Tiny Industries Association (TANSTIA) to set up a Business Development service provider—TANSTIA-FNF. These are all State-partnered private initiatives where the initiative of the Government triggered action from the private agencies. The exit route for the State Governments concerned also is carefully planned.

One of nodal officers-in-charge of setting up an Apparel Park narrated to me an experience that is also worthy to note.

Box 3.3: Setting up Apparel Export Park

The Nodal Officer narrated her experience in bringing up this Apparel Park during her tenure (1997-2002), which is worthy to record for the benefit of those implementing cluster development initiatives and infrastructure development programmes. "Government intervention for the development of such clusters of export-oriented apparel manufacturing units can yield positive results, if and only if, certain factors are in place. These factors vary from region to region depending on the nature of the cluster irrespective of the markets it caters to. Existence of an active local Industry Association can be very beneficial in order to build bridges of understanding between the Government agencies and the industry as also to build networks.

In the instant case, even though the land was to be made over by the State-owned Infrastructure Development Corporation, the procedures proved very cumbersome and led to a delay of nine months to one-year. Public intervention, was overly supply-oriented (infrastructure, training of manpower, attracting service providers, etc). Once the local industry representatives joined the Managing Committee of the Park, there was a shift towards demand-driven intervention. Thirdly, emergence of influential leaders among the local industry representatives helped resolving diverse issues by securing a consensus smoothly. Internal roads were not black topped for two years after setting up the Park and this had a deleterious

(*Contd.*)

Box 3.3 (*Contd.*)

impact on the manufacturing units—dusting the raw material, materials in process and finished goods whenever the trucks were passing through the Park. The lessons she drew from this experience are worthy of mention:

1. Engaging professional consulting team to conduct a diagnostic study before setting up the venture.
2. Involving them as a part of the Managing team would fix accountability for their suggestions and quickens the process for midcourse correction in project implementation.
3. Involve the Local Industry Association in the planning and designing of the project as also in execution. It is admissible to set up platforms to facilitate such consultations.
4. While there could be some pressure from the user groups to provide tax-holidays, cash incentives etc., it is better to distance such demands and instead the Government should directly involve in speeding up provision of electricity, telephone lines and other basic infrastructure at the doorstep of the entrepreneur, and to provide security during the period of construction.
5. It is essential to prepare a PERT-CPM charts and review them at prescribed intervals.
6. Developing sensitivity to local customs conventions and beliefs is very crucial (e.g., vastu—a science that deals with the manner and direction of structure and internal layout).
7. It is desirable to set up catalytic institutions in co-ordination with the Association, e.g. Institutes for manpower training and development, fashion designs, demand forecasting, Brand-building etc.
8. While initially the risks of public investment should be borne by the Government, as such of them cannot be absorbed by SMEs, transferring them later to the stakeholders should be planned and executed.
9. Special dispensations like making certain provisions of Labour laws inapplicable in the parks should be thought of in the beginning and should be legalised.
10. Exit of Government interventions and its timing are extremely critical in all cluster based initiatives.

Box 3.4: Industrial Policy of India

Features of the Industrial Policy of the Government of India Affecting the SMEs:

- Investment in Plant and Machinery for qualifying as tiny enterprises increased from Rs.0.2 mn to Rs.2.5 mn.
- Tiny enterprises are eligible for continuing support, including easier access to institutional finance, priority in the Government purchase programme and relaxation from certain provisions of the labour laws

(*Contd.*)

Box 3.4 (*Contd.*)

- Investment in Plant and Machinery Rs.50 mn qualify for being small enterprises as per the proposed SME Development Bill 2005.
- Investment in Plant and Machinery above Rs.50 mn but below Rs.100 mn qualify for being medium enterprises under the proposed SME Development Bill.
- Equity participation from the large-scale industries in a small-scale unit permitted up to a limit of 24 per cent of the total shareholding to promote sub-contracting and ancillarisation.
- Project limit for National Equity Fund raised to Rs. 5 mn with 30 per cent for tiny sector.
- Small Industries Development Bank of India establishes 'factoring' services to unlock the dues from the vendees.
- A scheme of integrated infrastructural development (including technology back-up services) to be introduced to facilitate locations of enterprises in rural and backward areas.
- Preferential allotment of indigenous raw materials to tiny and small-scale units.
- Turnover based simplified working capital assessment limit enhanced to Rs. 50 mn.
- The National Small Industries Corporation to concentrate on marketing, under a common brand name, mass consumption items produced by the small and tiny industries and to set up rating agency services to improve credit markets.
- SIDBI sets up Small and Medium Enterprise Rating Agency in September 2005.
- Threshold limit for Credit Linked Capital Subsidy scheme for technology upgradation increased from Rs. 4 mn to 10 mn and the related subsidy from 12 per cent to 15 per cent. The scheme extended up to March 31, 2007 and modified in terms of calculation of the capital subsidy to be calculated on the purchase price of the plant and machinery involved and not on the term loan disbursed to the beneficiary SSI Unit.
- Women enterprises redefined to include enterprises having majority shareholding and management control by women.
- Functional industrial estates to be established in areas where agricultural or horticultural produce is concentrated.
- Industries, services and business establishments are included within the ambit of tiny enterprises up to Rs.1 mn.
- Composite loans and working capital up to Rs. 5 mn to be extended through a single window.
- Upto Rs.2.5 mn without any collateral security under the GoI-SIDBI funded Credit Guarantee Fund Trust.
- Self-certification to reduce the statutory inspections under the Labour Laws and Factories Act.
- Market Development Assistance Scheme for SSIs introduced.

Cluster Development initiatives are implemented through two windows of Government of India, Development Commissioner (DC SSI) and Ministry of Industry & Commerce (MoI&C). While five are natural clusters that improved with UNIDO export during the last five years, the rest are in various stages of implementation and 90 per cent are in the beginning stage.

Competitiveness Constraints on SMEs

The law of comparative advantage provides that when markets are competitive, people will trade and will gain from trade when there are differences between countries in the relative costs of production.[12] In addition, many engaged in making trade policy have advocated export-oriented industrialisation for developing transitional economies, an approach that relies on an open economy that exports goods to industrialised nation and other markets. There has been a great deal of focus on the "East Asian Miracle" and on the extent to which the economies of South-east Asia relied on exports to generate significant rates of economic growth, often in the neighbourhood of 10 per cent per annum. In addition to the findings on export-led growth, there have been some significant findings in what has been termed "growth theory" on what makes countries develop. Classical growth models focused on capital accumulation, and particularly the increasing returns to capital that occurred in countries that were starved for capital. New growth theories, particularly endogenous growth theory, focus on technical change as the catalyst for economic growth. By technical change I mean decisions by individuals in an economy in response to market forces that result in growth because of technical change in three areas: (1) increasing returns to specialisation in an economy; (2) human capital accumulation through education and training; (3) "learning-by-doing" or growth through increased experience in business and production; and (4) research and development. All of these factors leading to growth complement the law of comparative advantage.[13]

- Noticeably, the policy thrust has been to enhance the capability of SMEs to perform effectively in this context. Structural changes in several economies have led to a shift in industrial production of SMEs from construction and low-tech manufacture to high-tech industries. SME presents a canvas with dull colours;
- No major shifts in employment pattern occurred despite sectoral

shift from the primary to the secondary and tertiary sectors over the last decade;

- Despite some initiatives on entrepreneurial promotion and development through specific institutional interlinkages, educated youth do not see setting up a new enterprise as a risk worth its while due to obsessive entry barriers, apparent inspector-raj, and uncertain markets;
- Technological obsolescence among the existing enterprises;
- Unwelcoming attitude of the financing institutions and banks in extending financial support for setting up new enterprises and for expansion;
- Lack of innovation in new financial products to suit the sector;
- Venture Capitals still hesitate to take risks in the manufacturing sector;
- Poor networking;
- Inadequate Market Support and Market Access Initiatives when compared to the developed nations.

SMEs are subject to a number of constraints on their competitiveness in public procurement markets. Many of the constraints result from the relative effects of fixed costs on SMEs to conduct public procurement relative to the effects of fixed costs on large businesses. Most important challenge is the fierce competition in costs and prices, improved designs, new product ranges, quick and on-time deliveries etc. The Town and Village Enterprises (TVEs) of China have aggressively moved from their domestic economy to the global arena and South Asia being in the immediate neighbourhood has to not merely take cognizance of it but should draw lessons from them. 'There was a dramatic reduction of rural poverty in China from about one-third of the rural population in 1978 to about 8 per cent in 1996.[14] Manufacturing TVEs constituted longest generator of employment opportunities accounting for around three-fifths of the total. TVEs contributed over 30 per cent of China's GDP and half of the value-added. In the year 2000 TVEs provided employment to nearly 25.8 per cent of workforce.[15] The sectors that are worthy of alliances within and outside the country among the SME firms are toys, RMG, electronic goods, and light engineering goods.

(v) Access to Finance: The Stress and Distress of SMEs

Limited access to finance has been identified as the major obstacle for growth in all the South Asian economies although their risk profile centres round obsolete technologies, limited marketing abilities, lack of adequate management skills, high share of intangible assets, poor accounting standards and the resultant low level of financial discipline. The net result is that they are discriminated against by the lending institutions. Impact of reforms on the small industries seems to have upset the apple cart and many small industries bemoan that the financing institutions distance them because of the fear of accumulation of non-performing assets. SMEs require greater attention from the lending institutions in terms of continuous follow-up and monitoring and this function increases the transaction cost for the Banks disproportionately. The post-liberalization era resulted in shedding manpower and uneconomic branches in the rural and semi-urban areas and therefore, has seen many banks turning to a market transaction function instead of lending function. In contrast, large enterprises, despite higher level of accumulation of non-performing loans, are accorded a privileged status by the banks and financial institutions. All the expert studies in India and elsewhere have confirmed that the small industries are denied adequate and timely credit facilities and therefore, losing their competitiveness.[16] Structural inadequacies and imperfections do not allow them to access equity markets.

(a) Start up Constraints

Most SMEs venture with small seed capital or promoters' contribution making them unattractive for the institutional lenders right at the start. Generally Banks insist 25-33 per cent of the total project cost as a contribution from the promoters. This requirement forces them to resort to finance from friends and relatives and even private sources at interest rates far higher than what the organized market generally offers. Another major constraint is that the gestation period allowed for the SMEs falls short by six months to a year compared to their elder brothers. These force SMEs generating resources for repaying their obligations far ahead of the commencement of commercial production. This places them in a debt trap.

(b) Debt Markets

The high risk, high return nature of SMEs suggest that excessive dependence on debt markets puts them to a high disadvantage particularly because they do not have cushion capital when the banker removes the umbrella either in a downpour or in hot Sun. Though National Equity Fund Scheme has been in operation in India for the last 20 years, the pick-up is very poor because the lending institutions did not propagate it and made the procedures for accessing it unfriendly to the entrepreneurs. Further, most enterprises being proprietary or family owned (90-98 per cent), pooling up required collaterals to access debt from time to time (banks are found to be seeking these collaterals with almost every request for additional quantum of loan) has later put them to partnership and family disputes. Whole nation suffers from this disability and therefore, we have to think of alternate ways of collateral-free financing. The overall thrust of the credit guarantee programmes should be to absorb part or some of the risks. Their success hinges on the ability to minimize the default rate. The premium and the extent of guarantee are the two essential parameters under such schemes and balancing them in a manner that would not upset the inherent incentivization effort seem critical to their success. Some efforts in Bangladesh, Pakistan and India are manifest lately though their impact has not been as significant as in the US, Japan, Germany, Italy, South Korea, other South East Asian economies in the post-East Asian crisis, China etc.[17] Started in Italy, Mutual Guarantee Programme has demonstrated its sustainability wherein the SME Associations promote own/mutual guarantee funds and leverage them with public or Government promoted guarantee funds for wider coverage. SIDBI's Guarantee Funds made a beginning in this direction. Among the Indian States, Rajasthan took the lead in this direction and the Government is putting in efforts to sensitize this sector.

(c) Export-Oriented SMEs

We view these enterprises on a different footing and created special zones for housing them (EPZs, SEZs, AEZs) where regulatory rigour is minimal and fiscal incentives are specially made attractive. The three instruments for the bank based trade financing system are: (a) transaction based, self-liquidating mechanisms for trade financing (including rediscounting mechanisms of the Central Bank); (b) Pre-shipment and post-shipment packing credit facilities; and institutions to deal with overseas buyers' non-payment risk, i.e., export credit insurance and guarantees Export Credit Guarantee Corporation (ECGC) (EXIM Bank of India both through direct financing and reinsuring mechanisms offers such facilities, to cite an instance).

(d) Micro-Finance

Micro-enterprises initiated by the self-help groups representing bottom of the pyramid farm and non-farm sectors' rural enterprises offer a faster growth prospect. Bank linkage programmes were evolved by the refinancing institutions like NABARD and SIDBI and are successfully implemented. Phenomenal growth of SHGs and micro-enterprises led to the need for and evolution of separate regulatory mechanisms to ensure their sustainability. Adoption of the systemic approach holds the key for sustainability of such micro-level interventions and maximizing the development impact. Enabling policy environment, promoting appropriate financial intermediaries, and financial infrastructure are essential ingredients of success as they delivered credit with extension and retail market support. Commercial Banks and Non-Banking Finance Companies reflect non-performing loans of just 1 to 2 per cent. One study conducted by APMAS reveals that hidden NPAs or rolled over credit unrelated to the activity requirements conceal the ills of the micro-enterprises *(See Notes 3)*.

(e) Venture Capital

Venture Capital (VC) is recognized as an important financial instrument to assist the high risk and innovative projects. However, the VCs are in infancy in South Asia. They have not opened their doors for the manufacturing industry and more specifically to the SMEs. All these countries are in transition and lack an exit route such as an organized public equity market at the lower end of investments. Republic of Korea and China made good in-roads and lately India is proving to be an attractive destination for the VCs. However, for the SMEs, VCs are still distant. SME-VCs should provide training to the new entrepreneurs in regulatory and financial management, and provide handholding support for at least two years. Even the SIDBI VC, which has been floated recently does not have these ingredients.

(f) Ailing SMEs

Prudential norms and reclassification of assets as a consequence of reforms in the financial sector in India and Pakistan have brought to the fore non-performing loans and in terms of number of accounts these abound in the SME and Farm sector loans. The *raison de'etre* of the classification of assets is the need for immediate attention to convert them to performing assets and in the absence provide for them as losses in proportion to the risk attached to them. The cause for the NPA will generally be not in the nature or value of the collateral but in the primary security the erosion to which may have taken place either on account of adverse selection or change in the complexion of risk. Moral hazard is not a function of collateral. NPA is, therefore, a ghost that may haunt but not a monster that would eat up the institution if the Banks through restructuring of debt or revitalizing the asset initiate corrective steps. Unfortunately, this did not happen. Absence of bankruptcy laws only accentuated the grief and burden of the 'sick' SMEs. Some of the State Governments under pressure from the Industry Associations and also with a concern of locked up monies and physical assets in sick small manufacturing enterprises initiated specific measures for rehabilitation (Gujarat, Andhra Pradesh, Kerala, Madhya Pradesh, Rajasthan and Tamil Nadu are cases in instance in India).

Supply side constraints do not border on lack of resources so much as on attitudinal disposition of the lending institutions as has been described above. Lack of transparency, undue obsession with collaterals, discriminatory application of lending norms under identical risk bearing portfolios in SMEs, absence of credit information exchanges, and almost non-existent credit rating for the SMEs, hostile attitude for upgrading the assets, etc., both from the commercial banks and specially set-up funding institutions like the SIDBI, State Finance Corporations in India, are some of the other supply side constraints that could be addressed easily. Demand side constraints largely arising from inexperience, lack of knowledge in systems and procedures, lack of adequate collaterals, pervasive default culture can be addressed through capacity-building efforts and with the involvement of Industry Associations and Chambers of Commerce and Industries. Recent experience suggests that a stringent law could effectively reduce the wilful default in the sector (Securitisation & Reconstruction of Financial Assets and Enforcement of Security Interest Act, 2003 has contributed to the net NPAs in Commercial Banks falling to a low of 2.4 per cent in 2004 from more than 7 per cent in 2000). There is a reprieve in the recent policy announcements of GoI and the RBI, which the institutions should set out to implement.

(g) Options Available

Attitudinal change is possible only when the Banks put in place incentives for better performance in SME sector-performance defined in terms of new enterprises financed, efforts at regular scaling up of assets and increase in the quantum of flow of credit and institutionalizing the risks by way of correcting information asymmetry, investing in capacity building, involving BDS providers and instituting internal rating mechanisms (it is worthwhile recording that the years 1993-2005 witnessed incentives for NPA management and many officers with a track record of showing consistent decline in the NPAs—not necessarily through cash recoveries—are today the CEOs of public sector banks).

(vi) Capacity, Capability and Competitiveness of SMEs

The 'traditional' way of doing things had its roots in the religion and native culture. In the case of Hindu society, caste and the joint family were generally considered hindrances to development of entrepreneurship that the new industrial growth poles centring on the SMEs.[18]

(a) Technology, R&D, Environment and ICT

SMEs generally suffer from technological obsolescence save the new generation enterprises and those in ICT and Bio-technology. This happens in the developing nations mainly on account of lack of information on emerging technologies, over-confidence on adaptation and reverse engineering processes, laggard approach of the financing institutions for technology investment proposals.

Technology divide does not have to follow income divide. It is created in response to market pressures—not the needs of the poor people who have little purchasing power.[19] Although technology modernization and upgradation is critical to both domestic and export competitiveness, it continues to be a casualty. SME sector in general is not capable of investments in R&D because of their cost intensity. Further, when technology is changing, enterprises have to invest in training workers to stay competitive. Smaller enterprises in particular can benefit from public policies that encourage coordination and economies of scale and partly subsidize their efforts. (Technology Information, Forecasting and Assessment Council—TIFAC, Technology Development and Modernization Fund of SIDBI are cases in instance).

TIFAC founded in 1988 had its task cut out clearly: It has to look ahead at the technologies emerging world-wide, and pick those technology trajectories, which were relevant for India and should be promoted. In its tasks, TIFAC networked various stakeholders: the Government, industries, users, scientific and technological institutions, financial institutions, and intellectuals.[20] It provides partial funding support to SMEs for demonstrating commercially viable new technology, products and processes. TIFAC has taken up 70 projects in ten years of which 15 have been commercialised and ten patented. Synergies between industry, academia and the Government are essential to develop a vibrant innovative eco-system much needed by the SMEs.

"The office of DC SSI in its latest publication 'SSI—An Engine of Growth' listed 20 industries as having proven advanced technologies. Their costs range from as low as Rs. 50,000 for dedicated test jigs in electronic industry to as high as Rs. 4 crore for bulk drugs and dyes industry. These twenty types are: (1) Foundry-Ferrous and Cast Iron; (2) Shoes Manufacture; (3) Glass Manufacture; (4) Rubber Processing; (5) Bulk Drugs; (6) Dyes; (7) Tissue culture for fruit plants, edible oil bearing plants, forest plants, medicinal and aromatic plants; (8) Tiles; (9) Electronics; (10) Transfer Manufacture; (11) Wires and Cables; (12) Lamps, bulbs and lighting service; (13) Multiple use hand tools; (14) Surgical equipment; (15) Bicycle parts; (16) Auto parts; (17) Machine tools; (18) Granite industry; (19) Diamond wiring; and (20) impregnated gang cutters, etc." (Raju, 2004).

A Government supported Regional Research Centre in Kerala and National Research Centre in Central India are coming up in the Food Processing Industry to promote R&D in the Food Processing Sector at a cost of Rs. 4000 million, for the benefit of SMEs. Commitments under TRIPS to promote technology transfer to developing countries remained only on paper. Either collective investments or public investments should spur the research and development activity. While India rides the wave of science and technology development in South Asia, it fades out into insignificance compared with NIEs viewed in terms of number of scientists and technicians engaged in R&D activity as also in terms of high-technology exports.

(b) Cluster Development Initiatives

A cluster is defined as a geographically bound concentration of similar, related or complementary businesses. A UNIDO study defines clusters as 100 registered small scale units.[21]

In India, clusters have been in existence for decades and they came up naturally. Some names that occur instantaneously are: Tirupur (Knitware), Muradabad (Brassware), Narsapur (Lace), Panipat (Cotton blankets), Sivakasi (Matches and fireworks), Agra (Foundry), Gujarat (Gems) and there are 350 such clusters. These naturally formed clusters could not, however, enter wider markets due to the absence of synergies that existed elsewhere in the world to acquire both product edge and cost competitiveness.

Effectively coordinating with the Union Ministries of Industry & Commerce and SSI, UNIDO put the Tirupur Cluster on international map firmly with quality exports jumping by more than 100 per cent during the last seven years to cross the US$11billion in the year 2003-04. Similar was the experience in Pune Food Processing Cluster, Jaipur (Textile Printing), Ludhiana, Panipat, Ambur (Leather), Ahmedabad (Pharmaceuticals) and Bangalore (Machine Tools). Swiss Agency for Development and Cooperation has also lent its support in the leather, pharmaceuticals and machine tool clusters. Several institutions that included export houses, common service centres, Non-governmental Organizations/Associations, have been revitalized and several Government Officials and promoting institutions have been sensitized to the initiatives. Planning Commission has included this as a major initiative in SME Development strategy and Employment-oriented Strategy. Two parallel initiatives are currently on stream-one, through the Ministry of Industry and Commerce and the other through the Ministry of SSI with the DCSSI as the nodal officer. SIDBI is also playing a major role in the development of textile clusters initiated by the Textiles Committee in its country-wide effort. A little over 1000 firms have benefited from this initiative so far. Five State Governments (Andhra Pradesh, Tamil Nadu, Karnataka, Madhya Pradesh and Rajasthan have adopted cluster development within the framework of their State Industrial Policies with UNIDO support. Numerous nation-wide support institutions joined, most prominent among them the State Bank of India, Development Commissioners—Handicrafts and Handlooms and the Chambers of Commerce. Innovative financial practices in the Consortia (Bangalore Machine Tool Cluster), Entrepreneurial development programme in Bagru (Jaipur), introduction of ICT in Ludhiana, building bridges of understanding between the regulators and the entrepreneurs in food processing sector in Pune on quality stipulations and legal requirements are some worthy interventions so far.

Clusters are found to have direct impact on the economic development of the region through enhancement of export potential, foreign direct investment, greater market penetration in export markets, quality standard assurance, higher income generation for SMEs, increasing employment opportunities in the cluster based services, productivity and competitiveness. The focus shifts to the capacities and capabilities of the SMEs to improve their competitiveness.

Competitiveness at the micro-level is functionally related to the skills and capabilities of the persons running the enterprises and the fuller utilization of the capacities installed in the enterprises. It, however, as of now measured and benchmarked at the industry, sectoral and national level (Porter, Sachs and McArthur, 2001: 21). It is also unfortunate that the data and information needed to monitor and compare the evolving capabilities and potential of the top players of SMEs in priority sectors and industries over time.[22] Such an exercise is crucial for developing an information bank on the core competencies of the Region.

(c) Enhancing SME Competitiveness through Cluster Development

Box 3.5: Enhancing SME Competitiveness through Cluster Development

Nature of Deficiencies in SMEs	*Remedies sought through Cluster Initiative*
Individualistic and non-sharing	Awareness and Trust-building interactions among the cluster participants.
Lack of collective thinking	Industry Associations are associated to bring about effective ideanization and participation; networking both within and outside the cluster.
High Costs	Consortium approach in procurement of raw materials, contracts and financing; sharing of costs of common services.
Low Quality	Common testing and R&D facilities; Awareness programmes on quality management; benchmarking.
Low Market Penetration	Product innovation; product diversification; market information and market diversification strategies; and market research.
Lack of poor brand image	Brand-building; Brand imaging; cobranding; and brand management.
Low capital, low debt and poor financial management	Improved access to debt markets; Skill-building in financial and accounting management; Build bridges of understanding between the financing institutions and entrepreneurs for better facilitation; Innovation of new financial products to suit the specific products and markets.
Lack of Institutional supports	Institution building and building strong public-private partnership in building product specific and general infrastructure.
Lack of enthusiasm and disinterested second generation entrepreneurs	Entrepreneurship Development programmes specifically designed for the cluster.

The necessity for training and capacity-building stems from the fact that Governmental efforts to reduce poverty through direct interventions have yielded mixed results and particularly so in the SME sector. Provision of services and capital is necessary but not sufficient condition for the growth of the sector in the region. Building social capital is extremely crucial. Social capital is defined as network of associations and organizations at community levels (McGregor, 1999).[23] The concept applies as much to SMEs as to the micro-enterprises. The lower factor productivity in SMEs is attributable in large part to lack of knowledge, knowledge of latest technologies and their application as also the quality stipulations and skills. There can be no denying the fact that a plethora of training institutions exist both in the Government and Non-government sector for entrepreneurship promotion and skill development. But the output from these institutions remains suspect because of the ground level results. For sure, the SMEs have been providing hands-on experience and skill, which had a spread-effect in human resource development. The other important facet is that owner-driven enterprises hardly had time to devote for self-development and many a time functioned in a world of their own that led to inefficiencies and lags in factor productivity. "A healthy, educated workforce is a productive workforce."[24] Investments in human resource development and entrepreneurship are crucial for private sector development that constitutes the core of SMEs.

Nearly four to five decades of existence of a few institutions engaged in entrepreneurship promotion and development reveals that only 10 to 11 per cent of the trained persons have set up the small enterprises in the country. Most of the cottage and small industries in the rural areas in our country sprang up under traditionally inherited skills or out of livelihood protection necessity. These enterprises continue to remain distant from the type of training required in the current market environment. They hardly possess knowledge relating to maintenance of accounts, preparation of business development plans and emerging technologies. In India, Entrepreneurship Development Institute of India (Ahmedabad) promoted jointly by leading financial institutions and Government of Gujarat, National Institute for Entrepreneurship and Small Business Development, National Institute of Small Industries Extension and Training—both promoted by the Government of India, several Universities in the country with the assistance of SIDBI have ventured into entrepreneurship promotion and development. These apart, some Banks like Syndicate Bank have set up Rural Entrepreneurship Development and Training (RUDISET). Realising that these efforts are proving inadequate, Industry Associations and Chambers have come up with their own ventures within the resources available to them. Bharatiya Yuvajana Seva Trust (CII), TANSTIA-FNF, FAPSIA-APSSI Centre (A joint venture of Federation of Andhra Pradesh Small Industries Association and Government of Andhra Pradesh) are cases in point. Leading financing institutions like the IDBI, SBI, OBC, Bank of Baroda, Punjab National Bank etc. (e.g., JNIDB, State Bank Staff College) have set up their own training networks to train their manpower in financing the SMEs through specially designed programmes on credit appraisal and risk management. However, there is no data and information on the evaluation of the programmes conducted at all these institutions in terms of the impact on the sector. The ground level perception is that there is a vast gap between what is required and what is imparted. There are a number of consulting institutions and individual consultants who are working in this area both through institutional interactions and handholding efforts. The National Small Industries Corporation has a programme to render mentor and advisory services to the SMEs through accredited individual Consultants. Their bureaucratic delays and apologetic cost subventions make the scheme almost a non-starter in several SME-dominant States.

Cluster and network development threw up challenges and opportunities in human resource development benefiting the SMEs. Small enterprises have come to increasingly realize that their investment in training is worth its while as they led to improving the value chain management and core competences.

Most Management schools and technology institutes in India (IIMs and IITs, Technology Universities) have their own curriculum based programmes for the SMEs though they just scratch the surface and their interaction with the SMEs is at a very low level as most of them consider that the SMEs cannot afford the salaries of their graduates at the entry point excepting the IT-related enterprises. Trainer training programmes and academia-industry interactions are interventions that deserve immediate attention of the policy-makers.

The poor pick up in HRD efforts arise from (a) structural deficiencies of the enterprises, and (b) high costs. The sector needs more handholding efforts through peripatetic training and consulting teams at affordable costs or at subsidized costs.

The saga of SMEs in India is a mixed bag of successes but has lessons for developing an integrative model built on cluster concept and sustainability, which is outlined here. Strategy outlined here will reduce spatial socio-economic disparities and enhances opportunities for enterprise and employment promotion.

SECTION-III

Sustainable Development of Rural Areas—Rural Entrepreneur Nucleus Estates (RENE)

The rural poor can be described as one of the most disadvantaged sections of the society in terms of not only their low socio-economic situation and access to information, resources and opportunities but also their degree of vulnerability to economic shocks. Vulnerability is a constant companion of the rural poor since they have precarious employment in the formal and informal sector.

> *Therefore, enhancing security by reducing the rural poor's vulnerability to risks and putting in place mechanisms and institutions to help them cope with adverse shocks is the key to reducing rural poverty.*

(i) The Unsustainability of Rural Areas

Innovative, workable, practical and functional paradigms of rural development are urgently required in the quest for sustainable rural communities—communities in which economic development is achieved in harmony with improvement in the quality of human life; and where a healthy economy supports the quality of both human and ecological systems. The majority of rural areas in India are vortices of unsustainability and social and economic distress. Communities seeking to become more sustainable must begin to view their towns and villages as living systems with solid economic foundations. Sustainable community planning in the past often did no more than add up the separate interests of stakeholders without ever achieving an integrated view of the community systems embedded in the local environment.

A new development perspective is needed, which provides fresh insights and tools for improving a community's industry, commerce, infrastructure, and household behaviour as a whole system. The goal should be to support business competitiveness and job creation through strategies that also improve quality of life in all dimensions.

The rural-urban connect emerges from the farm sector. Several of the small farmer families utilize their farm output for conversion into ready-to-eat foods although most of them do not conform to health and hygiene standards. While the earlier DWCRA and self-help groups endeavoured to empower these vulnerable groups and reasonably succeeded these have not proved to be sustainable. Another window of opportunity became available for MNCs to use these groups to take their products at least cost to rural markets while the MNCs did not make any effort to change the production pattern to suit the ever-demanding market standards.

(ii) Value Addition

The potential for achieving higher levels of value addition in Indian agriculture is substantial. At present, the value addition in our agricultural sector is estimated at about 10 per cent against a level of 25-30 per cent in China and the South East Asian countries and over 75-100 per cent in some of the developed countries. The 10 per cent value addition in Indian agriculture works out to US $ 114 billion (V. Venkateswarlu, 2001). It has further been estimated that an increase in value addition in agriculture by 50 per cent would result in a 1 per cent growth in the country's GDP.

Increasing value addition in the production and marketing of agricultural produce, therefore, should be a major thrust of future agricultural policy. There are several links in the value chain that links production on the farm to the last retail point that reaches the consumer. The production chain and the value chain extend from the farmer's field to the retailer's shelf through a set of links that include post-harvest technologies and marketing interventions. (See Chart). Policy attention needs to focus on the weak links in the chain for agricultural production to be made competitive and cost-efficient. A story put out by the World Development Report 2003 is worthy of quote: "A typical story of fisheries development starts with the discovery of new fish stock. As fishermen make good money, more capacity rushes in. Though scarcity and wasteful competition ensue, fishermen do not coordinate management. Over time, the catch falls, first per hour of effort, then in total with individual fishermen barely breaking even. This wasteful race to exploit the resource wastes labour and capital, since fish that should have been allowed to spawn and grow are caught. Government involvement can make matters worse, if subsidies distort signals and prolong over fishing. Contrast this story—of waste and no profit—with alternatives in which the value of the resource is maximized. A cooperative of fishermen could manage a resource well if it could control its members, if its members could control their organization, and if the cooperative could control entry into the industry—all big ifs." (WDR, 2003: 43).

The "ifs" can be realized through setting up institutions that respect accountable democratic processes. There is a broad realization that value addition to agriculture would take place only through agro industries and agribusinesses. The UPA Government has announced incentives for setting up agri-enterprises with the support from the banks and financial institutions. All these efforts are still in embryonic stages. Failures of earlier institutional mechanisms and the tardy progress in the new initiatives, the jobless growth in agriculture and industry wherever growth occurred, reflect the need for modern management and promotional ability. RURAL ENTREPRENEUR NUCLEUS ESTATES could provide the answer.

Traders having conflicting interests with the producers provide linkages of traditional village producers to the outside world. One approach that occurred to us is the development of nucleus estates in the rural hinterlands where all the existing schemes of assistance by both the farm and small and rural industry departments have a scope to converge. Briefly the nucleus estate approach involves the development of a processing and marketing centre linked to a number of satellite producers. Basically the idea is to provide a linkage between the traditional producers, such as rice-growers to rural domestic food product manufacturers like the rice-flakes, rice condiments, pryums, etc.

The goals of the nucleus estates project can be identified as follows:

1. To link mass of the population, tied to traditional means of production, into modern commercial sector of the economy;
2. To enhance the development of traditional agriculture by linking them to broader market and technological opportunities;
3. To bridge the gap between the development of the modern and the traditional sector of the economy; and
4. To stimulate the development of entrepreneurship based on health and hygiene.

It is imperative to realize that most farmers and women in the related households can only work and aspire towards what they can see, reach and understand. The cultural climate surrounding them has been characterized by its static and restrictive nature and their inability to see beyond their nose. The project unfolds a strategy to change their mindsets not by preaching but by handholding to make them see profit in their domestic enterprises while following hygienic practices in production of the same old goods they were used to producing in a different way.

The project helps to reduce the vulnerability of the rural poor and contribute to improvement in the quality of life of the disadvantaged communities living in rural areas of States like Andhra Pradesh, Karnataka, Tamil Nadu, Maharashtra, UP, West Bengal, Orissa, where crop diversity is taking roots, on a sustainable basis. By doing so, economic empowerment occurs through a *workable and practical model of integrated rural development* that would demonstrate best practices in community capacity-building and could also influence the donors, Government and other agencies for their promotion and replication. The success would depend upon the provision of necessary wherewithal within their existing domestic environs instead of shifting them to common work places. Agro-processing, therefore, assumes critical importance in this regard.

(iii) Agro-Processing

There are several agricultural products that can serve as raw material for the food processing industry as a wide variety of processed products can be made with their input. Rice, wheat, maize, sugarcane, soybean are some of the products that can enter into the value addition chain. Several types of fruits and vegetables have immense potential to be processed for domestic and global markets. Growth in demand is projected for: frozen/ dehydrated/canned fruits and vegetables, Fruit juices/concentrates/pulps, Pickles, Jams, Jellies, Squashes, etc. apart from ready to eat vegetables. A significant share of consumption in developed markets—almost 60 per cent in the US is for fruits and vegetables in processed form.

The constraints that the farmer faces in the agro-processing link of the supply/value chain are related to lack of adequate information about the possibilities of processing agricultural products, the importance of price and quality. Finance should not prove a constraint once the risks are properly addressed and insured.

(iv) Proposed Intervention

As mentioned above, several by-products of agricultural crops get processed into either intermediary products or ready-to-eat products in the small farmers' households. Women produce most such items. These household production centers in the food processing industry—broadly classified as Self-employed Non-farm Enterprises—do not generally follow hygienic ways of producing, mostly because of lack of awareness. They do not have what can be called a working platform in the household.

(a) In each village of a product cluster of, say, 10-15 villages, the block/mandal level Industries Extension Officers should be able to locate 20 promising households willing to go in for hygienic production given certain basic resources and training. With as little an investment as Rs.5,000 per household, a working platform can be built within the house and hand-gloves can be supplied at regular prescribed intervals to these households. Within a short period of 3-6 months of intensive supervision and extension, such units can be houses of quality production.

(b) For this cluster of 10-15 villages, a packaging unit can be established with very little investment of about Rs. 5 lakhs. Depending upon the quantum of production per day, a self-employed person can be commissioned for transporting the products to the nearby markets (either through a tempo or parcel van). The same person can bring in the raw materials and packaging materials.

(c) Such clusters, if they get developed in a block/*mandal*, say 10 to 15, then, a Brand supervisor can be employed to ensure that the standards of health and hygiene (HACCP) are appropriately taken care of. There can also be another small-scale unit for printing, labelling and computer coding. Thus, the whole food processing activity generated out of the farm produce closer to the farmstead can attain standards of excellence.

(d) Since the investments required for such interventions are very minimal, it is worthwhile to consider giving them by way of subsidy, mainly because such cost cannot be recovered through price by these tiny units. The village enterprises have the potential to become global players under such a scenario.

(e) Packing of fruits and vegetables can be done manually or by machines. Packaging material can consist of plastic/polythene material or aluminum foils, fibreboards or cardboard material of durable quality. The objective is ease of handling in loading and transit to the destination as well as minimizing loss or damage to the horticultural produce. Coordinating with the Regional office of the Indian Institute of Packaging will be quite helpful.

(f) Labelling of several products is being made mandatory with growing concern over genetically modified organisms in the food chain and concerns about threats to the environment through certain practices (marine food production, for example). A specific WTO Agreement—the Agreement on Technical Barriers to Trade (TBT)—deals with these issues and failure to meet the requirements set could have an adverse impact on exports of farm produce. Suitable interventions in this area of Packaging and Labelling can enable household producers to cater to the needs of the market in a better way. In short, the project empowers the rural poor not through just grants or subsidies but through establishing commercial viability of their household enterprises and through capacity-building. This is clearly illustrated in the charts appended.

(v) Funding Pattern

The project envisages utilization of existing funding mechanisms under SGSY, PMRY, CMEY, and Employment Guarantee Programmes dovetailed with the credit institutions' programmes. In effect the project aims at convergence of all the existing schemes for delivering through a single window—not as eleventh but as one. We would like to have a Special Project Vehicle for this purpose called RENE managed by a team of professional executives with ambition to work in the rural environment. I envisage the project to become self-generating and self-sustaining within a period of three years after which all the funding institutions would withdraw and the mentoring services would continue on pay-as-they-able basis by the user population. Animators of this project should be utilized for replicating this project in other regions with such modifications as the evaluation leads to.

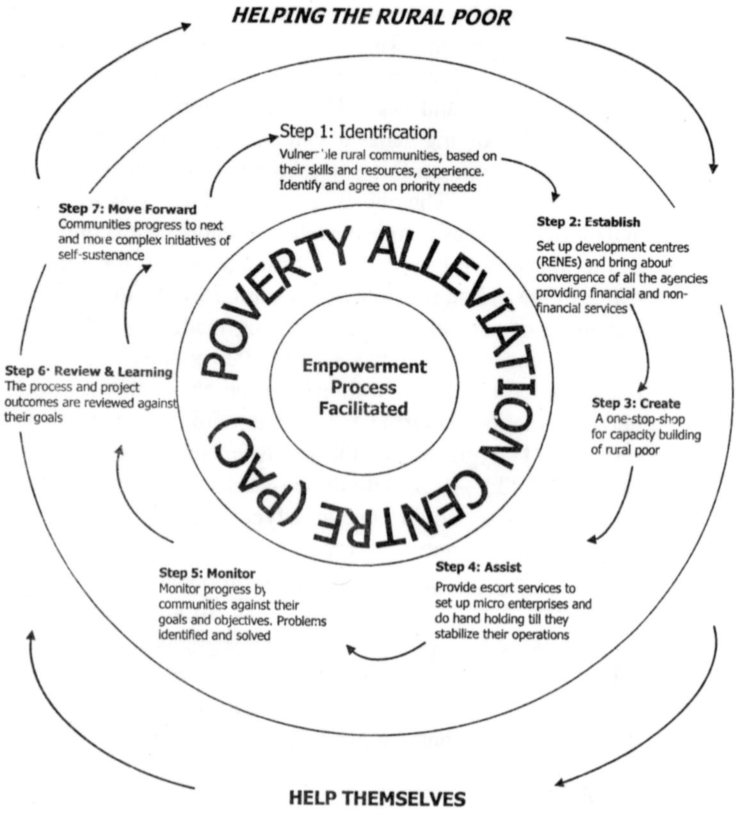

Fig. 3.1: The Cycle of Empowerment of Rural Poor

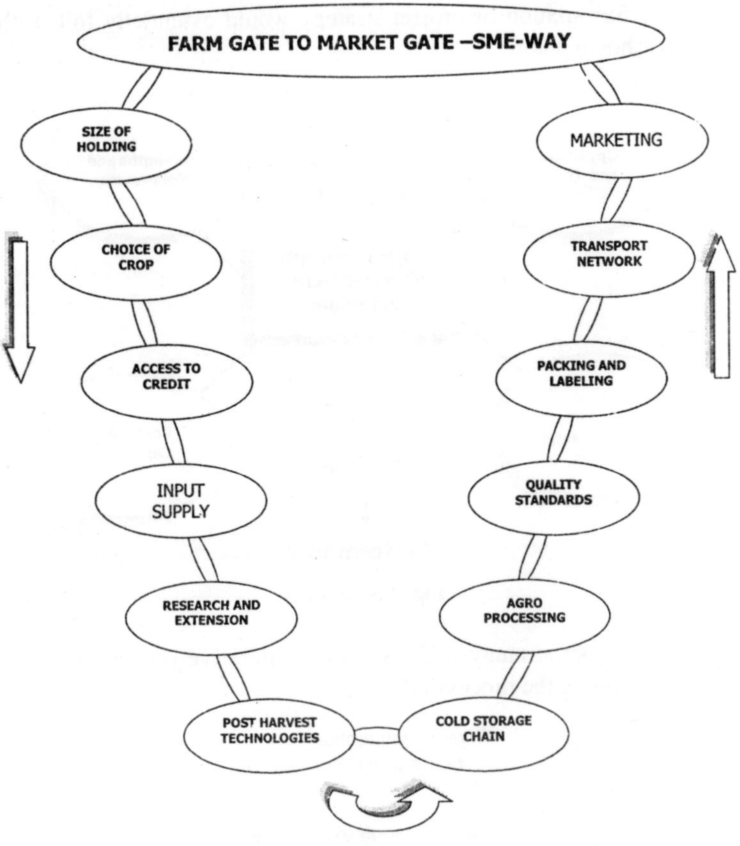

Sandilya Pvt. Ltd., Plot No. 143, Tarnaka, Hyderbad-500017

Fig. 3.2: Development of the Value Chain

The Rural Entrepreneur Nucleus Estates can act as focal points for convergence of all the programmes to bring about synergy between the various on-going poverty interventions. The main objective of these centres is to provide a 'one-stop-shop' for the local user population by strengthening every link in the value chain that leads to a commercial gain.

The formulation of project strategy would essentially follow the process shown below:

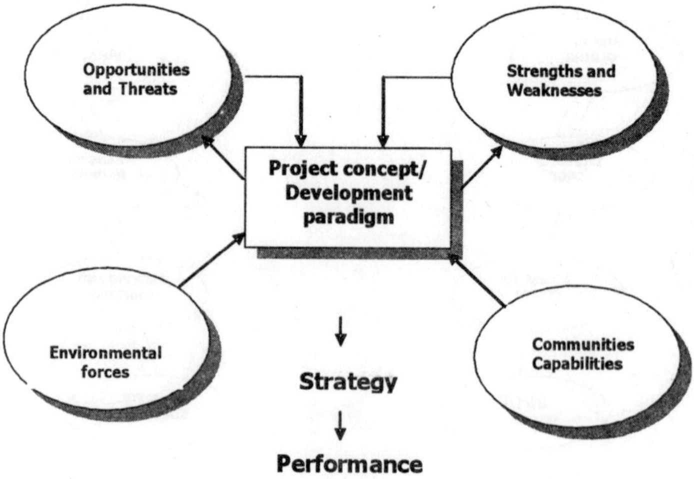

Fig. 3.3: Strategy

The project strategy is essentially an iterative process and will, therefore, follow the process given below:

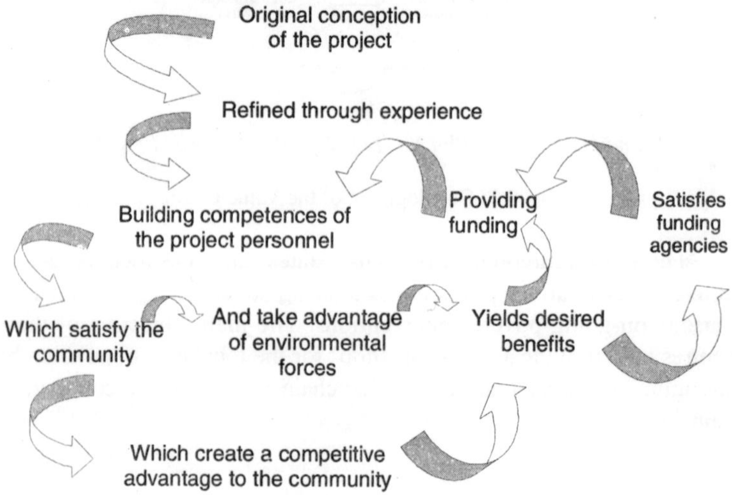

Fig. 3.4: Process

Institutional Framework: The overall functions of RENE are depicted below:

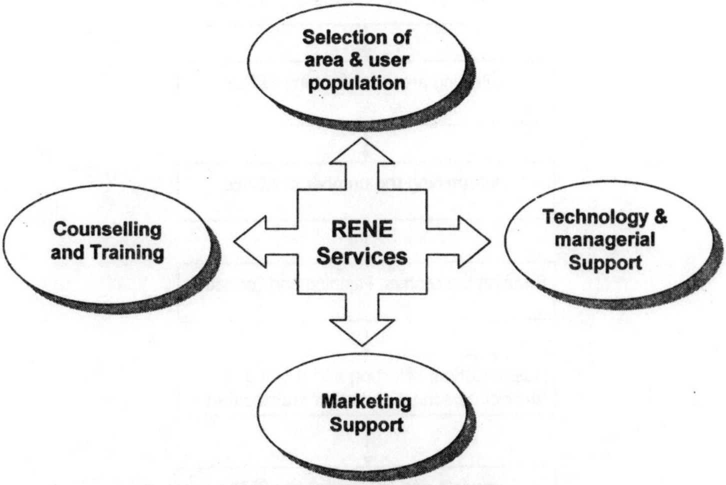

Fig. 3.5: Institutional Framework

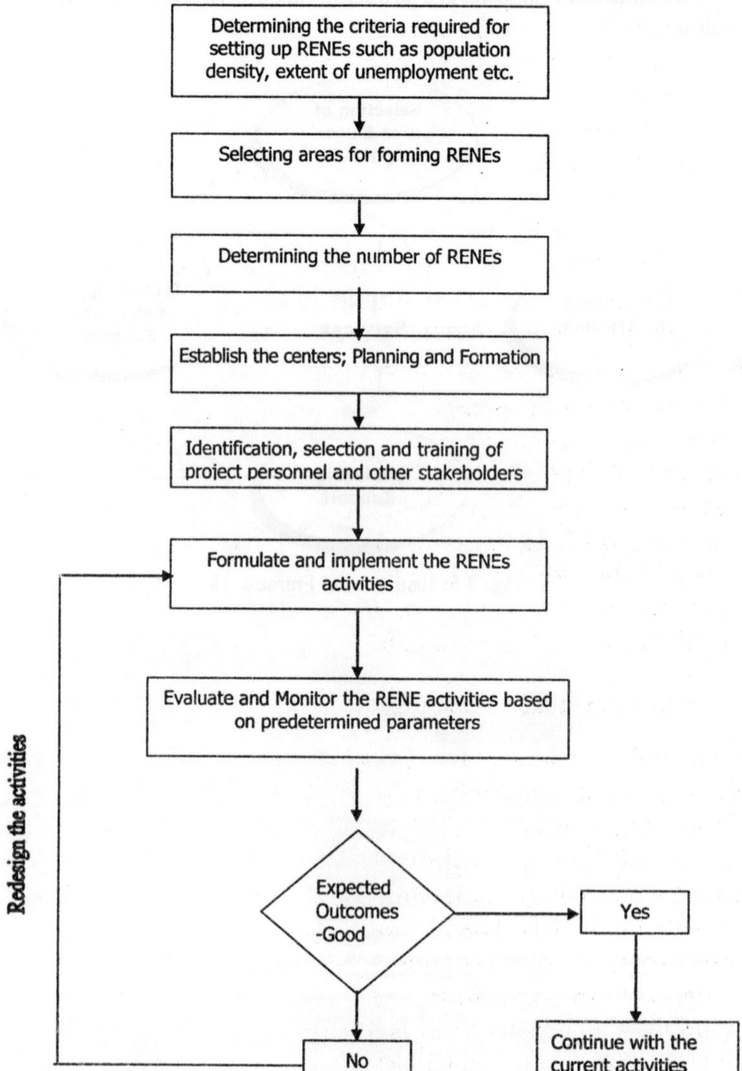

Fig. 3.6: Flow Process for Planning, Formation, Operationalizing and Maintaining the RENEs

Contextual Framework for Operationalising and Maintaining RENEs

The contextual framework in which the RENEs will operate is illustrated in the diagram below:

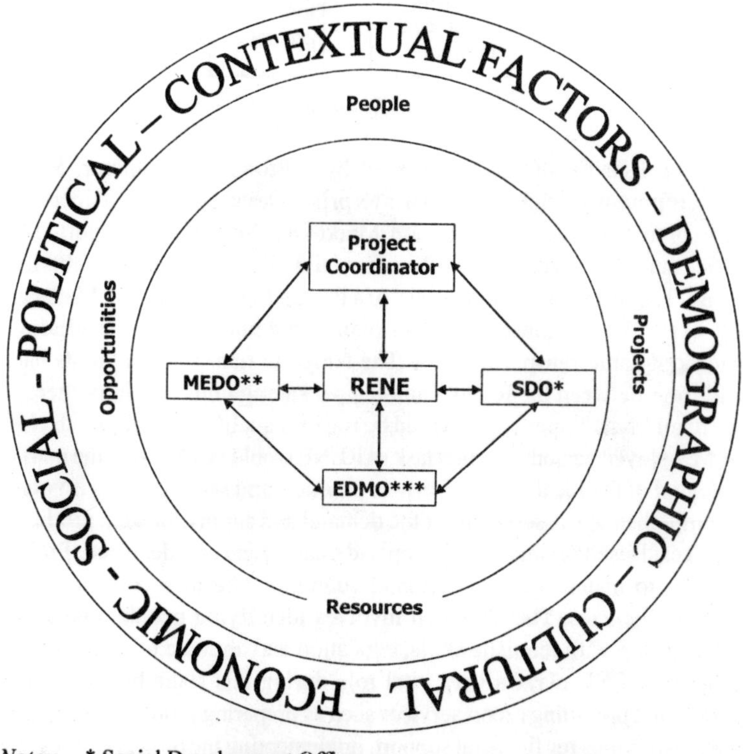

Notes: * Social Development Officer.

** Micro Enterprise Development Officer.

*** Entrepreneurial Development and Marketing Officer.

Fig. 3.7: Contextual Framework for Operationalising and Maintaining RENEs

The operationalzing and maintenance phase of RENEs involves setting in motion a chain of activities shown below. The cycle of empowerment of rural poor (shown above) illustrates these activities in a more holistic perspective.

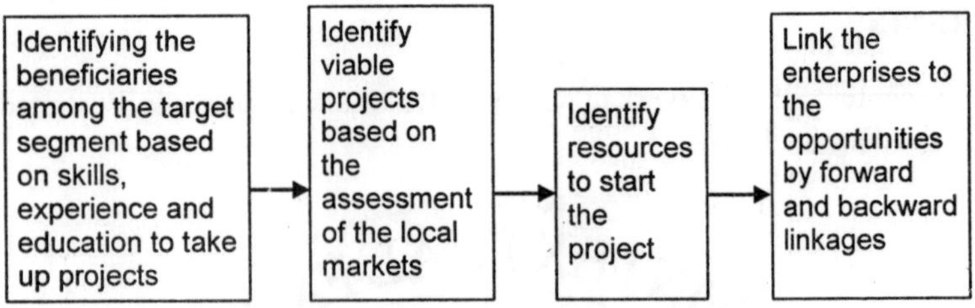

Fig. 3.8: Stages

The RENEs should be manned by a Project Coordinator, Social Development Officer, Micro-Enterprise Development Officer and Entrepreneurial Development and Marketing Officer and supported by office staff. The second tier officials would not be appointed in the first phase. The first phase would consist of a Project Manager, Quality Managers for every 10-15 households and plant personnel to run the grading, packaging and transporting unit. The tempo to transport the goods back and forth between the RENEs and the packaging units and from there to the retail Malls/Superbazars would be run by a qualified/trained tenth class self-employed person. The first task of RENE would be assessing the market potential of the local area for various products and services. This involves a comprehensive assessment of the demand and supply for such products/ services. Based on this study, supply-demand gaps are identified. Project profiles to match the local demand conditions are then collected from various agencies. The next step involves identifying people among the target group with requisite skills, education and aptitude to take up these projects. RENE plays a very vital role during this stage by identifying people and providing escort services such as preparing proposals, liasoning with institutions for financial support, implementing the projects, providing forward and backward linkages (procurement and marketing tie-ups) and training the entrepreneur and project personnel. It is extremely important that bureaucrats should not man RENE. There should not be a regulatory overhang on RENE. Effective monitoring and supervision should ensure quality and accounting standards in conducting RENE.

RENE's Vision

"Economic empowerment of the rural poor through effective value chain management."

The Four Components of Integrated Rural Development

Fig. 3.9: The Four Components of Integrated Rural Development

Stage I

The first stage in developing the village level development strategy is to do a comprehensive environmental scanning to understand the social, economic, cultural and political dimensions of the local area, which will impact and influence the project implementation and outcomes. In addition to this, an overall resource audit and mapping and identifying the resource linkages will have to be carried out. Based on these assessments, micro-levels plans need to be developed for the local community. However, for such plans to have legitimacy, the active involvement of all the local and state level stakeholders is necessary. At the same time, the plans cannot simply be the "wish list" of a given community. Hence, brokering the balance between the strategic and the local issues is the core concern of the entire project.

Stage II

Use the local map emerging from stage I as the basis for developing a local plan in consonance with the project goals of bringing out products based on health and hygiene.

Stage III

This stage would require identification of and establishment of market linkages for the products.

Stage IV

This stage involves the application of the usual monitoring and evaluation apparatus with feedback to the local implementation group (and also to strategy group). During this phase the lesson learnt from the project will be compiled, which will then be disseminated to all the stakeholders and various funding agencies for scaling-up the project or replicating it in other places.

Monitoring the Performance and Evaluating the Impact of this Project

The objective of the entire project is to ensure sustainable reduction in rural poverty and empower the rural poor to decide their own destiny. It is imperative to measure the performance of the project and, particularly, of RENEs against some preset milestones to evaluate and monitor the impact of the RENEs. The impact of the project can be broadly measured by the:

- Performance of the project,
- Impact on the rural micro-enterprises and livelihood options,
- Social and economic benefits from the project.

The performance can be more specifically measured by using a set of tangible and intangible indicators.

The tangible indicators include:

- Number of market enterprises established,
- Number of people gainfully employed,
- Number of people counselled,
- Number of people trained under various categories, skilled, unskilled—in-house/institutional,
- Number of clearance camps,
- Number of networks
- Quantum of marketing tie-ups,
- Number of marketing events,
- Number of job works provided.

The intangible indicators include:

- Social security,
- Capacity-building,
- Single-window mechanism (utility of financial and physical),

RENE has potential to become an independent brand for several household enterprises in the rural areas manufacturing agro-based products and ready-to-eat products within the project maturity period.

Some measurable benefits to the community include:

- Economic benefits through,
- Increased access to the market and linkages,
- Increased access to the investment and technology,
- Increased use of local resources,
- Better linkage between SHG/DWACRA with RENE,
- Value addition to the existing micro-enterprises.

Conclusion

India is a cynosure for investor's eyes both within and outside the country. UPA Government's new Common Minimum Programme, National Manufacturing Competitiveness Council's draft report unleashing an agenda for attaining global competitiveness of Indian firms, the new policies giving thrust to farm and SME sectors, the enactment of NREG and Right to Information Bills, the SME Development Bill waiting to be enacted, the capital market reforms and a host of other initiatives are responsible for upgrading the country's ranking by international rating agencies like the Moodys', Fishers' Standard & Poor etc. Implementation concerns, however, border on endemic issues like all-pervasive corruption, bureaucratic apathy to change, new regulatory institutions led not by professionals but bureaucrats of yester years (concept and not competence in question here) and the poor getting by-passed in the process. This contextual frame made me look at the mosaic of Rural India where 700 million of India's population still reside through the prism of PURA that A.P.J. Abdul Kalam, President of India unveiled before the nation in 2004.

One of the thrust areas for policy intervention is to facilitate increase in value addition in agriculture with a view to stimulate rural employment, generate higher incomes for rural households, promote exports and, at the same time, increase the knowledge and skill base of those living in rural areas. Opening up the agricultural sector to global markets within the framework of the WTO trade regime calls for innovative strategies to face changes in consumer demand patterns, meet international quality standards, and intense competition in the market place.

Issues of enterprise, employment and efficiency are viewed as critical to the growth of the economy. A strong SME sector well integrated with the large enterprise sector on one side and micro enterprises on the other has potential to unleash the energies of 'bottom of the pyramid' enterprises and employment-oriented growth of the economy. The Rural Entrepreneurship Nucleus Estates envisaged in this article would convert pro-poor initiatives into concretized action projects resulting in a paradigm shift. They carry the prospect of reducing spatial socio-economic disparities and enhance opportunities for private enterprises, employment and development taking the economy closer to attaining the Millennium Development Goals.

The enlightened among the rulers know where the dirt and stink lie. They also know that spraying a perfume or lighting an *Agarbati* would not remove them. It is courage of conviction, speed of action, and expending resources where they need most, subject to social audit, that would be the broad-spectrum antibiotic to the planning body politic.

Box 3.6: Major Poverty Alleviation, Employment Generation and Basic Services Programmes

- *National Food for Work Programme*: In line with the NCMP, National Food for Work Programme was launched on November 14, 2004 in 150 most backward districts of the country with the objective to intensify the generation of supplementary wage employment. The programme is open to all rural poor who are in need of wage employment and desire to do manual unskilled work. It is implemented as a 100 per cent centrally sponsored scheme and the foodgrains are provided to States free of cost. However, the transportation cost, handling charges and taxes on foodgrains are the responsibility of the States. The collector is the nodal officer at the district level and has the overall responsibility of planning, implementation, coordination, monitoring and supervision. For 2004-05, Rs.2020 crore have been allocated for the programme in addition to 20 lakh tonnes of foodgrains.
- *Swaranjayanti Gram Swarozgar Yojana (SGSY)*: SGSY, launched in April 1999, aims at bringing the assisted poor families (*Swarozgaris*) above the poverty line by organizing them into Self-Help Groups (SHGs) through a mix of Bank Credit and Government Subsidy.
- *Sampoorna Grameen Rozgar Yojana (SGRY)*: SGRY, launched in 2001, aims at providing additional wage employment in all rural areas and thereby food security and improve nutritional levels. The SGRY is open to all rural poor who are in need of wage employment and desire to do manual and unskilled work around the village/habitat. The programme is implemented through the Panchayati Raj Institutions (PRIs).

(*Contd.*)

Box 3.6 (*Contd.*)

- *Rural Housing—Indira Awaas Yojana (IAY)*: The Indira Awaas Yojana (IAY) operationalised from 1999-2000 is the major scheme for construction of houses for the poor, free of cost. The Ministry of Rural Development (MoRD) provides equity support to the Housing and Urban Development Corporation (HUDCO) for this purpose.
- *Pradhan Mantri Gramodaya Yojana (PMGY)*: PMGY, launched in 2000-01, envisages allocation of Additional Central Assistance (ACA) to the States and UTs for selected basic services such as primary health, primary education, rural shelter, rural drinking water, nutrition and rural electrification. For 2003-04 as well as 2004-05, the annual allocation of ACA for PMGY was Rs. 2,800 crore.
- *Rural Employment Generation Programme (REGP)*: REGP, launched in 1995 with the objective of creating self-employment opportunities in the rural areas and small towns, is being implemented by the Khadi and Village Industries Commission (KVIC). Under REGP, entrepreneurs can establish village industries by availing of margin money assistance from the KVIC and bank loans, for projects with a maximum cost of Rs.25 lakh. Since the inception of REGP, up to 31 March 2004, 1,86,252 projects have been financed and 22.75 lakh job opportunities created. A target of creating 25 lakh new jobs has been set for the REGP during the Tenth Plan. 8.32 lakh employment opportunities have already been created during 2003-04. For 2004-05, a target of creating 5.25 lakh job opportunities has been fixed.
- *Prime Minister's Rozgar Yojana (PMRY)*: PMRY started in 1993 with the objective of making available self-employment opportunities to the educated unemployed youth by assisting them in setting up any economically viable activity. So far, about 20 lakh units have been set up under the PMRY, creating 30.4 lakh additional employment opportunities. The targets for additional employment opportunities under the Tenth Plan and in 2004-05 are 16.50 lakh and 3.75 lakh, respectively. While the REGP is implemented in the rural areas and small towns (population up to 20,000) for setting up village industries without any cap on income, educational qualification or age of the beneficiary, PMRY is meant for educated unemployed youth with family income of up to Rs.40,000 per annum, in both urban and rural areas, for engaging in any economically viable activity.
- *Pradhan Mantri Gram Sadak Yojana (PMGSY)*: The PMGSY, launched in December 2000 as a 100 per cent Centrally Sponsored Scheme, aims at providing rural connectivity to unconnected habitations with population of 500 persons or more in the rural areas by the end of the Tenth Plan period. Augmenting and

(*Contd.*)

Box 3.6 (*Contd.*)

modernising rural roads has been included as an item of the NCMP. The programme is funded mainly from the accruals of diesel cess in the Central Road Fund. In addition, support of the multi-lateral funding agencies and the domestic financial institutions are being obtained to meet the financial requirements of the programme. Up to October, 2004, with an expenditure of Rs. 7,866 crore, total length of 60,024 km. of road works has been completed. The National Rural Roads Development Agency (NRRDA), an agency of the Ministry of Rural Development registered under the Societies Registration Act, provides operational and technical support for the programme.

- *Drought Prone Areas Programme (DPAP), Desert Development Programme (DDP) and Integrated Wastelands Development Programme (IWDP)*: DPAP, DDP and IWDP are being implemented for the development of wastelands/degraded lands. During 2004-05 allocation of Rs. 300 crore, Rs. 215 crore and Rs. 368 crore were provided for DPAP, DDP and IWDP, respectively. So far, during 2004-05, 2,550 projects covering 12.75 lakh hectares, 1,600 projects covering 8 lakh hectares and 165 projects covering 8.32 lakh hectares, have been sanctioned under DPAP, DDP and IWDP, respectively.
- *Antyodaya Anna Yojana (AAY)*: AAY launched in December 2000 provides foodgrains at a highly subsidized rate of Rs.2.00 per kg for wheat and Rs.3.00 per kg for rice to the poor families under the Targeted Public Distribution System (TPDS). The scale of issue, which was initially 25 kg per family per month, was increased to 35 kg per family per month from April 1, 2002. The scheme initially for one crore families was expanded in June 2003 by adding another 50 lakh BPL families. During 2003-04, under the AAY, against an allocation of 45.56 lakh tonnes of foodgrains, 41.65 tonnes were lifted by the State/UT Governments. Budget 2004-05 expanded the scheme further from August 1, 2004 by adding another 50 lakh BPL families. With this increase, 2 crore families have been covered under the AAY.
- *Swarna Jayanti Shahari Rozgar Yojana (SJSRY)*: The Urban Self-Employment Programme and the Urban Wage Employment Programme are the two special components of the SJSRY, which, in December 1997, substituted for various extant programmes implemented for urban poverty alleviation. SJSRY is funded on a 75:25 basis between the Centre and the States. The expenditure during 2003-04 was Rs.103 crore. For 2004-05, the allocation is Rs.103 crore, out of which Rs.90.38 crore were utilized by December 31, 2004.

(*Contd.*)

Box 3.6 (*Contd.*)

- *Valmiki Ambedkar Awas Yojana (VAMBAY)*: The VAMBAY launched in December 2001 facilitates the construction and upgradation of dwelling units for the slum dwellers and provides a healthy and enabling urban environment through community toilets under Nirmal Bharat Abhiyan, a component of the scheme. The Central Government provides a subsidy of 50 per cent, the balance 50 per cent being arranged by the State Government. Since its inception and up to December 31, 2004, Rs. 753 crore have been released as Government of India subsidy for the construction/ upgradation of 3,50,084 dwelling units and 49,312 toilet seats under the scheme. For the year 2004-05, out of the tentative Central Fund allocation of Rs.280.58 crore, up to December 31, 2004, an amount of Rs. 223.66 crore has been released covering 1,06,136 dwelling units and 20,139 toilet seats.

NOTES

1. The Millennium Development Goals

1. Eradicate extreme poverty and hunger: reduce by half the proportion of people living on less than $1 a day; reduce by half the proportion of people who suffer from hunger.
2. Achieve universal primary education: ensure that all boys and girls complete a full course of primary schooling.
3. Promote gender equality and empower women: eliminate gender disparity in primary and secondary education by 2005, and at all levels of education by 2015.
4. Reduce child mortality: reduce by two-thirds the mortality rate for children under five.
5. Improve maternal health: reduce by three quarters the maternal mortality ratio.
6. Combat HIV/AIDS, malaria and other diseases: halt and begin to reverse the spread of HIV/AIDS; halt and begin to reverse the incidence of malaria and other major diseases.
7. Ensure environmental sustainability: integrate the principles of sustainable development into country policies and programmes; reverse the loss of environmental resources; reduce by half the proportion of people without sustainable access to safe drinking water and basic sanitation; achieve significant improvement in the lives of at least 100 million slum dwellers by 2020.
8. Develop a global partnership for development.

2. ADB Report

The poor are still being left behind because of limited access to education, land, credit and infrastructure such as irrigation, roads and electricity, the bank said. Going forward, the ADB provided a number of scenarios for future poverty reduction.

In the best-case scenario, economic growth in Asia continues at the rates seen between 1999 and 2003. Extreme poverty would drop as a result, with the total number of people affected cut to 150 million by 2015, from 690 million in 2002. South Asia, however, would still be the most poverty stricken, accounting for 140 million of the total.

The worst-case scenario provides a more sobering picture. Growth in Asia would only have to dip by one percentage point below the 1999-2003 levels for the number of people living in extreme poverty to be 429 million by 2015.

The ADB report concludes that "policy-makers must focus on generating high rates of sustainable growth while ensuring that at a minimum the distribution of income does not worsen to any significant degree".

"The fight against poverty in Asia will be protracted."

Source: http://news.bbc.co.uk/2/hi/business/3600762.stm

3. "Who can make their way easily in Micro Finance institutions has been nicely presented at a seminar on Micro Finance 2008 held with DFID support by the APMAS"

- Organizations with relatively large outreach and large volumes.10 out of 94 organizations have access to market finance.
- Institutional Development (ID) partners are on the way to linkage but only after 3 to 4 years such linkage happens. Only 2 partners out of 16 ID POs linked to Financial market.
- To majority, fund is difficult to link due to low capital adequacy, negative ROA, inefficient MIS and poor information on market and products.
- Risk assessment is missing scenario. Even if it is there, allocation of fund to take care of it (by way of loan loss reserve) is seldom.
- Fund management skill is very poor when more than 2 funding sources are in the picture.
- Competent staff at accounting and operations are a boon for organisational effectiveness.

Source: Friends of Women's World Banking, India—2004

NOTES

1. Government of India, Ministry of Statistics and Programme Implementation *http://mospi.nic.in/mospi_iip.htm*

2 *Economic Survey* 2004-05.
3. *Support for this Digital Dividend "What Works":* Case Study Provided by The Microenterprise Development Division of the United States; Agency for International Development (USAID).
4. A.P.J. Abdul Kalam, *"Empowering Rural India"*, Express On-line, 25 August 2005.
5. *Business Standard,* Ahmedabad 18th April 2005.
6. *The DPR is a new, integrated analytic instrument that assesses a country's past development outcomes and identifies factors constraining better economic performance and faster poverty reduction.*
7. Report of UN on *Unleashing Entrepreneurship*, 2004.
8. Rangarajan, 2000.
9. This definition is largely based on Fisher, Thomas and Mahajan, Vijay (1996): *The Forgotten Sector: Rural Non-Farm Employment and Enterprises in Rural India*, Oxford and IBH, New Delhi, ITDG, UK.
10. See SEDF.
11. DC SSI, GOI recently carried out Third Census Survey and hence more statistical detail is available.
12. Kenan 1994.
13. Francois and Shiells 1993.
14. C.H.Hanumantha Rao, 2005 Rural Transformation in India, p. 31.
15. C.H. Hanumantha Rao, 2005.
16. The Reserve Bank of India appointed from time-to-time a number of committees to examine the problems in meeting the financial requirements of the small industries. The prominent among them are: Puri Committee (1980), Tewari Committee (1985), Nayak Committee (1993), Kapoor Committee (1999) etc. The Banks on their own and a few State Governments have also addressed the issues relating to sickness in the SSI sector: SBI Study; ASCI Study sponsored by the Government of Andhra Pradesh (1997) and another by the Government of Jammu and Kashmir (1998). Government of India also initiated two major studies in this regard: Expert Committees with Abid Hussain as Chairman (1998), S.P. Gupta, Member of the Planning Commission as Chairman (2001) and Ashok Ganguly, former Chairman, HLL and presently Director, RBI as Chairman (2004). All these studies were unanimous in their opinion relating to the inadequacies and inefficiencies in the lending systems of the financing institutions and banks in meeting the requirements of the small industries. A specialized institution set up for meeting the requirements of this sector, viz., Small Industries Development Bank of India (SIDBI) did not also go far enough to redress the grievances of the small industries in this regard. Therefore, all the efforts put in for reducing the supply side constraints proved a damp squib. It is but necessary to look at these supply side constraints, as we have to seek solutions in the problem arena and not elsewhere. (See Raju 2001:RGICS WP33 for detailed analysis of the Indian Financial System vs. SMEs). The

Internal Working Group on SMEs, 2005 of the RBI have made certain recommendations that found merit with the GoI resulting in a renewed thrust on this sector from the banks and financial institutions vide the Policy Announcement of Finance Minister dated August 10, 2005.

17. Deposit Insurance and Credit Guarantee Corporation of India, an extended arm of the RBI (1962-97) and a similar Credit Guarantee Corporation in Sri Lanka were wound up. SIDBI, at the instance of and with the support of Government of India started in the year 2000 a Credit Guarantee Fund Trust with participation from the Industry Associations. It took nearly an year for enthusing the public sector banks to become members of this scheme. None of the private banks has become a member which speaks of their interest in catering to the requirements of the really small enterprises whose working capital requirements are below Rs.2.5 million.
18. Mario Rutten and Carol Upadhya (Ed), *Small Busines Entrepreneurs in Asia and Europe—Towards a Comparative Perspective*, Sage Publications, New Delhi, 1997.
19. *Human Development Report, 2001*, Oxford University Press, New Delhi p. 3.
20. A.P.J. Abdul Kalam & Rajan 1998: 47.
21. Gulati, Mukesh (1998), SME Development Cluster Development Programme--Prospects, Problems and Strategies in Small Industries in India, Yerram Raju, 2004.
22. (Ms) Thitapha Wattanapruttipaisan, Promoting SME Development: Some Issues and Suggestions for Policy Consideration; Bulletin on *Asia-Pacific Perspectives* 2002/03, pp. 57-67.
23. McGregor Allister (1999), "Micro-finance and Social Capital: A Social Resource Perspective", Paper presented at the Asia Microfinance Workshop, Action Aid, December 1999, Hyderabad.
24. Unieashing Entrepreneurship 2004 UNDP, Geneva.

REFERENCES

"Annual Report 2004-05", Reserve Bank of India.

"Cluster Development and Sectoral Programmes—Achievements and New Initiatives" by Ministry of SSI.

"Food Insecurity Atlas of Rural India"—M.S. Swaminathan Research Foundation.

"India five years of Stabilization and Reform and the challengers Ahead". A World Bank Country Study.

"India's Economic and Social Development National and International Perspective"—International Conference in Honour of Professor C.H. Hanumantha Rao 16-17, November, 2004.

"Small Scale Industries in India—An Engine of Growth" (2002)—Office of the Development Commissioner, Small Scale Industries, Ministry of SSI, New Delhi.

"World Economic Outlook (2004). The Global Demographic Transition". Published by International Monetary Fund.

Abid Hussain Committee (1997). *Expert Committee Report on Small Enterprises* p. 21.

Bibek Debroy, Shyam Babu, D. (2004). *The India Mosaic Search for an Identity* Published by Rajiv Gandhi Institute for Contemporary Studies

Carl Dahlman, Anuja Utz (2005) *India and the Knowledge Economy—Leveraging Strengths and Opportunities*. Published by World Bank Institute.

C.H. Hanumantha Rao, 2005 *Rural Transformation in India*, p. 31.

Coordinating Group for Studies on South Asian Perspectives; Perspectives on South Asian Cooperation.

Country Strategy and Programme Update (2002-04) July 2001 Bangladesh, Asian Development Bank.

Developing SME Policy in Pakistan, SME Issues Paper—for Deliberation by SME Task Force—Ministry of Industry, Government of Pakistan, 2002.

Final Results: Third All India Census of Small Scale Industries 2001-02, Development Commissioner (SSI) Ministry of Small Scale Industries, Government of India, New Delhi.

Financial Express, September 28, 2005.

Fisher, Thomas and Mahajan, Vijay (1996). *The Forgotten Sector: Rural Non-Farm Employment and Enterprises in Rural India*, Oxford and IBH, New Delhi, ITDG, UK.

Friedrich Ebert Stiftung, (1994). *Coordinating Group for Studies on South Asian Persectives—Perspectives on South Asian Cooperation*, Published by Pakistan Office of the German.

Friedrich Elbert Stiftung 1994.

Friends of Women's World Banking, India, 2004.

http://pmindia.nic.in/speech/content.asp?id=98;

http://www.ers.usda.gov/Briefing/India/issuesandanalysis.htm.

http://www.dmaran.nic.in/speechdisplay.php?id=83.

Human Development Report 2004, Published by Oxford University Press.

Human Development Report, 2001, Oxford University Press, New Delhi p. 3.

India Development Report 2004-05 (Ed). Kirit S.Parikh, R. Radhakrishna Indira Gandhi Institute of Development Research Oxford University Press 2005.

Industrial Policy, Privatisation and SME, National Policy Review Forum 2003 Bangladesh.

Joseph Stiglitz, (2002). *Globalisation and its Discontents*, Penguin Books, New Delhi.

Kalam, A.P.J. with Y.S. Rajan, (1998). *'India 2020—A Vision for the New Millennium*, Viking by Penguin Books India (P) Ltd.

Kim Seung Jin, Suh Jang won, (1992). *Co-operation in Small and Medium-scale Industries in ASEAN*, Asian and Pacific Development Centre, Malaysia.

Lavish Bhandari and Bibek Debroy, (2004). *Small-scale Industry in India*

Academic Foundation in collaboration with Friedrich Naumann Stiftung and Rajiv Gandhi Institute for Contemporary Studies, New Delhi.

Mario Rutten and Carol Upadhya (Ed.), (1997). *Small Business Entrepreneurs in Asia and Europe—Towards a Comparative Perspective*, Sage Publications, New Delhi.

Mario Rutten and Carol Upadhya, (1997). *Small Business Entrepreneurs in Asia and Europe*, Sage Publications, New Delhi.

Omkar Goswami and Rama Bijapurkar in *Business Standard,* July16, 2005.

Porter, M.E.J.D. Sachs and J.W. Mcarthur (2001). "Executive Summary: Competitiveness and Stages of Economic Deveopment", www.cid.harvard.edu/cr/pdf/GCR0102percent20Execpercent20 Summary.pdf

Prahlad C.K. (2004). *The Fortune at the Bottom of the Pyramid: Eradicating Poverty through Profits*, Wharton School Publishing.

Prime Minister Inaugurates Agriculture Summit, April 9, 2005.

Promoting Business and Technology Incubation for Improved Competitiveness of Small and Medium-Sized Industries Through Application of Modern and Efficient Technologies in Bangladesh Momtaz Uddin Ahmed, Member, General Economics Division, Planning Commission, Dhaka, Bangladesh 2004.

Promoting Business and Technology Incubation for Improved Competitiveness of Small and Medium-Sized Industries Through Application of Modern and Efficient Technologies in Nepal Mr. Chiranjibi Nepal, Director, Mr. Bishwa Raj Karki, Marketing Manager, Industrial Enterprise Development Institute, Kathmandu.

Raghbendra Jha (2000). *Growth Inequality and Poverty in India: Spatial and Temporal Characteristics*. Published by Indira Gandhi Institute of Development Research.

Rangarajan, C. (2000). *Perspectives of Indian Economy—A Collection of Essays*. UBS Publishers Distributors Ltd., New Delhi.

Rohini Nayyar and Alakh N. Sharma (2005). *Rural Transformation in India—The Role of Non-farm Sector*.

S.S. Bhalla and Tirthanmoy Das: *Why be afraid of the truth? Poverty, Inequality and Growth in India, 1983-2000*, December 6, 2004.

SEDF, Dhaka, Bangladesh 2002.

Special Group on Targetting Ten Million Employment Opportunities per Year—Employment Generating Growth. (2002)—Government of India, Planning Commission, New Delhi.

Swaminathan M.S. "Vision of Rural Agro Industrial Complexes" Agricultures, and Food Industry, February, 2004 pp. 51-52.

Technical Assistance (Financed by the Government of Denmark) to the Kingdom of Bhutan for Small and Medium Enterprise Development Asian Development Bank Tar: Bhu 37166 October 2004.

Technology Transfer in SMEs: Problems and Issues in the Context of Nepal

Chiranjibi Nepal, Bishwa Raj Karki And Kabya Prasad Niraula Industrial Enterprise Development Institute (IEDI), Kathmandu.

The Small and Medium Enterprises (SMEs) In Bangladesh: An Overview of the Current Status Dr. Momtaz Uddin Ahmed, Vice-Chancellor, State University of Bangladesh 2004.

United Nations 2000 Report on Positive Agenda and Future Trade Negotiations.

Unleashing Entrepreneurship, 2004 Report to the Secretary-General of the United Nations, UNDP.

Vijay Mahajan and Yerram Raju B. (2004). Institutional Framework and Enabling Environment for Rural Non-farm Sector in *Rural Transformation in India: The Role of Non-farm Sector* Edited by Rohini Nayyar and Alakh N. Sharma, Institute for Human Development, New Delhi.

Wojciech Hubner (2000). SME Development in Countries of Central Asia (Kazakhstan, Kyrgystan and Uzbekistan): Constraints, Cultural Aspects and Role of International Assistance UNIDO 2000.

World Development Report 2002.

World Development Report 2003.

World Development Report 2005, p. 85.

World Economic Outlook, September 2004.

Yerram Raju, B. (2005). "Occasional Paper—36; Commodity Study on Diversification of Indian Agriculture". Department of Economic Analysis and Research, National Bank for Agriculture and Rural Development.

Yerram Raju, B. (2001). RGICS WP33 for detailed analysis of the Indian Financial System Vs SMEs.

Yerram Raju, B. *Sustainability of Agriculture: Issues and Choices.*

Yerram Raju, *Marketing Holds the Future for Rural Industrialization*, Vol. 24(2) April-June 2005—National Institute of Rural Development, Hyderabad.

Yerram Raju B. (2004). Small Industries in India; Policies and Perspectives in *the Emerging Context, GIFT EXCEL Series*, New Delhi.

4

Sisal: Potential for Employment Generation and Rural Development

MOHINI SAXENA, SHIRAMDAS MURALI, MOGALI J. NANDAN and N. RAMAKRISHNAN

Introduction

Utilization of plant fibres for generating employment in rural sector is cost-effective and ecologically sustainable. India has a vast resource for different natural fibres viz., jute, *sisal*, banana, coir etc., which are abundantly available in many parts. Presently, the production of natural fibres in India is more than 400 million tonnes (Amit Rai and Jha, 2004). Among others, *sisal* has many advantages like thriving in wastelands and yielding superior fibre continuously for 6-8 years with least management input.

The Portuguese introduced *agave* in India in the 15th century. They are now found throughout the country. The major species of *agave* available in India are *A. sisalana, A. mexicana, A. americana, A.cantala,* and *A. veracruz*. Among the agaves, the most prevalent is *Agave sisalana,* commonly known as *sisal*, which is a xerophyte and can survive on poor soils in drought prone tropical regions. *Sisal* occupies 6th place among fibre plants, representing 2 per cent of the world's production of plant fibres (Rehm and Espig, 1991). The leaves of *sisal* plants yield a strong fibre, which is traditionally used for making ropes, cordage and twines. Apart from the traditional applications *sisal* finds its way in environment friendly engineering materials due to low density, high specific strength and bio-degradability. *Sisal* and allied activities like cultivation, fibre extraction, product making etc. are labour intensive, low-tech and has high potential for employment generation in rural and semi-urban sectors. The paper discusses various aspects of *sisal*, its applications and possibilities for sustainable rural livelihood.

Cultivation of *Sisal*

Sisal is a perennial hardy plant, which unlike the other fibres is not a seasonal crop. It can establish and easily grow in all States of India covering sub-humid to arid and semi-arid regions, which cover major parts of India. It can also survive in almost all soil types and its input costs are least for its survival, regeneration and maintenance on sustainable basis. *Sisal* tolerates prolonged droughts and high temperatures also. It yields parallel hard fibres.

The *sisal* plant has a stalk on which the succulent leaves are arranged spirally. Its dimensions are about 1 to 2 m in height, with a diameter of ~ 20 cm. The lance-shaped leaves, growing out from the stalk in a dense rosette, are fleshy and rigid, with dark green colour. Each is 0.8 to 1.5 m long, 7 to 8 cm wide at the base, and 10 to 15 cm across at the widest portion, terminating in a sharp spine. Cutting usually starts after 2½ years when the plant has about 100 leaves. The plant matures fully in 4 to 8 years after planting (Fig. 4.1).

Fig. 4.1: *Sisal* (Agave Sisalana) Plant

A monocarpic perennial *sisal* plant yields 220-250 linear, lanceolate leaves with a terminal spine in the 6-9 years before it poles (flowers) and dies. *Sisal* is stoloniferous plant, which produces shoots from the stolons, known as suckers or bulbils, which can be used for propagation. The suckers/bulbils are grown on in nurseries until they are about 50 to 70 cm high and then planted in the main field (Wienk, 1969).

Production Scenario of *Sisal*

In India, *sisal* is not cultivated and the sector is unorganized. However, considering the employment and income generation potential, the District Rural Development Agency (DRDA), Koraput, Orissa implemented an integrated "*Sisal* plantation, Fiber extraction and Rope making" programme under Jawahar Rojgar Yojana (JRY) in the year 1995 (Purandare and Sambi Reddy, 2001). Sisal is currently found on embankments, bunds and roadsides, serving the purpose of soil conservation and protection as hedge plantation. Presently *sisal* fibres are collected and utilized for conventional purposes like ropes, anchors, cordage and handicrafts. They are not optimally utilized and commercially exploited with respect to their abundant availability, superior characteristics/quality and wide applicability. A high yielding sisal hybrid (*Leela*) has been developed at Sisal Research Station, NIRJAFT (ICAR), Bamra in 1985 which has a yield potential up to 25 q/ha (http://www.crijaf.org/about/aicrp.html).

Extraction and Properties of *Sisal* Fibre

Sisal fibre is traditionally extracted by retting, a biodegradation process involving microbial decomposition of *sisal* leaves, which separates the fibre from pith. The fibres are washed and processed further. This process takes 15-21 days for a single cycle of extraction and degrades the quality of fibre. The retting process is water intensive, unhygienic and not eco-friendly. The other methods available for the extraction of fibre are chemical treatment and mechanical extraction. The mechanical extraction is done with the help of Raspador machine, developed by RRL, Bhopal. It is a semi-automatic machine suitable for small-scale operations (Fig. 4.2).

Sisal fibre consists of 66-72 per cent cellulose, 12 per cent hemicellulose and 10-14 per cent lignin. The superior engineering properties (diameter 50-200 mm; microfibril angle 10-22°, Ultimate Tensile strength of 468-640 Mpa; Modulus of 9.40–15.80 Gpa and elongation of 3–7%) makes it as an excellent material for manufacturing high strength textile and reinforcement in composites for various applications (Navin Chand and Rohatgi, 1994; Soumitra Biswas *et al.*, 2005).

Fig. 4.2: *Sisal* Fibre Extraction with Raspador

Uses of *Sisal*

The leaves of *sisal* yield a strong fibre (Fig. 4.3), which is traditionally used for making ropes, cordage and twines. It is also being used to manufacture coarse fabrics, rugs, carpets, handicrafts, mats, fishing nets etc. The *sisal* pulp left after fibre extraction (which is about 95%) is used for making paper/paperboards, hecogenin (a cortico steroid), wax, biogas etc. (CFC, 2000).

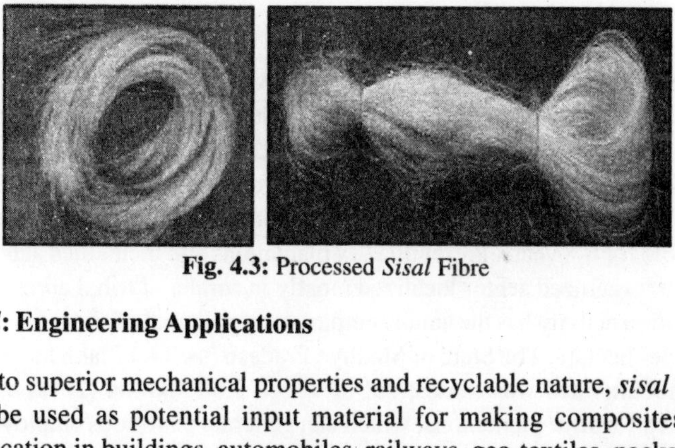

Fig. 4.3: Processed *Sisal* Fibre

***Sisal*: Engineering Applications**

Due to superior mechanical properties and recyclable nature, *sisal* fibre can be used as potential input material for making composites for application in buildings, automobiles, railways, geo-textiles, packaging industries etc. *Sisal* fibre reinforced composite building materials like;

wood substitute products, panels, doors, corrugated roofing sheets and instant houses suitable for disaster (floods/tsunami/earthquake) prone areas would be made which attract prospective entrepreneurs and stake-holders due to its durability and cost effectiveness. Asbestos fibres being carcinogenic, *sisal* fibre cement corrugated roofing sheets, which are eco-friendly, can be an effective alternative.

Present scenario indicates that the use of plant fibre (*sisal*/flax/hemp etc.) based automobile parts like trim parts, various panels, seat backs, shelves, brake shoes etc., are picking up momentum world-wide. Reduction in weight (10%), energy of production (80%) and cost of the components (5%) as experienced elsewhere, attracts the automobile industry to employ *sisal* fibre composite parts in India. The conservative estimates indicate that about 6,000 TPA plant fibre based composite parts can find their way into passenger cars and multi-utility vehicles (OSEC, 2004). Railways are also a potential application area where in it is estimated that about 350 TPA fibre composites is required for manufacture of doors, luggage racks, panels, partitions, seating etc. (Nangia *et al.*, 2005).

Packaging materials for bags, boxes, crates, containers, which is now made up of wood, can be replaced by cost-effective *sisal* reinforced composites. Boats can be made by replacing the conventional polymer composite fibres with *sisal* as reinforcement. The market potential of geotextiles for roads, paved road networks and railways applications in India is estimated to be 2,72,500 tonnes, of which a considerable portion can be earmarked for *sisal*-based textiles (Vibrant Gujarat, 2005).

Sisal: Employment Generation

Since *sisal* is a xerophyte, its establishment in major parts in India is not a constraint and its cultivation involves least water and crop management during its survival period of about 10 years. They can be raised along the bunds, hedges, forest and wastelands for soil conservation. Once established, an assured production of about 2.5 tonnes/hectare/year is possible for 6-8 years. Presently *sisal* plantations and their allied activities is an unorganized sector localized mostly in rural and tribal areas. *Sisal* plantation activity has the annual employment potential of about 113 man-days per hectare. The State of Madhya Pradesh has 14.17 lakh ha. barren and uncultivated wastelands, out of which 3.00 lakh ha (20%) can be targeted for *sisal* plantations, which can generate enormous employment potential for 1.20 lakh people for cultivation and fibre extraction. The *sisal* fibre is an input material for various rural applications involving cordage and ropes, mats, handicrafts and other utility items, which can augment the employment opportunities for income generation in rural sector.

Sisal: Spin Offs

It is discernible that there is a huge quantity of sisal waste (95% of the leaves by weight) can be effectively utilized for composting, vermi-composting and extracting other valuable products. The pilot scale demonstrations show that it is a valuable feedstock for biogas plants to cater to the local and rural energy requirements. One tonne of *sisal* pulp can generate 54.3 m^3 of biogas by the methanogenesis (Leo Oudshoorn, 1995). *Sisal* leaves are known to contain wax (0.38%) and hecogenin (0.10%). These expected spin offs in the form of secondary/allied activities which are labour intensive and have applications in rural and small-scale industrial sectors.

Sisal: RRL Expertise

Regional Research Laboratory (RRL), Bhopal is involved in R & D activities on *sisal* fibre extraction, processing and product design. It has developed a versatile machine (Raspador) for the mechanical extraction of *sisal* fibre (Fig. 4.2). The laboratory has been imparting training to the rural masses, especially the women folk in making handicrafts (Fig. 4.4) from *sisal* fibre.

Fig. 4.4: *Sisal* Fibre Handicrafts and Training Programme on *Sisal* Fibre Handicrafts Organized by RRL, Bhopal

RRL has also developed the vermi-composting technology by utilizing *sisal* leaf residue to produce eco-friendly manure. Based on the expertise the laboratory has developed *sisal* fibre reinforced corrugated roofing sheets (Fig. 4.5) and *sisal* fibre-polymer composites, which are alternate building materials superior in quality, eco-friendly, non-carcinogenic and cost-effective. Basic Research and Design need to be carried out for the development of crimp property of the fibre for making yarn, various surface treatments and processes to overcome the hydrophilic nature, improving the interfacial bonding and mechanical properties of the fibre. The 'Building Materials Characterization and Testing Centre' and 'Technology Enabling Centre' for R-WOOD are the additional expertise and facilities available to substantiate.

Fig. 4.5: *Sisal* Fibre Reinforced Corrugated Roofing Sheets

The *sisal* fibre and its allied activities like cultivation; fibre extraction, processing and making value added products are proven sources of employment opportunity and income generation. The present experience indicates that trained artisans and women are able to sustain with the income generation through the *sisal* related activities. The engineering applications of *sisal* fibre like buildings, automotives, railways etc. are the prospective areas where it can generate employment potential. Easy to raise on wastelands in a wide range of agro-climates, sustained fibre yield for a considerable long duration, versatility of the fibre as a potential input material for various applications makes it a viable option for employment generation and rural development.

REFERENCES

Amit Rai and C.N. Jha (2004). Natural fibre composites and its potential as building materials. *Express Textile*, 25th Nov. 2004.

CFC (2000). Alternative Applications for Sisal and Henequen—Tech. Paper No. 14, Proc. Seminar by the FAO and the Common Fund for Commodities (CFC), FAO, Rome, 13th Dec. 2000.

http://www.crijaf.org/about/aicrp.html

Leo Oudshoorn (1995). Biogas from Sisal Waste—A new Opportunity for the Sisal Industry in Tanzania, Energy for Sustainable Development, Vol. II, No.4, Nov. 1995.

Nangia Sangeeta, A. Mittal, G. Srikanth and S. Biswas (2005). Composites in Railways, *News and Views*, TIFAC, DST, New Delhi.

Navin Chand and P.K. Rohatgi (1994).

Natural Fibres and their Composites.

OSEC (2004). India Automotive Components Industry Swiss Business Hub India, OSEC Business Network, Switzerland.

Purandare, A.P. and B. Sambi Reddy (2001). Integrated Sisal Plantatiòn based activities in Koraput District, Orissa. In: Experiences in Implementation of Innovative Employment Programmes, pp. 341-64.

Rehm, S. and G. Espig (1991). *The Cultivated Plants of the Tropics and Sub-tropics.* (http://www.nnfcc.co.uk/crops/info/sisal.html)

Soumitra Biswas, G. Srikanth and Sangeeta Nangia (2005). Development of Natural Fibre Composites in India, *News & Views*, TIFAC, DST, New Delhi.

Vaish, A.K. (1994). Research on the utilization of sisal leaf waste. Proc. Technical Workshop on Waste Utilization, NML, Jamshedpur, 12th Sept. 1994.

Vibrant Gujarat (2004). Project Proposals: Global Investors Summit—2004, Government of Gujarat.

Wienk, J.F. (1969). Long Agave Fibres, in Ferwerda, F. P., and Wit F. (eds), *Outlines of Perennial Crop Breeding in the Tropics*. H. Veenman and Zonen, N.V. Wageningen, The Netherlands, pp. 1-21.

5

Rural Tsunami: A Famine in 21st Century

A Case Study of Kalidhindi Mandal of Andhra Pradesh in India

ANAND SAGAR K. and VIJAYANAND KOMMALURI

Introduction

Blindly aping the World Bank model of agriculture, Andhra Pradesh had pumped in huge finances to push an industry-driven agriculture that has not only exacerbated the crisis leading to an environmental catastrophe but also destroyed millions of rural livelihoods, which echoes to "Rural disaster". As a result, the State had turned into capital of shame for farmers' distress, visible more through the increasing rate of suicides in the rural areas.

The ground realities are far removed from the rhetoric and the statistics that have bred immunity against compassion. We are all part of a global food system, which perpetuates poverty and deprivation. The claims of improved technology for agriculture ignore the stark realities like increasing indebtedness, growing poverty, resulting in human suffering and hunger.

The State Government insists on using the term "drought conditions" and not famine. The farmers' suicides were due to man-made disaster rather than natural disaster due to failure of the State and its machinery. The State Government should be aware of the fact that while drought conditions are caused by the vagaries of nature, a famine is not a natural phenomenon. Most of the agricultural labourers have lost opportunities for gainful employment and small and medium farmers have been forced to leave their land fallow for want of water.

Remote Sensing and GIS plays an important role in rural development and drought preparedness and response. Planning and implementation of complex development goals and drought response efforts often require spatial analysis and map creation capabilities. Data for these efforts is often in a variety of formats like maps, databases, spreadsheets, images etc. from multiple agencies and multiple sources. The regionally differentiated strategy for drought preparedness and response in integration of GIS and decision support system helps to access the information in a spatial and visual medium and allows improving decision-making and promoting better disaster management.

Different Stages in Disaster Management Lifecycle

There are five important phases of disaster management, namely disaster prevention, disaster mitigation, disaster preparedness, disaster response, and disaster recovery. Of them, disaster prevention, disaster mitigation, and disaster preparedness constitute the pre-disaster planning phase. Pre-disaster planning is the process of preparing in advance, to meet a future disaster. Disaster prevention is the action taken to eliminate or avoid harmful natural phenomena and their effects. Disaster mitigation is the action that deals with reducing human suffering and property loss. Disaster preparedness encompasses those actions, which are taken to limit the impact of natural phenomena by structuring response and establishing a mechanism for effecting a quick and orderly reaction.

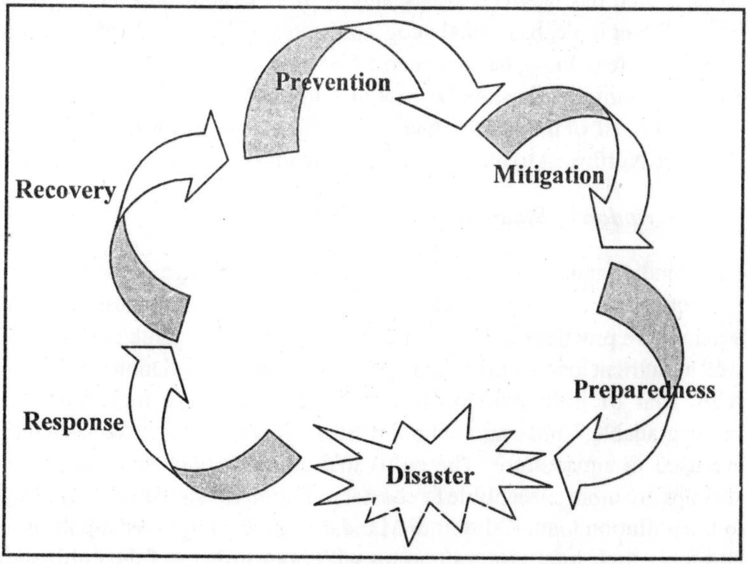

Fig. 5.1

Disaster response covers responding to disasters by various organizations, providing many services that need to be mobilized on a moment's notice, and functioning for an indeterminate period in a coordinated manner under stressful and difficult circumstances, and may be demobilized after the emergency has abated. The ability of an agency, or a group of agencies, to manage emergencies, rather than just react to crises, is critically dependent on the availability and flow of real time and archived information from monitoring systems, thematic databases, and decision support systems that are linked through national networks. Disaster recovery is the last phase of disaster management and is concerned with providing relief after the disaster has struck. It deals with providing food and shelter to the disaster victims, restoring normal conditions and providing financial and technical assistance to rebuild.

Case Study

In this article, Kalidhindi Mandal in Krishna district of Andhra Pradesh State in India has taken as a case study as shown in Fig. 5.1.

(i) Geographical Area

India, which lies between 84° N and 37°6' N latitude and 68°7' E and 97°25' E longitude, has a total geographic area of 32,87,782 km^2. Andhra Pradesh State in India, lies between 12°41' and 22°N latitude and 77° and 84°40'E longitude. Krishna District in Andhra Pradesh lies between 15° 43' and 17° 10' of the Northern latitudes and 81° 33' of Eastern latitudes. From the Northwest to the Southeast it extends over an area of 8734 km^2.

(ii) Description of Study Area

Kalidhindi Mandal, which is adjacent to the Bay of Bengal, is well known for Aquaculture farming in Krishna District of Andhra Pradesh. Coastal aquaculture practices in Kalidhindi Mandal lead to organic pollution as well as nutrient loading in coastal waters. This form of pollution is different from that of industrial pollution, since the waste materials are bio-degradable. Unlike agricultural wastes no toxic material like pesticides are used in aquaculture. The cultivated aquatic organisms, especially shrimps are more susceptible to coastal pollution rather than contributing to the pollution load. Kalidhindi Mandal is affected by coastal pollution and loses their total crop or end up with poor growth of the cultivated organisms due to the poor quality of water. Increased siltation and sedimentation of coastal water in consequence of deforestation, mining and inappropriate agricultural practices in this area causes to degradation of soil, depletion of water level, which reflects the drought conditions in this Mandal.

Study Area Map

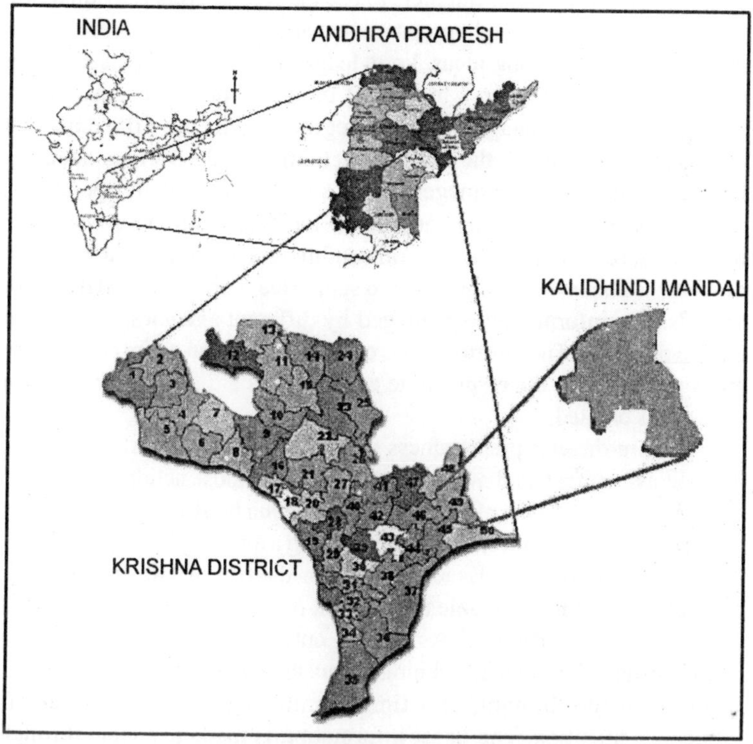

Fig 5.2: Study Area of Kalidhindi Mandal of Krishna District in Andhra Pradesh State, India

Web-based GIS for Drought Preparedness and Response

The Web-based GIS can be used in many ways for effective disaster management. Also, the applications of the Web-based GIS can range over the entire disaster management cycle. Integration of the GIS, Web technology and the Decision Support System (DSS) technique such as Analytic Hierarchy Process (AHP) with regionally differentiated approach taking into consideration of soil conditions, rainfall intensity, ground water conditions, demographic data, infrastructure data etc. can be used to significantly increase the usage and accessibility of the spatial data, which is a key requirement before, during and after any disaster. The application functions very much like a "Whiteboard in the Sky" in which a number of agencies can share event information via the web. The approach allows several agencies operating on different technology platforms and using different communication channels to use the web in order to collaborate while managing the drought preparedness and response etc. It provides a platform for exchanging ideas, knowledge and the latest update during the event of any disaster, which is of utmost importance. Also, Web-based workflow management is integrated with the GIS data and applications to manage the multiple activities of the various agencies involved in the disaster management. In a community's daily operations, a Web GIS collects, maintain and store vital map information related to infrastructure, rainfall intensity, soil conditions and ground water etc., all of which is displayed graphicallv and supported with associated database records. The information is managed by different agencies and used in daily workflows. The information is collected across an entire community and shared among the responsible entities, thus a disaster management network is created.

The pre-disaster preparedness activities like drought identification, drought assessment, and awareness are aimed at those actions, which are taken to limit the impact of a natural phenomenon by structuring response, and for establishing a mechanism for ensuring a quick and orderly reaction. This is organized very efficiently through the use of GIS on Web, for example, by making available the drought maps, as well as the "do's and don'ts" on the Web. It also allows carrying out some basic "what-if" scenario analyses and helps the local administrators to take effective measures.

During the drought, real time monitoring and response needs immediate attention. The latest information is made available through Web giving a detailed picture of the event tracking, forecast of the affected region, the alternative measures, availability of food resources, movement of various agencies like NGOs, Health Care Supervisors etc. Thus, they'll take the measures to create psychological awareness in order to prevent the death toll from suicides.

The post-disaster activities like empowerment of rural people with providing food and shelter to the disaster victims, restoring normal conditions and providing financial and technical assistance to rebuild can be effectively coordinated by using Web-based GIS. This includes making the latest information available regarding the spatial coordinates of the affected people and sources of providing relief and rescue, the regional extent of the calamity, and the geopositioning of the "lifelines" like water, food supply and transportation network etc. The use of Web GIS provides an effective solution for this purpose. With the help of Web GIS, the latest information of affected areas, the demographics of the affected area is posted on the web and thus it facilitates all agencies to coordinate their efforts in a more effective manner to empower the rural people and giving a helping hand to save the death toll from suicides.

Conclusion

A web-based GIS integrated with decision support system helps to manage the entire range of spatial, as well as non-spatial data to carry out various governmental responsibilities and services. Data exchange at local, regional, city, state and national levels can be a continuous and cohesive process, resulting in better management and control of resources.

The State Government insists on using the term "drought conditions" and not famine. Those at the helm, after all, should be aware of the fact that while drought conditions are caused by the vagaries of nature, a famine is not a natural phenomenon and suggests the failure of the State and its machinery. The provided 'solutions' are really the causes for the problems in the first place; and behaving like an ostrich is not going to eclipse hunger and death from the politico-economic radar screens. Unless, the State Government realizes the facts and figures regarding rainfall, losses in terms of damage to crops and the extent of land left uncultivated and thrive to formulate the strategies for rural empowerment, the State Economic policies like free electricity and enhancing bank credit will remain as proverbial Emperor's clothes.

Policy, makers, agricultural scientists, social scientists, academicians and even the civil society groups should recognize the fundamental flaw that forces farmers to their deaths. The idea of industrial farming should keep away since the majority of the population in India makes their livelihood with small portions of land and gainful employment from agricultural sector.

I look forward to "Another World..." where the people can sustain their lives without hunger, greed and inequity.

REFERENCES

Anand, Sagar K., Vijayanand K. (2004). "Facets of Disaster Management: Recipe to Relief", *Proceedings of Map Asia 2004*, Beijing, China.

Anand, Sagar K. (2003). "Agmarknet: Rural Empowerment", *Geospatial Today*, July-August 2003, Vol. 2, Issue-2.

Anand, Sagar K. (2004). "Rural Banking: Changing Contours of Rural India", *Proceedings of ICTD Conference*, NIRD, Hyderabad.

Anand, Sagar K., Vijayanand (2003). "Role of GIS in Good Governance", GIS India, Posted on GIS Portal at http://www.gisindia.com/article_read, Site last Visited 08/03/2005.

Anand, Sagar K. Vijayanand K. Satyajith Mohanty (2004). "Shaping the Earth for Sustainable Development: A Multi linguistic Approach for India", *Proceedings of Map Asia 2004*, Beijing, China.

Densham, P.J. and M.P. Armstrong (1994). "A Heterogeneous Processing Approach to Spatial Decision Support Systems", Advances in GIS Research, Proceedings of T.C. Waugh and R.G. Healey (Eds.), *Sixth International Symposium on Spatial Data Handling*, 1, 2945.

Singh, R.K. (2004). "National System for Disaster Management", *Proceedings of India-United States Conference on Space Science*, Applications and Commerce.

Symon Francis, (2003). "GIS and Risk Assessment—Decision Made Easy", *Geospatial Today*, July-August 2003, Vol. 2, Issue-2.

6

Can Andhra Pradesh Achieve Millennium Development Goals?

S. MAHENDRA DEV

Introduction

During the last decade there has been an increasing realization that measures of progress of society or a Nation should go beyond those based on money-metric. While income remains the predominant *means* indicator for assessing well-being, emphasis is placed more on basic functionings such as hunger and malnutrition. That the relation between income poverty and non-income poverty is not necessarily one-to-one is evident from the recent experiences of several countries. For instance, the incidence of child malnutrition in India is greater than the incidence of poverty. The highest incidence of malnutrition is in the poorest states and the lowest is not in the richest, but in one of the middle-income states. The emphasis on 'non-income' concerns of well-being acquired greater space in the policy arena. The Millennium Development Goals were formulated out of these concerns.

At the Millennium Summit in September 2000, the world leaders adopted United Nations Millennium declaration to reaffirm their commitment the goals of achieving sustained development and eliminating poverty. The Millennium Development Goals (MDG) that emerged from the declaration are a set of numerical and time bound targets that express key elements of human development. They include halving income poverty and hunger, achieving universal primary education and gender equality, reducing under-five and maternal mortality rates by two-thirds, reversing the spread of HIV/AIDS and halving the proportion of population without access to safe drinking water and sanitation facilities (Box 6.1). For each goal there are one or more indicators to track its progress. For most of the targets, the goal is set for 2015 with a benchmark at 1990. The MDGs become a widely accepted benchmark for measuring development progress at National and sub-National level. The goals are mutually reinforcing and are directed at reducing poverty in all its forms. The issues that MDGs address are crucial for sustainable human development and draw priority attention in policy-making, especially in the era of globalization. If the Human Development Reports of UNDP have succeeded in bringing 'people' to the centre of development debate, the MDG declaration has enrolled the international community and the Governments to commit to work towards this. India is one of the signatories to the MDG declaration.

Alleviation of Poverty in all its forms has been a major goal of development policy in India since the beginning of planning. Recognizing that the pressure for investments in the development of infrastructure and production sectors would leave relatively small allocations for social services, the Government of India in the V Plan launched the minimum programme (MNP). The MNP concerns with providing services to the poor in the areas of education, health, water, housing etc. Since MNP, the planning in India focused on these sectors in one form or others. The Tenth Five Year Plan (2002–2007) identified a set of monitorable targets for different indicators of human development, which are considered central to the attainment of the plan targets (Box 6.2). The issues addressed in MDGs and the X Plan overlap to a great extent.

India achieved high rate of economic growth in the post-reform period and reduced the number of people below the poverty line considerably. However, its performance with respect to human development has been mixed. Literacy rates continue to rise, but still about less than half of women cannot read or write. Health continues to remain an area of neglect. Malnutrition and mortality rates are alarmingly high, especially among women and children. The data indicate that more children and women suffer from malnutrition than from income poverty. But India is a country of continental proportions. There are large variations across the states in the performance. The present study is focused on Andhra Pradesh.

BOX 6.1: Millennium Development Goal

1. Eradicate Extreme Poverty and Hunger,
2. Achieve Universal Primary Education,
3. Promote Gender Equality and Empower Women,
4. Reduce Child Mortality,
5. Improve Maternal Health,
6. Combat HIV/AIDS, Malaria and Other Diseases,
7. Ensure Environmental Sustainability,
8. Develop Global Partnership.

BOX 6.2: Monitorable Targets for the Tenth Plan and Beyond

1. Reduction of poverty ratio by 5 percentage points by 2007 and by 15 percentage points by 2012,
2. Providing gainful and high-quality employment at least to addition to the labour force over the Tenth Plan period,
3. All children in school by 2003; all children to complete 5 years of schooling by 2007,
4. Reduction in gender gaps in literacy and wage rates by at least 50 per cent by 2007,
5. Reduction in the decadal rate of population growth between 2001 and 2011 to 16.2 per cent,
6. Increase in Literacy rates to 75 per cent within the Plan period,
7. Reduction of Infant mortality rate (IMR) to 45 per 1000 live births by 2007 and to 28 by 2012,
8. Reduction of Maternal mortality ratio (MMR) to 2 per 1000 live births by 2007 and to 1 by 2012,
9. Increase in forest and tree cover to 25 per cent by 2007 and 33 per cent by 2012,
10. All villages to have sustained access to potable drinking water within the Plan period,
11. Cleaning of all major polluted rivers by 2007 and other notified stretches by 2012.

Context of MDG in Andhra Pradesh

Andhra Pradesh has been vigorously pursuing economic reforms since the mid-1990s to step up economic growth and alleviating poverty. Reforms were underway in the power and irrigation sectors with an aim to enhance their efficiency. Significant strides have been made in the areas of participatory managements of infrastructure, resources like water, land and forests. Measures for fiscal correction were launched to wriggle the state out of the financial crises. The State has undertaken promotion of women self-help groups in a big way[1]. These institutions have formed the central element of poverty alleviation strategies of the State. During the last decade the State of Andhra Pradesh earned a name as one in the forefront in the economic reforms and raised the expectations within the country and abroad about its performance.

The State of Andhra Pradesh has always been ranking in the middle among the States of India with respect to many economic and human development indicators. If budgetary allocations are taken as indicators, the social sector does not seem to have the state's attention as it should have. The public annual expenditures on social sector services constituted 6-8 per cent of GDP during the eighties dropped below 6 per cent in the nineties. Similarly, expenditure on rural development dropped from 1-2 per cent of GDP in the eighties to less than 1per cent of GDP in the last decade. A recent study indicates that the position of the State has not improved significantly during the last decade (Rao and Mahendra Dev, 2003). The annual growth rate of GSDP in the nineties was 5.3 per cent pa. lower than all India growth rate of 6.13 per cent p.a. The trends show that employment growth in A.P. declined drastically in the 1990s as compared to the decade of 1983 to 1993-94. In rural A.P., it declined from 2.40 per cent per annum during 1983 to 1993-94 to 0.29 per cent during 1993-94 to 1999-2000. The corresponding numbers for urban areas were 4.28 per cent and 0.01 respectively. Real wages in the State increased in the 1990s, but at lower rates than were observed in the 1980s for both men and women. The trend rate of growth for males declined from 5.74 per cent in the 1980s to 1.53 per cent in the 1990s. Similarly for women, the trend rate of growth in real wages has declined from 4.76 per cent in the 1980s to 1.94 per cent in the 1990s. Growing rural wages in real terms was one of the key determinants of poverty reduction in India in the 1980s.

The head count ratio of poverty, estimated as alternative to the official ratios, which were disputed under several counts show that about 26 per cent of the rural population is still poor. The indicators of other human development show mixed trends. Literacy levels in AP are lower than the levels of all India. Life expectancy is close to all India average and shows better performance with respect to other health indicators like malnutrition, infant and maternal mortality. However, among the southern States AP ranks low with respect to several indicators. The human development index of AP has been lower than all India and what is worrisome is that its rank slipped from 8th in 1981 to 10th in 2001.

Objectives

Achieving the MDG is a major challenge ahead for Andhra Pradesh. As the target year is closing in, it is important to examine if the progress towards MDGs is on track to identify the challenges involved. This forms the central objective of the present exercise. A good assessment of the progress and the challenges towards the MDGs should go beyond the averages. The average trends often mask the wider dimensions of equity and justice. Failure to do so would leave behind several deprived sections of population and laggard regions, although at mean level the MDGs are met. Therefore, the analysis is carried out at disaggregated level. The level of disaggregation of the analysis is severely restricted by lack of availability of reliable data. Keeping this in view, the analysis is confined to the following level of disaggregation:

I. *Social*: (1) Scheduled Castes, (2) Scheduled Tribes, (3) Other Backwards Castes, (4) Other Castes.
II. *Regional*: (1) Costal Andhra, (2) Rayalaseema, (3) Telangana.
III. *Occupational*: (1) Self-Employed in Non-Agriculture, (2) Self-Employed in Agriculture, (3) Agricultural Labour, and (4) Others

The objectives of the study are set as:

(a) Following the present trajectory, can AP achieve the MDGs.
(b) Where will *geographical pockets (regions) of poverty* remain in AP in 2009 and 2015?
(c) How *social groups* are likely to fare?
(d) Which are the *occupational groups* likely to lag behind?
(e) Draw up a combined picture of where the poor are likely to be against the MDGs when compared with the general population.
(f) What are the challenges ahead?

The focus of the study is rural Andhra Pradesh. Projections were also made for aggregate rural and urban areas and for combined Andhra Pradesh. The indicators for the study are presented in Table 6.1.

Data and Methodology

The study is extensively based on secondary data. The two major sources of the data for the analysis are:

1. Sample of Surveys conducted by National Sample Survey Organisation (NSSO).
2. National Family Health Survey conducted by International Institute of Population Sciences (NFHS).

Table 6.1: Projected Poverty Head Count Ratios

Groups	Poverty Ratio			
	1994	2000	2009	2015
Occupational Groups				
Self-employed in non-Agriculture	20.39	23.12	9.50	4.45
Agricultural labour	40.34	34.04	17.11	8.45
Other labour	33.03	25.66	11.21	5.09
Self-employed in agriculture	20.43	21.48	9.28	4.56
Others	23.29	13.11	5.85	2.72
Social Groups				
Scheduled Tribe	40.71	42.94	23.92	13.88
Scheduled Caste	42.43	35.95	19.25	9.85
OBC	—	27.35	13.77	6.64
Others	23.61	12.71	6.20	2.85
Regions				
Coastal Andhra	30.77	18.00	9.40	4.47
Rayalaseema	30.44	47.72	23.08	13.61
Telangana	25.73	27.56	13.73	7.01
All				
Rural	28.83	26.67	13.36	6.89
Urban	17.13	10.90	7.31	4.33
Andhra Pradesh	25.67	22.38	11.68	6.17

The NSSO conducts large-scale sample surveys nation-wide on different topics relevant to socio-economic planning and policy-making. Two of the topics on which the surveys are conducted regularly every year are Household Consumer Expenditure (HCE) and Employment and Unemployment. The household consumption data is useful for estimating levels of poverty, and other distributional parameters. The employment and unemployment data contains educational and schooling information. The data of these surveys conducted in 1993-94 (50th round) and 1999-2000 (55th round) were used for projecting poverty and educational indicators. The data of HCE is also used in conjunction with NFHS data for making projections of other non-income indicators.

The NFHS data forms a rich source of data for various health and demographic parameters. So far, two rounds of NFHS were conducted in India, viz. 1992-93 and 1998-99. The periods of survey of NFHS roughly correspond to those of NSSO. The NFHS gives rich data on several health indicators.

Methodology

The forecast of the MDG indicators were made using a simple model.

Broadly the methodology consists of anchoring the progress of indicators to growth in income and to make future projections. The model assumes that all the other determinants of the indicators are held constant and projects the trajectory based on growth. Once the trajectory is determined and the future challenges are identified the 'other determinants' play crucial role in meeting accelerating the progress to meet the goals.

The projection procedure consists of following steps:

(i) Aggregate income is projected based on a pre-assumed growth rate. For making assumption about growth, we examined the historical trends. We find that the aggregate expenditure in Andhra Pradesh was growing at 2.94 during 1972-73 and 1999-2000. If we exclude the nineties, the growth rate would be about 3.5 per cent pa. This indicates that there was a 'slowdown' in the growth during nineties most of which is due to the dampened growth in the rural areas[2]. With the recent spurt in rural investments, especially in rural areas we assume that the growth rate of pre-nineties will be restored to the pre-nineties level during the period 2000-09. We expect further acceleration in growth rate during 2009-15. However, we also use a scenario wherein the growth rate remains as observed during the recent period, i.e. 3.0 per cent p.a. for the period 2000--09 and accelerates further to 3.5 per cent p.a. for the rest of the period.

(ii) Using the projected populations, the aggregated incomes are converted into per capita terms. The population projections given by GoI (1996) were modified using the growth rates observed in the last decade. The next step of the procedure involves splitting the overall per capita income into rural and urban per capita incomes.

(iii) We assume that the rural-urban differential follows the historical trend. In Andhra Pradesh, the PCE has been growing at a higher rate than the urban PCE, as a result the rural-urban differential narrowed over time. We estimated a trend equation for the rural-urban differential and projected the differential ratio for the forecast period.

(iv) Once we have the growth rate for rural areas, the next step is to project growth rates for each of social and occupational groups and regions. For this we estimated elasticities of group level growth with respect to overall growth.

(v) From the projected growth rates of per capita income for each group (social/occupational/regional) we can predict the MDG indicators.

(vi) For income poverty we make use of the fact that the per capita income follows log normal distribution. For this we need to make assumptions about the income inequalities for the projection period. Historically, the NSSO data do not show any historical trends in Andhra Pradesh. We therefore assume the G will remain constant for the forecast period. We also use a scenario assuming that the Gini coefficient increases by 1 per cent p.a. for the forecast period.

(vii) For non-income MDGs, we used elasticities of these indicators with respect to expenditure. We estimated the elasticities separately for gender, region and social and occupational groups. For this we make use of both NSS and NFHS data.

(viii) All the results are presented for the reference scenario, which assumes growth rate of 3.5 per cent p.a. for 2000-09 and 4 per cent for 2009-2015 and constant inequalities. As low growth scenario we assume 3 per cent and 3.5 per cent for the periods 2000-09 and 2009-2015. Under a third scenario (inequality rising scenario) we assume growth rate as in scenario I but with increasing Gini coefficient.

Projected Trajectory of Indicators of MDGs

Extreme Poverty and Hunger

Indicator 1: Poverty Head Count Ratio (MDG: halve the ratio by 2015)

There has been a debate and controversy around the poverty estimates based on the recent 55th round NSS Consumer Expenditure Survey[3]. This is due to the changes in design of the survey instrument with regard to recall periods used to collect the data on consumption of various food and non-food items. As a result of these changes the consumption aggregates derived from 55th round were found to be incomparable to those of the earlier surveys. The 55th round survey design departs from the earlier surveys in two ways, viz., simultaneous use of one-week and one-month reference and change of reference period from one month to one year for certain non-food items. Fortunately, several scholars have addressed this issue of comparability of the 55th round data and suggested innovative methods to overcome it. Deaton and others have made use of the fact that for certain commodity groups the reference period in the 55th round was the same as was in the earlier rounds. They derive the probability distribution of income conditional on the consumption of this comparable group for the 50th round and use the same for 55th to estimate poverty levels. This approach assumes that Engel relation between the consumption of comparable group and total expenditure is stable over the two periods. Further, this approach assumes that relative prices are constant over time[4]. A second approach consists of making adjustments to the data of 50th round, which could be adjusted to make them comparable to 55th round. This approach gives poverty estimates comparable between 50th and 55th rounds. For the present exercise we use Deaton adjustments to forecast poverty ratio.

Table 6.1 presents the poverty estimates for the forecast period under scenario I along with the targets that each social and occupational group/region has to meet. The results indicate the following:

1. Poverty ratio for AP will come down to 11.68 in 2009 and further down to 6.17 by 2015. With a base of 25.67, AP will meet the target of halving the poverty by 2015.
2. All the social and occupational groups and regions will reduce the poverty beyond the target of 13 per cent, except Scheduled Tribes among the social groups and Rayalaseema among the regions, which falls short by about one percentage point.
3. But the gap between poverty ratio at State average and those of the groups remain large for some of the social groups and regions. In case of STs and Rayalaseema, the poverty ratio would be almost double the average of AP. Poverty among Scheduled Castes and Agricultural labour will be 3-4 percentage points higher than the average.
4. However, in terms of absolute numbers poverty remains a challenge. By 2015 about 5.5 million people would continue to remain poor. Almost of eighty per cent of them would be in rural areas.
5. 1.2 millions Scheduled Castes and 0.8 millions of Scheduled Tribes would be poor even by the end of 2015. Among OBCs, who constitute highest proportion in the State's population, almost 2 million will remain poor.
6. Among the Occupational groups, agricultural labour will have 3.4 millions poor.

7. Higher growth benefits all, but with variations. An additional half per cent growth will benefits the laggards more.
8. Rising Inequalities will dampen the gains. The impact of inequalities is more in case of groups and regions with higher poverty.

Indicator 2: Poverty Gap Ratio

This indicator is to be monitored along with the poverty head count ratio. There is no target set for this, but it is a useful indicator for measuring the depth of poverty. The estimated poverty Gap Ratios are presented in Table 6.2. The poverty gap ratios fall faster compared to the poverty ratio for all groups. From the table it can be concluded that:

1. Poverty Gap Ratios fall faster than the poverty ratios. The rural poverty gap is always lower than the urban poverty ratio. This indicates that among the poor the proportion of very poor is lower in rural compared to urban areas.
2. The poverty gap is high for Rayalaseema, SC, ST and agricultural labour.

Table 6.2: Projected Poverty Gap Ratios

Region/Groups	Year		
	2000	2009	2015
Rural	3.26	1.46	0.6
Urban	5.84	2.74	1.18
Andhra Pradesh	3.89	1.69	0.67
Coastal Andhra	3.33	1.56	0.68
Rayalaseema	2.4	1.06	0.45
Telangana	8.83	4.82	2.47
ST	6.6	3.21	1.43
SC	4.21	1.93	0.82
OBC	2.3	1	0.42
OC	3.43	1.55	0.66
Self-employed in Non-Agriculture	8.52	4.73	2.47
Self-employed in Agriculture	4.78	2.32	1.05
Agricultural Labour	4.84	2.37	1.08
Other Labour	2.55	1.47	0.81

Indicator 3: Percentage of Dietary Deficient Population

This indicator does not have a target. It gives the proportion of population whose dietary intake is less than the recommended levels. Although an important indicator of food security, it is embroiled in controversies in India[5]. Taking 2700 as the requirement per consumer unit per diem, we estimated the percentage of households suffering from the dietary deficiency. The estimates are presented in Table 6.3. As can be seen, the estimates present somewhat different picture from the poverty estimates. Even by 2015, 40 per cent of rural households suffer from dietary deficiency. But what is puzzling is that the estimates do not show many variations across social and occupational groups and regions. In fact the proportions of dietary deficient households are marginally higher for the OCs, Coastal Andhra and Self-employed in agriculture. These are the groups, which have farewell in terms of poverty. The dichotomy between poverty and energy deficiency is because of the diversification that is taking in the Indian diet and the relevance of recommended levels of intake. It is observed that the prominence of energy-intensive foods like cereals is decreasing in the Indian consumption basket. The per capita monthly intake of cereals in rural India declined from 15.25 kgs. to 13.27 kgs. during the last three decade (Table 6.3.1).

Table 6.3: Percentage of Dietary Deficient Population

Region/Groups	*Year*			
	1993-94	*1999-2000*	*2009*	*2015*
Self-employed in non-agriculture		70.30	58.40	53.34
Agricultural labour		72.89	60.50	55.24
Other labour		76.32	63.13	57.54
Self-employed in agriculture		61.42	50.58	46.00
Others		54.27	43.19	38.63
ST		72.24	59.37	53.95
SC		75.15	62.38	56.95
BC		69.72	57.77	52.70
OC		57.83	47.27	42.84
Coastal Andhra		62.18	50.94	46.22
Rayalaseema		72.78	59.93	54.51
Telangana		72.87	60.42	55.13
Rural	65.5	68.14	56.00	50.89
Urban	70.8	73.65	61.36	56.30

The decline is faster among the top 30 per cent and middle 40 per cent of the population. As a result, the calorie intake declined from 3151 Kcal to 2692 Kcal for the 30 per cent and from 2345 to 1968 Kcal for the middle 40 per cent of the population (Table 6.3.2). The trends show that higher the diversification higher is the decline in energy intake. Since the diversification is taking place faster among the richer groups, large proportions among the rich end up calorie deficient.

Table 6.3.1: Cereals Per Capita Per Month (in kgs.)

	CEREALS (Kgs. per capita per month)					
Decilne	*1972-73*	*1977-78*	*1983*	*1987-88*	*1993-94*	*1999-00*
Bottom 30	11.25	12.73	10.53	12.51	11.25	10.89
Middle 40	15.20	15.85	15.39	14.25	13.24	12.81
Top 30	19.31	18.97	20.19	16.32	15.33	14.21
All	15.25	15.85	15.37	14.35	13.27	12.66

Table 6.3.2: Calorie Per Capita Per Diem

	1972-73	*1977-78*	*1983*	*1987-88*	*1993-94*	*1999-00*
	CALORIE (Kcal per capita per diem)					
Bottom 30	1612	1764	1690	1499	1443	1400
Middle 40	2345	2536	2306	2032	1968	1841
Top 30	3151	3557	3118	2776	2692	2614
All	2367	2611	2365	2095	2028	1980

Indicator 4: Percentage of Malnourished (MDG: Half the Ratio by 2015)

The MDG indicator is defined for children of age 0-5 years. In India the data on anthropometric measures are available for 0-3 years in the NFHS (1998-99) and 0-4 years in NFHS (1992-93). Therefore, we confine the indicator to the age group 0-3 years[6]. *The estimated percentage of children under-weight in 1992-93 was 45.68, implying that the goal is to reduce the malnutrition to 22.84 per cent by 2015.* Among the measures of nutrition, those based on body measurements are more preferable as they are highly sensitive to even minor levels of malnutrition. The indicators used in MDG relates to underweight, which reflect both chronic and acute malnutrition[7]. The measure gives the percentage of children in the age group of 0-3 years whose weight are 3σ limits below from the median weight of the reference population. This includes both severe (2σ limits below median) and moderate (between 2σ and median) forms of malnutrition. The projected malnutrition ratios are presented in Table 6.4. It can be inferred from the table that:

1. The malnutrition rates are much higher than the poverty head count ratios for all groups.
2. Overall, the trends indicate sluggish decline in the malnutrition rates. For AP, the malnutrition rate will come down from 45.68 per cent to 27.54 per cent. In Rural areas one-third of the children of 0-3 years continue to suffer from malnutrition in 2015.

Table 6.4: Projected Percentage of Underweight Children (0-3)

Region/Groups	*Years*				*Gap*	*Gap as % 2015*
	1992	*1999*	*2009*	*2015*		
Rural	47.96	40.70	33.87	29.14	6.30	21.61
Urban	39.52	28.60	25.70	23.47	0.63	2.70
Andhra Pradesh	45.68	37.40	31.60	27.54	4.70	17.06
ST	50.08	44.40	37.00	31.87	9.03	28.33
SC	48.58	45.50	38.53	33.62	10.78	32.06
BC	46.52	40.40	34.34	30.07	7.23	24.04
OC	44.46	34.40	28.25	24.04	1.20	4.98
Self-employed in non-agriculture	44.87	39.18	32.67	28.14	5.30	18.84
Agricultural labour	47.61	41.65	34.70	29.89	7.05	23.58
Other labour	45.10	40.63	34.34	29.92	7.08	23.67
Self-employed in agriculture	45.26	39.10	32.92	28.60	5.76	20.13
Others	44.96	38.61	31.54	26.72	3.88	14.52
Coastal Andhra	38.26	35.22	28.13	23.39	0.55	2.37
Rayalaseema	56.14	51.70	44.88	39.98	17.14	42.87
Telangana	39.72	42.70	34.59	29.10	6.26	21.51

3. The State will miss the target of reducing malnutrition rate to 23 per cent in 2015 by about 4.7 percentage points falls short by 17 per cent. This is entirely due to the performance of the rural areas in this regard. The urban area miss the target by 0.6 percentage points and the rural by 6.3 percentage points. In other words the malnutrition level in 2015 in urban area would fall short by 3 per cent and in rural by 17 per cent. Considering the fact that the rate of decline of malnutrition is about 2 per cent per annum, this means that the target of 23 per cent would be met 7-9 years later.
4. The trends across social/occupational and regions show large variations. SCs and STs miss the targets by as much as 9-10 percentage points implying that they fall about 30 per cent short of the target and would take another 10-15 years to achieve the goal.
5. Similarly agricultural and other labour children will have almost 7 per cent more children than what is targeted. The gap will take another 10 years to disappear.
6. Rayalaseema turns out to be worst performing region in the State. The malnutrition rate in this region would be 17 percentage points below the target in 2015 and would need another three decades to accomplish the goal.

7. The projected trends indicate that 2 million children of AP will be under weight in 2015—a one-third million more than what achievement of MDG would have meant.

The malnutrition rates are higher than the poverty rates and are less sensitive to income growth. This is not a phenomenon specific to AP. The malnutrition for 20 fractile classes formed on the basis of standard of living index, also revealed that malnutrition declines slowly with improvements in the standard of living, and persists even among the top fractile classes. For example, among the highest fractile class, 28 per cent of rural children and 21 per cent of urban children suffer from malnutrition. Their results further show that the probability of a child falling into malnutrition decreases with improvement in mother's nutrition status, mother's education, mother's age and ante-natal visits but, increases when the mother is working. The adverse effect of the working status of the mother may be a characteristic among the poor households. The risk of malnutrition also decreases with the standard of living of the household and increases with household size.

Table 6.4.1 shows Ante-natal Care practices by five quintiles in rural Andhra Pradesh. Percentage of mothers having ante-natal is as high 93 per cent for the State. It ranges from 85 per cent in the lowest quintile to almost 99 per cent in the top. However, the quality of ante-natal care in terms of timing of first visit and number of visits show substantial variations across the quintile group. In the State only about 57 per cent of mothers had their first ANC visit at three months of pregnancy and a third had less than three check-up visits. There are large gaps between the poor and rich. Among the poorest quintile, 40 per cent of mothers had first visit before the third month of pregnancy and 50 per cent after fifth month. About 60 per cent of these poor pay less than three visits for ante-natal check up. In contrast among the richest quintiles 80 per cent of mothers have early (before third month) antenatal checkups and two-thirds have more than 4 visits of ANC. Similar disparities can be observed with respect to institutional deliveries. In 1998-99 the institutional deliveries constituted only 50 per cent—much lower among the poor. Thus, though large numbers of mothers access ante-natal care, the quality of car-even in the limited sense-needs to be improved substantially.

Malnutrition among the children is also inter-generational. The NFHS data show that Mothers who suffer from Chronic Energy Deficiency (with body mass index of less than 18.5kg/M^2) have 13 per cent more malnourished children than health mothers have.

Table 6.4.1: Ante-natal Care Practices During 1998-99

Quinties	Any ANC?	Timing of First Visit				Number of Visits		
		1-3 months	4	5	> 6	≤ 3	4-5	>6
1.00	84.95	40.34	14.20	45.45	15.91	58.74	18.45	22.82
2.00	88.39	44.72	12.06	43.22	8.54	45.09	26.34	28.57
3.00	93.70	53.94	9.06	37.01	6.69	39.63	23.70	36.67
4.00	98.21	64.55	10.45	25.00	6.36	20.98	24.55	54.46
5.00	99.01	80.00	6.00	14.00	2.00	11.39	17.82	70.79
All	92.90	57.10	10.20	32.70	7.63	35.44	22.38	42.18

Education

Indicator 1: Net Primary Enrolment Rates (MDG: Universal Primary Education)

This is the ratio of the number of children of official school age who are enrolled in primary school to the total population of children of official school age. It is common in India (a) for children younger than primary age to be enrolled in primary school, (b) for children older than primary age to be kept back as repeaters, and (c) for some primary age children to be enrolled at other levels of education. This is why the gross enrolment ratio can be more than 100 per cent. The official age for primary education in Andhra Pradesh is 5-10 years. Recent official statistics indicate that the gross enrolment ratios in almost all the districts have crossed 100 per cent[8]. Therefore, in this study we have taken what is known as the 'net enrolment ratio', which is the ratio of the number of children of primary age to total number of primary age children. The net primary ratio is always less than 100 per cent. The projections were made separately for boys and girls.

Tables 6.5 and 6.6 present the projected enrolment ratios for the boys and girls. The projections indicate that near universal enrolment of boys would be achieved by most of the social, occupational and regions. By 2015, every boy of 5-10 years of age in urban areas would be enrolled in school. Similarly, all the boys belonging to 'Other Castes' or Self-employed in non-agriculture would be enrolled in school. Among the regions Coastal Andhra would achieve 100 per cent enrolment ratio. Scheduled Tribes and agricultural labour lag behind the 100 per cent enrolment target by more than ten percentage points. In case of girls (Table 6.6), Scheduled Tribes and Agricultural labour lag behind the 100 per cent target by nearly 26 and 17 percentage points respectively.

Table 6.5: Projected Primary Enrolment Ratios (Boys)

Region/Groups	*Year*			
	1993	*2000*	*2009*	*2015*
Rural	74.17	89.37	93.23	97.28
Urban	75.76	94.97	98.43	100.00
Andhra Pradesh	74.55	90.72	94.48	97.93
ST	65.16	73.55	79.76	86.89
SC	70.77	87.44	91.58	95.19
OBC	73.56	85.47	88.55	92.50
OC	76.35	97.13	99.40	100.00
Self-employed in Non-agriculture	76.52	93.89	98.32	100.00
Self-employed in Agriculture	77.63	88.36	93.37	98.43
Agricultural Labour	66.96	82.46	85.92	88.62
Other Labour	88.76	90.66	94.73	98.12
Coastal Andhra	74.07	92.50	96.86	100.00
Rayalaseema	79.36	90.71	94.50	97.76
Telangana	71.96	85.60	90.51	95.51

Among the regions, Rayalaseema and Telangana fail the target by more than 10 percentage points (Table 6.7). The combined picture of boys and girls indicate that the target of 100 per cent enrolment eludes the rural areas by 6 percentage points. The laggards are STs, followed by SCs and OBCs among the social groups, Agricultural Labour and cultivators among the occupational groups and Rayalaseema and Telangana among the regions. However, the gap is less than 10 percentage points for most.

Table 6.6: Projected Primary Enrolment Ratios (Girls)

Region/Groups	*Year*			
	1993	*2000*	*2009*	*2015*
Rural	61.41	83.20	87.69	91.50
Urban	75.00	88.87	92.80	96.22
Andhra Pradesh	64.26	84.39	88.76	92.49
ST	46.37	60.48	67.23	73.24
SC	53.53	82.67	86.72	90.15
OBC	59.37	83.55	88.18	92.12
OC	65.21	93.32	97.52	100.00
Self-employed in Non-agriculture	69.97	92.54	98.75	100.00
Self-employed in Agriculture	61.90	83.89	89.53	94.38
Agricultural Labour	54.47	77.11	80.12	82.64
Others	81.60	91.73	95.81	99.24
Coastal Andhra	64.50	90.73	95.86	100.00
Rayalaseema	66.87	79.56	82.96	85.82
Telangana	55.59	77.70	83.04	87.62

Table 6.7: Projected Primary Enrolment Ratios (All)

Region/Groups	*Year*			
	1993	*2000*	*2009*	*2015*
Rural	67.92	86.35	90.52	94.45
Urban	75.42	92.25	95.92	98.32
Andhra Pradesh	69.62	87.68	91.74	95.32
ST	54.72	66.28	72.79	79.30
SC	62.52	85.16	89.25	92.78
OBC	66.94	84.57	88.38	92.32
OC	70.53	95.14	98.42	100.00
Self-employed in Non-agriculture	73.34	93.24	98.53	100.00
Self-employed in Agriculture	70.25	86.26	91.57	96.53
Agricultural Labour	60.63	79.75	82.98	85.59
Others	85.39	91.17	95.24	98.65
Coastal Andhra	69.38	91.63	96.37	100.00
Rayalaseema	73.31	85.30	88.91	91.97
Telangana	63.93	81.72	86.84	91.64

However, it is equally important to ensure that every enrolled child is retained in school to complete the primary stage of education. In Andhra Pradesh, it observed that one-third of enrolled children in the primary stage drop-out before completing the primary education. The drop-out rates vary substantially across the districts—60 per cent in the backward districts like Mahabubnagar and Adilabad to about 20 per cent in Guntur and Krishna. Therefore, it is important to monitor whether the enrolled children complete primary stage to enter the next stage of education. The next indicator relates to this issue.

Indicator 2: Primary Completion Rates (MDG: 100% completion rates)

Primary completion rate is defined as ratio of the number of children completing the primary education in a year to the total number of children official graduation age. The ratio is computed differently in different sources depending on the availability. The age groups used vary from 15-19 years to 12-13 years. In AP the level next to primary stage is called Middle school (grades VI-VII), the official age for this is 12-13 years. Children in the middle school age are supposed to complete the primary level of education. We estimate the ratio for the age group 12-13 years.

The projected completion rates for boys are presented in Tables 6.8 to 6.10. As can be seen the projected completion rates fall far below the 100 per cent target in 2015 for all the groups and regions. For the State as a whole, the projections indicate that about 25 per cent of the children

Table 6.8: Projected Completion Rates for Boys

Region/Groups	*Year*			
	1993	*2000*	*2009*	*2015*
Rural	61.33	66.53	74.22	80.38
Urban	69.91	77.82	82.69	86.42
Andhra Pradesh	64.14	70.23	76.99	82.36
ST	27.99	42.16	47.53	51.87
SC	61.09	68.13	71.41	73.90
OBC	63.10	64.20	66.23	67.75
OC	65.12	76.13	85.17	92.42
Self-employed in Non-agriculture	56.23	71.52	72.79	73.73
Self-employed in Agriculture	68.24	69.43	76.06	81.29
Agricultural Labour	50.21	57.58	60.63	62.95
Others	63.81	59.80	68.63	75.88
Coastal Andhra	57.89	66.12	70.89	74.58
Rayalaseema	65.83	75.60	80.29	84.01
Telangana	62.78	62.69	62.69	62.69

Table 6.9: Projected Completion Rates (Girls)

Region/Groups	*Year*			
	1993	*2000*	*2009*	*2015*
Rural	46.70	49.86	54.93	`58.95
Urban	68.90	75.74	80.59	84.44
Andhra Pradesh	54.04	58.42	63.42	67.39
ST	24.93	29.85	36.36	41.97
SC	47.96	40.55	44.92	48.40
OBC	47.68	51.83	56.50	60.16
OC	47.39	67.14	69.14	70.65
Self-employed in Non-agriculture	47.01	57.74	65.23	71.28
Self-employed in Agriculture	51.44	58.23	70.03	80.09
Agricultural Labour	34.66	40.26	43.51	46.04
Other Labour	47.02	25.53	27.64	29.28
Coastal Andhra	44.75	52.64	60.71	67.36
Rayalaseema	45.21	50.62	59.10	66.16
Telangana	49.88	45.64	54.69	62.39

fail to complete the primary stage of education. Unchecked, the current trajectory will take another three decades to reach the goal of 100 per cent completion rate. All the social and occupational groups and regions have shares in this failure. Even among the OC children, the completion rate reaches 82 per cent by 2015. About half of ST children and one-third of SC and OBC children drop-out from primary stage of education without completing it. The projected achievements of Labour households are equally poor. It can be seen that lower performance of completion rates of girls considerably has blunted the overall progress in achieving 100 per cent completion rates. The completion rates of girls in 2015 fall short of the target of 100 per cent by more than half in case of STs, SCs, and Agricultural Labour. For these groups achieving the MDG is a distant dream.

Table 6.10: Projected Completion Rates (All)

Region/Groups	*Year*			
	1993	*2000*	*2009*	*2015*
Rural	54.17	58.37	64.78	69.89
Urban	69.41	76.80	81.65	85.45
Andhra Pradesh	59.19	64.44	70.34	75.01
ST	26.38	35.70	41.67	46.68
SC	55.27	55.90	59.67	62.59
OBC	55.44	58.06	61.40	63.98
OC	56.85	71.94	77.69	82.26
Self-employed in Non-agriculture	49.18	60.98	67.01	71.86
Self-employed in Agriculture	59.97	63.92	73.09	80.70
Agricultural Labour	42.86	49.39	52.53	54.95
Other Labour	56.95	45.79	51.87	56.83
Coastal Andhra	51.17	59.22	65.68	70.89
Rayalaseema	56.31	64.06	70.51	75.77
Telangana	56.67	54.62	58.90	62.55

Indicator 3: Ratio of Boys and Girls in Primary, Secondary education (MDG: Remove Gender Disparity)

This indicator reflects the gender disparity in the education. The projected ratios indicate that the State will fall short of the target by 15 percentage points (Table 6.11).

Table 6.11: Projected Ratio of Boys and Girls in Primary, Middle and Secondary Schools

Groups/Region	*Years*			
	1993	*2000*	*2009*	*2015*
Rural	67.3	74.9	78.5	84.9
Urban	79.4	79.4	87.3	97.0
Andhra Pradesh	71.5	76.6	79.4	85.0
Scheduled Tribe	49.4	85.8	88.2	92.6
Scheduled Caste	61.0	62.7	64.2	67.0
Other Backward Castes	65.7	73.9	74.6	75.9
Self-employed in Non-agriculture	70.1	80.1	84.2	91.8
Agricultural Labour	69.6	70.9	76.9	88.3
Other Labour	60.3	85.5	85.1	84.4
Self-employed in Agriculture	64.4	75.7	69.1	58.9
Others	72.1	64.9	64.6	64.1
Coastal Andhra	68.0	76.5	78.6	82.4
Rayalaseema	71.7	69.8	74.0	81.9
Telangana	64.2	75.9	81.1	91.1

Health

Indicator 1: Infant and Under Five Mortality (MDG: Reduce by two-thirds)

The infant mortality rate is typically defined as the number of infants dying before reaching the age of one per 1000 live births in a given year. If data are drawn from vital statistics registration systems, this is considered to be more robust than the under-five rate. The data constraints do not permit to do systematic analysis at disaggregated level. We have taken the SRS data to examine the trends. For group level estimates, we used NFHS data. The base level IMR for groups is projected using the growth rates observed in SRS data. During the eighties IMR declined at 2.2 per cent and 1.1 per cent per annum respectively in rural and urban areas. At aggregate level it declined at 2 per cent. But in the nineties the IMR was almost stagnant without showing any trend in rural areas. On the other hand, in urban areas it declined faster at 3 per cent pa. At aggregate level the rate of decline fell to less than one per cent p.a. (0.7%).

The goal of reducing the IMR by two-thirds is not achievable at the present trend, based on which the projections were made (Table 6.12). The gap varies from 21 percentage points for STs, 24 for SCs to almost 18 for Agricultural Labour.

Table 6.12: Projected IMR

Groups/Region	*1993*	*1999*	*2009*	*2015*	*Target*
Rural	77.00	78.60	68.17	62.00	50.82
Urban	61.90	46.40	38.22	33.58	40.85
Andhra Pradesh	73.20	66.00	57.24	52.06	48.31
ST	99.30	88.65	76.89	69.93	65.54
SC	89.83	92.79	80.47	73.18	59.29
BC	81.52	76.52	66.37	60.36	53.80
OC	73.20	63.96	55.47	50.45	48.31
Self-employed in Non-Agrl.	62.61	72.09	62.52	56.86	41.32
Agrl. Labour	69.29	83.49	72.41	65.85	45.73
Self-Emp. in Agrl.	85.12	84.59	73.37	66.72	56.18
Others	77.07	64.01	55.52	50.49	50.86
Coastal Andhra	78.26	79.89	69.28	63.01	51.65
Rayalaseema	109.62	82.32	71.39	64.93	72.35
Telangana	88.28	75.67	65.63	59.68	58.27

For under five mortality rates we used NFHS data. Elasticities estimated at aggregate level were used for projecting the under five mortality rates for the groups and regions. Similar to IMR, the goal on under five mortality is not achievable (Table 6.13). Large gaps for SCs and STs among the social groups, Rayalaseema among the regions and Agricultural Labour among the Occupation groups.

Table 6.13: Projected Under Five Mortality Rates

Region/Groups	*1993*	*1999*	*2009*	*2015*
Rural	110.10	102.10	91.67	85.93
Urban	72.60	56.90	40.17	32.60
AP	100.60	91.30	81.97	76.84
ST*	165.58	103.39	92.83	87.02
SC	137.27	121.60	109.17	102.34
BC	117.41	99.13	89.00	83.43
OC	97.54	85.78	77.02	72.19
Self-Employed in Non-Agrl.	98.39	91.25	81.92	76.79
Agrl. Labour	120.29	111.19	99.83	93.58
Self-Emp. in Agrl.	115.20	107.01	96.08	90.07
Others	92.40	86.77	77.90	73.03
Coastal Andhra	118.26	100.90	90.59	84.92
Rayalaseema	111.43	124.33	111.63	104.64
Telangana	100.22	92.69	83.22	78.01

The target for safe deliveries is 100 per cent. By 2015, 70 per cent of the deliveries in rural areas would be attended by professionals (Table 6.14). There are large variations across regions, social and Occupational Groups. Between 2012-2015, about 1.5 million deliveries will take place without

Professional Attendance. The entire unsafe deliveries would take place only in rural areas.

Table 6.14: Percentage of Deliveries Attended by Professionals:Safe Deliveries

Groups/Region	*1993*	*1998*	*2009*	*2015*
Rural	39.70	58.49	67.32	75.49
Urban	78.30	85.29	91.45	97.01
All	50.21	65.79	74.02	81.56
ST	49.30	27.69	31.88	35.75
SC	32.15	58.94	67.37	75.13
OBC	38.63	59.73	68.35	76.30
OC	45.12	67.72	78.25	88.03
Self-employed in Non-agriculture	44.27	65.22	74.51	83.04
Self-employed in Agriculture	38.20	56.28	64.69	72.46
Agricultural Labour	40.90	60.25	68.87	76.79
Others	42.84	63.11	74.32	84.89
Coastal Andhra	49.58	65.66	75.75	85.10
Rayalaseema	24.40	46.20	53.11	59.49
Telangana	37.76	57.82	66.14	73.80

Regarding safe drinking water, if we take Tap and Bore-well (public/private)—more than three-fourths will have access to safe drinking water. All Regions/Social and Occupational Groups will meet goal (Table 6.15). If we exclude public tap, the access comes down to 45 per cent. The entire population without access to safe drinking water will be located in Rural Andhra Pradesh. Nearly one-fourth of Rural population remains without access to safe drinking water.

Table 6.15: Percentage of Households without Safe Drinking Water Source

Groups/Region	*1992*	*1998*	*2009*	*2015*
Rural	56.64	26.84	24.59	22.70
Urban	17.22	4.88	4.31	0.00
Andhra Pradesh	47.21	21.35	19.52	17.03
Self-employed in Non-agriculture	55.37	24.69	22.50	20.66
Agricultural Labour	60.73	33.73	31.80	30.18
Other Labour	58.61	30.16	28.07	26.33
Self-employed in Agriculture	56.49	26.58	24.35	22.47
Others	57.32	27.98	25.40	23.22
ST	53.85	20.68	18.24	16.20
SC	69.66	27.10	24.97	23.19
OBC	62.21	27.04	24.89	23.09
OC	54.77	27.90	25.62	23.71
Coastal Andhra	66.70	41.95	40.13	38.61
Rayalaseema	67.99	7.21	4.39	2.02
Telangana	33.57	18.26	15.86	13.85

Table 6.16

Group/Region	Poverty	Underweight	Enrolment Boys	Enrolment Girls	Enrolment All	Completion Rate Boys	Completion Rate Girls	Completion Rate All	Gender Ratio	IMR	U5M	Safe Deliveries	Drinking Water	Sanitation
Occupational Groups														
Self-employed in non-agriculture	●	●	●	●		●	●	●	●	●	●	●	●	●
Self-employed in agriculture ?	●	●	●	●		●	●	●	●	●	●	●	●	●
Agricultural Labour	●	●	●	●		●	●	●	●	●	●	●	●	●
Other Labour	●	●	●	●		●	●	●	●	●	●	●	●	●
Others	●	●	●	●		●	●	●	●	●	●	●	●	●
Social Groups														
Scheduled Tribe	●	●	●	●		●	●	●	●	●	●	●	●	●
Scheduled Caste	●	●	●	●		●	●	●	●	●	●	●	●	●
OBC	●	●	●	●		●	●	●	●	●	●	●	●	●
Others	●	●	●	●		●	●	●	●	●	●	●	●	●
Regions														
Coastal Andhra	●	●	●	●		●	●	●	●	●	●	●	●	●
Rayalaseema	●	●	●	●		●	●	●	●	●	●	●	●	●
Telangana	●	●	●	●		●	●	●	●	●	●	●	●	●
All														
Rural	●	●	●	●		●	●	●	●	●	●	●	●	●
Urban	●	●	●	●		●	●	●	●	●	●	●	●	●
Andhra Pradesh	●	●	●	●		●	●	●	●	●	●	●	●	●

In the study, we have considered 14 indicators. It is depressing to note that Andhra Pradesh will not meet MDGs in 10 out of 14 indicators (Table 6.16). Thus, except in poverty, enrolment of boys and girls and drinking water, Andhra Pradesh may not achieve millennium development goals in crucial indicators of education, health and sanitation at current rates of progress. Similarly, out of 14 indicators, 11 for Telangana, 9 for Coastal Andhra, 12 for STs, 11 for SCs, 12 for agricultural labourers will not meet MDGs by 2015 (Table 6.17). There are significant rural and urban disparities. In the case of rural areas 11 out of 14 will not meet MDGs while in the case of urban areas only 4 indicators will not meet the goal. These results on MDGs thus show that there is a cause for concern for Andhra Pradesh.

Table 6.17

	Off-track
Occupational Groups	
Self-employed in non Agriculture	9
Self-employed in agriculture	9
Agricultural Labour	12
Other Labour	9
Others	8
Social Groups	
Scheduled Tribe	12
Scheduled Caste	11
OBC	12
Others	8
Regions	
Coastal Andhra	9
Rayalaseema	10
Telangana	11
All	
Rural	11
Urban	4
Andhra Pradesh	10

What is Needed to Improve MDGs for Andhra Pradesh?

There is a need to concentrate on five areas. First, economic growth particularly pro-poor growth in terms of concentrating on agriculture and employment are important in order to reduce poverty. However, economic growth alone will not be sufficient to lift some people above poverty. Despite achieving the MDG on poverty, a large number of SCs, STs will remain poor even if the growth rate is hiked. Women and children constitute significant proportions of poor. The paradigm shift in poverty eradication through women empowerment can provide a solution to this problem of poverty. There is a need to link between livelihoods and social sector. The ongoing women empowerment models in AP may be of high value. The recent study conducted to evaluate the contribution of this ongoing programme indicated that this programme has contributed to rapid decline in poverty and inequality, expansion of economic resource at household level. The ongoing *Velugu* programmes can be sharply focused towards the women belonging to the vulnerable sections. Under which they can be provided with land. The land distribution can be in the form of distribution of government land purchase of land from private individuals and providing clear entitlements to the earlier distributed lands and keeping the entitlements in the name of the women.

Second, the study also shows that although one can achieve at State level, there are significant disparities across regions, groups, gender etc. Reduction in gender disparity itself will achieve the relevant MD goal. Similarly, there is high level of geographical concentration of poverty and low human development in many districts of North Coastal Andhra, Telangana and Rayalaseema. Within developed districts also, there are many backward mandals. It indicates the need for geographical targeting of resources for poverty reduction. For example, improving the MDG indicators in the above poor regions will increase levels for AP. Similarly, there is a need to concentrate on social groups such as SCs and STs to improve their MDG indicators.

Third, there is a need to increase public expenditure on health and education. Effectiveness of these expenditures has to be improved.

Fourth, carefully targeted, sector specific interventions (particularly on health and education) are needed to achieve MDGs. For example, expanded child and material immunization, ante-natal care coverage, nutritional supplementation (including promotion of exclusive breast feeding) and home-based neo-natal services (including treatment of pneumonia) is likely to bring about significant reduction in both infant mortality and child malnutrition. In HIV/AIDS is another problem to be tackled in AP.

Last, development of institutions and strengthening the present institutions of service delivery are important. Several institutions seem to have failed in delivering better services particularly in health and education in rural areas. Institutions seem to be responsive when women are empowered. Decentralization in terms of strengthening PRIs has to be improved in AP in order to have better delivery systems.

To conclude, it is depressing to note that if Andhra Pradesh follows the current rate of progress, it cannot achieve millennium development goals in 10 out of 14 indicators. In order to improve human development indicators and attain MDGs, more efforts are needed than now. There is a need to concentrate on backward regions of all the three regions particularly in Telangana, Rayalaseema, North Coastal Andhra, socially disadvantaged groups and women. Sector specific interventions and strengthening of Government, civil society institutions and PRIs are needed for better delivery systems particularly in rural areas. To achieve this, empowerment, participatory and rights-based approaches are needed.

NOTES

1. DWCRA, SAPAP, APDPIP, APRPRP.
2. The recent experience in India and in many States was that the growth in rural areas has slowed down considerable during the nineties.
3. See Radhakrishna and Ravi, 2004, Deaton and Dreze, 2002, Tendulkar, 2003, Sen and Himanshu, 2004.
4. See Deaton and Dreze, 2002 for price adjustments.
5. See Mahendra Dev (2005).
6. For 1992-93, estimates are made for the age group 0-3 years from the unit level data.
7. Stunting (Height-to-age) reflects chronic malnutrition while wasting (Weight-for-height) indicates acute malnutrition.
8. GoAP(2003).

REFERENCES

Deaton, Angus and Jean Dreze (2002). 'Poverty and Inequality in India: A Re-examination', *Economic and Political Weekly,* September 7, Vol.37, No.36.

Dev, S. Mahendra (2005). 'Calorie Norms and Poverty', *Economic and Political Weekly*, February 19, 2005.

GoAP, (2003), *Economic Survey 2002-03*, Planning Department, Government of Andhra Pradesh.

Radhakrishna, R. and C. Ravi (2004). 'Measurement of Changes in Economic Welfare in India: 1970-2001', *Journal of Quantitative Economics,* Vol. 2, No.2 (New Series), July, pp.58-75.

Rao, C.H.H. and Dev, S. Mahendra (2003). Andhra Pradesh Development: Economic Reforms and Challenges Ahead, Centre for Economic and Social Studies, Hyderabad.

Sen, A. and Himanshu (2004). 'Poverty and Inequality in India-I and II', *Economic and Political Weekly,* Vol. 39, No. 38.

Sundaram, K. and S. D. Tendulkar (2003). 'Poverty in India in the 1990s: Revised Results for all India and 15 major States for 1993-94', *Economic and Political Weekly*, Vol.38, No. 46, November 15.

7

Self-Sustaining Development in Rural India

A Study of Technological, Geographic and Social Changes in Three Villages of Sawai Madhopur District of Eastern Rajasthan

V. NATH*

Introduction

The study, the results of which are described in the following pages, was done first in 1961, when Dr. Nath was the Collector of the district. Mrs. Nath participated in the study. Mrs. Nath and three researchers have done the repeat study in 2001. The two studies show changes in production techniques, increase in output of cereals and commercial crops; major increase in incomes and consumption of people of all castes, reduction in social disparities and increase in education, improvement in health and major changes in aspirations of the peoples of the area. It will be shown that the changes are the result of facilities provided by the State Government and their intelligent use by the people themselves. Furthermore, the technological and economic changes have not been at the expense of the natural resources of soil, water, pasture and woodland. On the contrary, physical texture of the cropland is better and its productivity is higher as a result of the technological changes. The livestock, mainly milch buffaloes, kept in the villages are stall-fed and graze in the fields, which are fallow after the crop has been harvested. The grazing checks growth of weeds; the dung of the animals adds organic

* This article was written in 2001, since then Dr. V. Nath is no more and the article stands as he left it. Mrs. Kamla Nath has done some editorial changes only.

manure to the fields. Ploughing in of roots of sorghum, millets, mustard and groundnut further add to the fertility of the cropland.

The District

Sawai Madhopur district had an area of 5000 sq. kilometres and a population of 875,000 in 1991. These figures relate however to the bifurcated southern part of the district. The bifurcation was done in the 1980s for administrative convenience. Sawai Madhopur town and its suburb Maantown, in which the district offices are located in this half. The northern half was constituted into a separate district at that time.

Seventeen per cent of the population in 1991 was urban and the remaining 83 per cent was rural. More than 40 per cent of the population consists of scheduled castes and scheduled tribes. In the villages studied, however, they form the majority of the population. The *Meenas*, who comprise most of the farmers, are classified as a scheduled tribe while the agricultural labourers or small cultivators-*cum*-labourers belong to the scheduled castes. The high castes—Brahmins, Rajputs and Mahajans—and the non-scheduled tribe cultivators, *Gujars* and *Jats*, form about one-third of the population. The district has five community development blocks which are coterminus with panchayats samitis. There are 197 panchayats for almost 900 villages. Annual rainfall averages 62 cms and is quite dependable. Most of it is received in the three months July to September of the monsoon season. Diesel operated pumps are the principal means of irrigation in all the three villages studied. There are no factory industries. However, there are small industrial units, concerned with processing of crops, wheat and oil seeds. In some villages rolling of *beedies* and making glass bangles are traditional cottage industries.

Self-Sustaining Development

Self-sustaining development has two meanings in our view. First it should not make undue demands on the natural resources of soil, water, pastures or woodlands. Second while the State Government should provide the catalyst for change-advice and inputs for use of improved agricultural techniques, e.g. use of dwarf high yielding varieties of wheat, chemical fertilizers and pesticide, however, the cultivators themselves have, realizing their value, bought tractors for cultivation and diesel operated pumps for irrigating their croplands. Medium and long-term loans for buying these have been given by the Land Mortgage Bank and the nationalized banks, branches of which are established in Maantown market and also in two villages—Maanpur and Raipur.

The crop rotation—sorghum, millets and groundnuts in the monsoon season and wheat, gram and mustard in the cool season—is such that the nutrients removed by sowing wheat are replaced by the organic matter left in the soil by the roots of sorghum, millets, gram and the foliage of groundnuts and mustard. Furthermore, the dung produced by buffaloes provides organic manure to the soil. Double cropping of the cropland means that the land has crops through the year and are not lying waste subject to wind and water erosion after the monsoon crops have been harvested.

Meena and other farmers and the scheduled castes did not send their daughters to schools. They were afraid that educated girls would not work on the farm or do unskilled labour. Now, they send their daughters to schools. They realize that educated girls make better housewives and mothers. They look after their children better and keep their homes clean. Furthermore, there is less work in the farm because of use of tractors and other machines such as mechanical harvesters—threshers, winnowers—for sorghum, millets and wheat. Because of cleaner homes and access to safe drinking water from tubewells and well-kept wells, the health of women and that of their children and men is much better. Furthermore, facilities for medical care are available from two hospitals, four *vaidyas* and a medical practitioner in Maanpur, all located within easy reach of the people of these villages.

Earlier the small Meena farmers and scheduled caste farmers-*cum*-labourers thanked their stars if they had enough cereals to eat through the year and avoid hunger during the lean season of July to September, i.e. before the monsoon crop of sorghum or millets became available. Now they have enough cereals to eat and milk, vegetables and fruit, which they could not afford in 1961, are a regular part of their diet. Earlier, the worldview of the Meena and scheduled castes was linked to their immediate environment of a few villages. They did not expect their children to do better than themselves. Now the better-off Meena and scheduled castes are hoping that their sons would become white-collar workers after education.

Increase in agricultural output has had the effect of major increases in income and consumption of most of the people of the area. Increase in education of boys and girls, and in literacy and awareness of the people along with improvement in health has resulted in remarkable change in their attitudes and aspirations.

Self-sustaining development also means that rise in income and improvement in health, reduction in illiteracy or ignorance among males or females; increase in education of boys and girls should have been shared by people of all castes and not just confined to the high castes and/or the cultivators. In other words, disparities in income and consumption, health status and education should not have increased. From all available evidence, this would appear to be the case in these villages. Most of the people of the three high castes—Brahmins, Rajputs and Mahajans in Maanpur and Raipur the two multi-caste villages had higher income and levels of consumption in 2001 than they had in 1961. Their sons have got educated and have become white-collar workers—school teachers, secretaries of cooperative societies or panchayats, extension officers of various departments. Their daughters have been married to white-collar workers; some have after education became school teachers.

The landowners among them have higher income from farming thanks to greatly increased production of cereals and cash crops due to cultivation by tractors, irrigation of most of the cropland by diesel oil operated pumps; use of dwarf high yielding varieties of wheat, chemical fertilizers, pesticides, etc. has increased yields per hectare 2-3 times.

Most of the *Meena*, *Mali* and *Gujar* cultivators, all except those whose land holdings are too small for self-cultivation also have higher income due to these improvements. The State has provided facilities for increasing agricultural output, i.e. extension officers who advise on use of chemical fertilizers and pesticides, cooperative societies provide short- term credit for crop production, land mortgage and other banks provide loans for buying tractors, pump sets and other machines. The farmers have accepted the facilities and used them.

The people of all the three villages have contributed large sums amounting to 4 million rupees in each of the two large villages for construction of overhead tanks which supply piped water to a large number of houses. The State has assisted by providing the tubewells and the operators to pump water to the tanks. The people have not only invested in having piped water in their houses but paved the main streets and constructed side drains for drainage of waste water in areas in which the houses of the high castes and most of the Meena cultivators in Raipur are located.

The *Meenas* of Jenapur have constructed the overhead water tank, the tubewell and piped water supply in their houses and concrete paved the main street, without assistance from the State Government. These developments indicate the value these people attach to adequate and clean drinking water.

With increase in agricultural output and income, they have invested in improvement of their houses—a large percentage of them live in brick and mortar houses. They have contributed funds through the panchayats for construction of buildings for primary, middle and secondary schools for boys and girls, offices of panchayats and cooperative societies, residences of VLWs and stores for chemical fertilizers and pesticides. The state has provided the hospitals and their staff, Ayurvedic dispensaries and their staff; family planning advisers and other such facilities. The people are using these services in increasing numbers. There are private schools in the two large villages. The standard of teaching in them is higher than in the schools run by the Department of Education, so that they have is a larger number of students (boys and girls) despite much higher fees. There is in Maanpur a qualified physician and two *vaidyas* in the villages who have remunerative practices. A trained auxiliary nurse mid-wife also has a remunerative private practice. A multi-facility hospital, with 40 beds is located within a kilometre of Raipur.

The small farmers-*cum*-agricultural labourers or non-agricultural labourers of *Chamar* and other scheduled castes have benefited as much as or even more from the social and economic changes of the last five decades as the high caste or the Meenas. By 1961, they had already given up their traditional obligations to the *jagirdar* and had become small farmers or small farmers-*cum*-agricultural or non-agricultural labourers. They had constructed a primary school in their own hamlet and were sending their sons to it. The Panchayat of Raipur had improved the drinking water well of the *Chamars* to provide them with adequate and safe drinking water. The primary school in the hamlet of *Chamars* of Maanpur has now become an 8-grade middle school; the school which was located in a temple then has its own building. The building has been constructed and is being maintained by the villagers themselves. A number of boys of both Maanpur and Raipur are attending the higher secondary school of Maantown. The Meenas of the hamlet of Maanpur have constructed 8-room building so that the primary school in their hamlet can be upgraded to a middle school.

Their aspirations have changed significantly. In 1961, the high castes especially the Brahmins and Mahajans whose landholdings were small, were satisfied if their sons could become Patwaris, VLWs, school teachers, secretaries of cooperative societies or panchayats or clerks in the cement factory or some office. Now they want that their sons should get better white-collar jobs after education. One or two sons of our informants in both Raipur and Maanpur are working in States other than Rajasthan. They occupy positions of a status that their grand parents could not even dream of for them. The Meenas and the *Chamars* are also keen that their sons should get white-collar jobs—clerks, school teachers, secretaries of cooperatives or panchayats and others. These jobs are valued for security of service, fairly high wages and the 'power' that they confer on the incumbents.

Caste and Occupational Mobility: A Rajasthan Village

Caste played and continues to play a very important role in occupational change from the traditional village economy to a market economy in which the economic motive is at least as important as the traditional caste base. By 1961, a large number of Brahmin young men had become *Patwaries*, VLWs, secretaries of *Panchayats* and cooperative societies, teachers in primary schools, *vaidyas* (practitioners of ancient Ayurvedic system of medicine), clerks in offices of the Rajasthan Government and the cement factory. After getting educated up to the middle school or higher secondary school examination, some of the Brahmin and Mahajan girls had also become teachers in primary schools. These jobs paid well and since the young men were posted within 30 kilometres of Raipur, they could visit their families on weekends and could keep in touch with them through letters. However, no Brahmin had taken to trading or become an artisan such as carpenter, mason, smith, driver of motor vehicles, etc. despite the high level of earnings from these activities. These occupations were literally closed to them.

The Mahajans of Raipur could become traders in the Maantown market or like the Brahmins white-collar workers after passing the necessary examinations. However, they preferred to remain trader-moneylender-*cum*-non-cultivating owners of their land within their own villages. Besides, relatively high earnings, prestige was a major consideration in staying in the village.

The Rajputs of Maanpur were satisfied with remaining non-cultivating owners of their lands that they could retain for self-cultivation. They employed *Chamars*, *Meena* and *Malies* as agricultural labourers or sharecroppers for managing cultivation of their lands. Their aspiration was that one or more of their sons should get educated and become white-collar workers. However, one son must remain to look after the land of the family. This gave prestige to the family and also provided sizeable income. They would marry their daughters to educated Rajput young men who were in the Army or were white-collar workers or were managing family farms.

The *Meenas* were satisfied with managing their farms better to increase agricultural production. Those with small holdings supplemented their income with unskilled work such as repair of the railway tracks or the roads or work in the cement factory. They were interested primarily in producing field crops (sorghum, millet, wheat and oil seeds) but were not interested in producing vegetables, fruit or milk for sale. Producing fruit and vegetables for sale was the job of the *Malies*. In addition, they produced wheat, sorghum and oilseeds. The *Gujars* were helping to meet the demands for milk. Similarly, the large demand for meat was met by Hindu *Khotiks* and the Moslems.

The *Chamars* although interested in being recognized as a cultivating caste, the equal of the *Meenas*, had like the *Meenas* become unskilled workers in the railway and roads. Some of them had become carpenters, masons, smiths, and so on and found well paid jobs in Maantown market and building houses in Maantown. The teachers in schools and doctors in hospital could be Brahmins or Mahajans while nurses and mid-wives in hospitals could be mainly *Chamars* or Moslems; that is, the women who performed these tasks by tradition. The great increase in demand for meat in Maantown was being met by Hindu *Khotiks* or Moslems. The high demand for meat had not led any Hindu caste except the *Gujars* to increase the size of herds. The meat sold in the Maantown market was all goat meat. Beef or meat of camels could not be sold because of the sensibilities of the people in this predominantly Hindu area. Cattle were reared in the woodlands to produce bullocks. Meat of sheep and goats was also sold by *Gujars* who grazed their herds on the stubble of the cultivated fields and in the forests. But this was done primarily to earn some cash and to keep size of their herds within the availability of the pasture. Thus, even in 2001, the economic motive had not become dominant. Caste considerations remained the important factor in the choice of new economic activities in the new market economy.

Change in values: People of all castes including the *Meenas*, *Malies* and *Chamars* had accepted the fact that education of girls to the primary school level was a good thing. The *Meenas*, *Malies* or *Gujars* did not feel that girls educated to primary schools would not work in the fields. Moreover, cultivation of land by tractors and use of other equipment such as mechanical shellers of sorghum and millets had reduced the women's work in the field. They could, therefore, devote their time to their work as mothers and homemakers. The men realized also that educated women would get prompt treatment for themselves and their children in case of communicable or other diseases. Furthermore, clean homes and access to clean and adequate drinking water supply from wells located in the villages or piped water from overhead tanks meant less disease among their men, children and themselves.

Most of the *Meenas* still said that they did not accept any bride price, but gave dowry on marriage of their girls. However, the truth of this statement could not be verified. Child marriages had been reduced greatly. The Brahmins, Mahajans and Rajputs married their daughters only when they were 16 or 18 years of age and the young men were 22-25 years old. By this time, both were mature and would not fall victim to communicable diseases. Widows had always been a burden among the high castes who wanted to keep their number to the minimum. Even the *Meena*, *Mali* and other cultivators wanted that their daughters should be married when they were old enough to look after their homes and children. Although, widow remarriage was allowed among the *Meenas*, *Malies*, *Chamars* and all other castes, except the high castes the widows did not have a high social status. Even the Moslems did not accord a high status to their widows.

A few Moslem widows of Raipur had become tailors so that they could earn some money while sitting at home. A Delhi based NGO Dastkar, had opened a centre in Raipur where Moslem women made quilted bed sheets for export to Europe and embroidered dresses for sale to the tourists. Women of no other caste had learnt this trade despite the high earnings. Brahmin widows carried milk from Raipur to the Maantown market on their heads to earn a little money to support themselves. They were looked down upon by the rest of the community for being so poor. But these supposedly enlightened people were so stupid that they would not allow their widows to learn a remunerative trade like tailoring. The Brahmin women just did not do this skilled work. On the other hand, Brahmin cooks who used to cook for marriages in the village and went to Mumbai to work as cooks in Marwari homes, some of them still cook for marriages in Raipur and have also opened food and sweet meat sellers' shops in Maantown market and were doing very well.

In 1961, almost all the district level officers were Brahmins, Rajputs, Mahajans or Kayasthas. All the clerks in the Collector's Office belonged to these castes. Almost all the policemen were either Rajputs or Jats. The president of the Bar Association was a Brahmin lawyer who was also the chairman of the Sawai Madhopur Municipal Committee. These jobs preferences have changed only to the extent that many *Meenas* have become BDOs and *Chamars* have become clerks or junior officers.

Thus, most of the jobs in the new economy had been taken by those for whom the change to these jobs was permitted.

Monetary Corruption: A certain amount of monetary corruption was also widespread in the 1960s. Its scale was limited however. The collector previous to the author Mr. Mool Singh was reported to have used his position to enrich himself and his business in Makrana to which he belonged. He used his position to get favours done to him by the cement factory.

The clerks in the office and the junior officers talked about it. It was also known that the clerk in the Collector's office would take a bribe of Rs. 20 or more from a cultivator to give a decision in his favour. There was no stigma against it. However, there were also a few officers or clerks who were scrupulously honest and very competent. One of them was an old Brahmin who was a superb draftsman in Hindi. I could trust him to interpret the cases of the cultivators (I could not understand their language—*Haroti*—well and after a few attempts to write my judgements in Hindi, gave it up because he could write them better than I could). The SDO, Sawai Madhopur a Brahmin was also scrupulously honest; but two SDOs of Gangapur and Karauli were not.

There was much greater corruption at the State level also. One leader of the Jats from western Rajasthan had about 300 people eating at his official residence in Jaipur every night. He was a wealthy farmer but his main source of income was his position as chairman of the Rajasthan Cooperative Federation which controlled distribution of improved seeds, chemical fertilizers, pesticides and improved implements to the farmers through the cooperative societies.

We could not obtain any data on the scale of monetary corruption during our revisit but it was well known among the cultivators that in order to get any work done in the BDO's office one had to pay a bribe. The villagers recognized this as a routine official practice.

Could the Development be consider Sustainable: Could the improvement in the economic position of all the farmers, agricultural labourers and other rural people be considered sustainable or could it lead in one or more decades to a decline in income and consumption due to depletion of the available resources of land, water and forests.

We consider it sustainable in more than one way. Increase in agricultural production was due to better cultivation of land by tractors, larger area under irrigation by wells fitted with diesel pumps. The cultivated area had not increased; on the other hand the increase in agricultural output through use of dwarf high yielding wheat and fertilizers, pesticides other grains (sorghum, millets and oilseeds) in the fields in which there was more retained moisture was evident. The larger production of fruit and vegetables added further to the volume of agricultural output. This production was however in small areas by *Malies* who had this as a traditional highly specialized resource.

The forest resource was being preserved in order to provide a suitable habitat for wildlife; its destruction would have greatly reduced the income of the State Government from this remunerative activity.

The development was self-sustaining in another sense also. Women educated even up to primary school level could look after themselves, their men and their children better. This meant better health and better capacity to work than with the people, who used to fall victims to various communicable diseases. The female literacy rate has doubled during the last few decades. Its effect on growth rate of population will undoubtedly be visible. The increase in per capita income in Rajasthan by 83 per cent during the last two decades is more than the national average. Steadily from a position of extreme backwardness due to the critical lack of social services, the state has seen remarkable improvement.

Leadership in Village and Panchayati Raj Institutions

In 1961, as well as in 2001, there are two conditions which are essential for leadership in the village and these institutions. First, the leaders should belong to a family with a large landholding or an established position of affluence or prestige in the village. In 1961, late Hari Singh a Rajput with large landholding and who had been the principal functionary of the former *Jagirdar* of the village was the *Sarpanch* of Manpur *Panchayat*. He had been *sarpanch* since the *Panchayat* was established in 1952. His position was unquestioned for two reasons. First, he had the qualities which kept the leaders of various communities (*Rajputs*, *Mahajans*, *Meenas*, *Malies* and the scheduled castes) satisfied and he had also established good relations with officials. In 2001, the *Sarpanch* is a successful *Mahajan* businessman. But late Hari Singh's younger brother Mr. Suraj Bhan Singh an ex-*Sarpanch* of the village himself, is consulted by leaders of various castes in matters relating to the welfare of all the villages.

The leaders of the hamlets of the *Chamar* and the *Meenas* are also cultivators of more than average landholdings.

The second essential condition for leadership is that he should be convincingly vocal about the needs and problems of the village to the officials and at the same time convince the villagers that he can put forward the demands of all the villagers as well as or better than any one else. Late Hari Singh and Suraj Bhan Singh has these qualities.

In Raipur, the two most respected leaders in 1961 were (i) a middle aged *Meena* with a large landholding. He had been a member of the first *Panchayat* established in 1952. The second leader was a middle aged affluent *Mahajan*. The family owned a sizeable landholding and besides had money-lending as a principal economic activity. He too had been a member and for many years *Sarpanch* of the *Panchayat* established in 1952. The two leaders had mobilized funds, labour and material contributions from the villagers for constructing the buildings of a middle school for boys and a primary school for girls, repair of the approach road to the village, pavement of streets in the habitation area of the village and construction of buildings for the *Panchayat* and the cooperative society- *cum*-store for improved seed, fertilizers and other inputs. They had obtained grants or loans from the community development administration for construction of all these works and also for construction and repair of wells in the hamlets of the *Chamars* and *Gujars*, located on the outskirts of the village.

Mr. Sham Lal Sharma, the *Sarpanch* of Raipur in 1961 was very critical of the regime of these leaders; he accused them of corruption a charge which they reciprocated. But he recognised that he could not mobilise the funds for the works mentioned above that the traditional leaders had done. The maximum that he could do was to maintain these works and at the same time ensure that the *Panchayat* had enough funds for maintaining clean drains in the habitations of the high castes. The villagers did not allow him to raise taxes to undertake any new development works. Although he was a well paid clerk in the cement factory, he came from a poor family. Thus he lacked one essential condition for leadership. Thus in 1961, there were two essential conditions for position of leadership. First, the leader should be well off enough not to appear to utilize his leadership to enrich himself from the grants or loans that were available for development works in the village. Second, he should be convincingly vocal about his demands and the difficulties of the villagers to the officials.

The *Pradhan* (chairman) of the Sawai Madhopur *Panchayat Samiti* in 1961 was a middle aged *Meena* from the central part of the tehsil which was the stronghold of the *Meenas*. His family had a large landholding and the *Meena community* had chosen him to be a convincingly vocal representative of the *Meena* caste. Furthermore, he had such personal qualities that the decisions of the *Panchayat Samiti* were seldom overturned by the District Development Officer or the Collector.

In 1961, the MLA of the Sawai Madhopur (unreserved) constituency was a reasonably successful Moslem lawyer. His brother was also a successful lawyer in the Collector's Office. Furthermore, he had the backing of the mostly Moslem forest contractors who mobilised men and money for his campaign in return for his ability to get permission to cut forest and other favours from the Collector and the State Government.

He was succeeded in the election of 1962 by a Rajput ex-*jagirdar*, a traditional leader from a reasonably affluent family. Furthermore, he had the ability to approach the *Meena*, *Malies* and the *Chamars* and to convince them that he would further their interests with the State Government.

These conditions have remained unchanged in 2001, no matter who occupies the various positions in the Panchayat Raj Institutions. The *Sarpanch* of Raipur in 2001 is a woman of washerman caste, the lowest caste in the caste hierarchy, in an extremely caste conscious village. She is completely ineffective however. She is accused also of corruption because her husband drinks. She was elected because the seat had been reserved for a member of the scheduled caste and the scheduled caste men could not decide on a better candidate.

However, the real leaders of the community remain a reasonably affluent *Meena* and two or more reasonably well off *Mahajans*.

In Manpur while the *sarpanch* is a Mahajan, Suraj Bhan Singh, the ex-*Sarpanch* is consulted in various matters for his traditional position in the village. The leaders of the hamlets of the *Chamars* and the *Meenas* are also farmers with relatively large landholdings who are convincingly vocal about the demands of their constituents.

Village Caste and Community

The community development programme, a programme of integrated rural development launched by the Government of India in 1952 was based on the concept that the tightly nucleated rural settlement characteristic of north western and north central India including Punjab, Haryana, western U.P., Rajasthan, Gujarat and north west Madhya Pradesh was a community. The village is in fact a fairly large administrative unit. The revenue administration includes in it besides the main village one or more hamlets each of which had a particular caste as its main residents. Thus, Manpur and Raipur are with their two hamlets each, relatively largely multi-caste villages.

The community development administrator and later the Ministry of Community Development which gave loans and grants to these communities for construction of drinking water wells, renovation and improvement of other drinking water services, construction of approach roads, opening primary schools for boys and girls, higher primary (middle) or higher secondary schools in larger villages, provision of facilities for preventive and medical care within easy reach of villagers and providing the facilities to increase agricultural output by giving loans to the farmers to buy improved seeds, chemical fertilizers, pesticides and improved implements in order to increase agricultural production regarded them as communities. The administrator and later the Ministry was very satisfied when the villagers provided through their *Panchayat* or otherwise funds, labour and construction materials required to build approach roads, drinking water wells, particularly for the poor, clean and improve tanks or step-wells, and construct buildings for primary or secondary schools. The farmers response to increased output with use of improved seed and other inputs was also adequate or unexpectedly enthusiastic.

The only regret of the administration and later the Ministry was that funds were available for only a few villages and a large number of villages had to remain uncovered. The funds have been progressively increased over the years and an increasing number of villages have been provided these facilities. However, maintenance of the facilities is variable depending upon the urgency of the needs of the villagers or in case of drinking water wells, of the group that uses it.

Some of them like the school buildings, the drinking water wells and pavement of streets and drains in them inside the villages were well maintained. However, approach roads to the villages were not so well maintained unless the road formed part of a network road of the state PWD.

Are the Villages Communities?

The single caste *Meena* village is a community. It has in its Sarpanch a leader who can speak for the whole village. However, the multi-caste Raipur and Manpur are not communities, although they responded enthusiastically to the officials of the community development administration to obtain funds for construction of middle school for boys and girls, buildings for the *Panchayats*, the cooperative society, the store for improved implements and a residence of the VLW with store for improved seed, fertilizers and pesticides.

Over the years, all the farm inputs are being sold by traders in the market who also mobilise funds for them.

These multi-caste villages are not community however and do not have one spokesman. In Manpur, there are two leaders one each for the hamlets of the *Chamars* and the *Meenas* besides one or more spokesmen of the main village. The *sarpanch* of the Manpur *panchayat* is a *Meena* because the seat has been reserved for a *Meena* but for any work requiring contributions from the villagers, the villagers will need to consult Suraj Bhan Singh and others whom they trust and recognise as their leaders.

Is each caste a community? Yes, for some purposes the Rajputs of Manpur have more kinship with Rajputs of adjoining villages of Sawai Madhopur taluk or even Janpur district than they have with the Brahmins or Mahajans of their own village. They would marry with other Rajputs and present a solid front to the elections for protection of their rights.

What Next

Economic development is a tumultuous process; its course is never standard. During the last four decades there have been major increases in income, consumption and level of living of these people due mainly to increase in agricultural output as a result of major technological changes. The farmers are educated, are healthier and are much better informed thanks to the presence of radio and T.V. receivers in their homes which create a keen desire for living 'better', i.e. like the urban middle class. Would these improvements continue during the next two decades or would there be a retrogression. It is difficult to give a definitive answer to this question, just as it was difficult to imagine that the changes that have occurred would occur. Although there does not appear to be any new growth pole in the economy, agricultural output could continue to increase somewhat by still better utilization of improved seeds of wheat and mustard the principal cash crops. There are no minerals which could be processed to locate industries. Reopening of the cement factory would depend upon a steep increase in the demand for cement for which there is no immediate prospect. The only other industries would be wheat flour mills and oil mills to process mustard. It would provide employment to a small number of workers. The income of farmers would increase further if supply of electricity is improved because using electricity to operate irrigation pumps would be much cheaper than use of diesel. Use of electric lights at night and fans during the day would reduce expenditure on kerosene oil and increase comfort.

Young men of the high castes would after graduation from higher school or college, migrate to Jaipur, Kota, Delhi or elsewhere to get white-collar jobs. The more affluent parents would finance education of their sons through degree and professional colleges so that they can become professionals and earn high incomes. This has happened in case of sons of our informants in both Raipur and Manpur; they are professionals who are working at long distances from their homes.

Competition for white-collar jobs in offices of the district would increase greatly so much so that most of these jobs would have to be given to young men of *Meenas* and the scheduled caste. Most of the female school teachers could however continue to be young Brahmin or Mahajan women.

Although there is no visible new growth pole in the economy, it is unlikely that with the improvement in income, education, health and awareness of the people, there should be any decline in their income, education and health in the next two decades. But it is difficult to predict the ways in which further improvement will be achieved just as it was difficult to imagine in 1961 that incomes and level of living, education, health, awareness and aspirations of these people would increase during these four decades, as much as they have. In 1961, most of them were poor, ignorant and in poor health and were resigned to their fate. Now most of them can be classified as being in the lower middle income. They are better educated, and better informed; their health is better. They aspire to raise their economic status further to middle income. It is difficult to predict whether or how they will achieve their aspirations.

Recommendations

1. The supply of electricity by the Rajasthan State Electricity Board should be improved so that the farmers can use electrically operated pumps, instead of diesel-operated pumps for irrigation. This will reduce greatly the cost of irrigation because diesel is very expensive. The reduction in cost of irrigation will also mean significant increase in income of farmers. Furthermore, they can use fans and air coolers in their homes during the day and electric light instead of kerosene oil or petromax light at night which would add to the comfort of the residents.

2. The road to Jaipur *via* Tonk should be kept in good repair so that tourists motoring the Delhi-Jaipur-Agra-Delhi route in three days can add a visit to this game sanctuary in one day. This will not only add to the income of the farmers but will give some employment to landless labourers in hotels taking care of the tourists.
3. Advantages of the reservation of posts for their sons in services of the district administration. In 1961, these posts were the exclusive preserve of the higher castes. Earlier, the small farmers and the scheduled castes had to pay a bride price to get married. By marrying, the man acquired a valuable worker on the farm. Bride price has been replaced by dowry. The amounts of dowry are not small either. A Patwari, a school teacher in a primary school or a clerk in an office demand and get dowry ranging between Rs. 150,000 to Rs. 3,000,000.

8

Gender Equality and Women's Empowerment

KAMLA NATH

Introduction

According to a CIDA Report "*Gender Equality* means that women and men have equal conditions for realizing their full human rights and potential to contribute to national, political, economic, social and cultural development and benefit equally from the results.....

Attaining gender equality demands a recognition that current social, economic, cultural, and political systems are gendered; that women's unequal status is systemic and that it is necessary to incorporate women's specificity, priorities and values into all major social institutions." (CIDA)

Because of current disparities, however, equal treatment of women and men is insufficient as a strategy for gender equality. Equal treatment in the context of inequalities can mean the perpetuation of disparities. Achieving gender equality will require changes in institutional practices and social relations through which disparities are reinforced and sustained. It also requires specific action, which will lead to positive discrimination in favour of the depressed gender.... Since women are the depressed gender it will require a strong voice for women in shaping their societies.

Gender Equality is the process of being fair to women and men. To ensure fairness, measures must often be available to compensate for historical and social disadvantages that prevent women and men from otherwise operating on a level playing field, measures that will empower women and achieve political, social, economic, cultural and environmental security (OECD, 1998).

Equality between women and men should be promoted in ways that are appropriate to each particular context. This begins with an understanding of the current status of women and men, their relations to each other and the interconnections between gender and other factors. In order to understand the quantitative disparities in the current status of men and women it is important to undertake gender analysis through sex segregated data of important social and economic sectors as well as women's position in decision-making. *Gender analysis* can provide an understanding of the social relations and how inequality can be addressed.

Both men and women have a stake in building a more just society where all people are equally valued for their contributions. Yet, gender equality has often been seen as a 'women's issue' separate from men and 'gender' often been confused to mean women only. This is part of a tendency to consider men's characteristics as the norm and women's characteristics as different from the norm. Yet, both women and men are influenced by their gender and thus, ideas about the status of women are interrelated with ideas about the status of men. In short, change for women means change for men and *vice versa*. However, since men are considered superior to women, gender equality is approached from the point of empowering women to become equal to men.

Gender Equality is not only a social/gender justice issue and a measure of equal participation, it is an important component for economic development. Because inequality of access to development resources and to decision-making power between women and men results in lower contribution to development by women, who are half of the labour force, than that of men. No society can exploit its full potential for development if half of its labour force is less developed than the other half. The United Nations identified this as a major reason for slow economic development of developing countries as early as at the end of its First Development Decade (1960s) and emphasized on the member countries to identify the status of women in their respective countries to identify gender economic and social inequalities and evolve strategies to reduce the inequalities. All member states of the UN are committed to gender equality. Research in this area revealed, however, that gender inequalities exist in all societies—developed and underdeveloped in different forms and degrees. This is when women of the world got united to raise a voice against inequality, which is reflected not only in lower economic and social positions of women but is also the major cause of gender violence.

The Human Development Report of the UN published every year to measure human development, which according to it is a major contributor to development. The Report also stresses on the need for gender equality and empowerment of women in order to achieve equality. It has developed indices for measuring improvement in these indicators. Equality is considered by the UN as a fundamental input in human development and subsequent economic development.

Gender Equality as a Development Goal

Gender Equality is both an essential element of any development process and a result of sustainable development. Development efforts should not only aim to advance equality between women and men as an outcome, but should also advance equality throughout by applying gender analysis in planning, consultations, design, implementation and assessment.

Women's Empowerment

Since it is women who predominantly experience inequality (inferiority), empowerment supports women to become more aware of the unfair power relations they face. Empowerment allows both men and women to take control over their lives by gaining skills, setting their own agendas and building self-confidence and self-reliance. It also means that women have increased control over their lives by acquiring a voice to challenge and overcome the inequality they experience. Empowerment is both a process and a result that cannot be achieved by an individual or organization for someone else. A woman can only empower herself, although organizations and agencies around the world can play a role in supporting this journey and providing an enabling environment. Gender equality is both an essential element of any development process and a result of sustainable development. Development efforts should not only aim to advance equality between women and men as an outcome, but should also advance equality throughout by applying gender analysis in planning, consultations, design, implementation and assessment.

It is one of the goals of Millennium Development (4), which proposes to Eliminate gender disparity in primary and secondary education, preferably by 2005, and to all levels of education no later than 2015. The Indicators for Achievement of Millennium Development Goal 4 are Equality in:

- Ratio of girls to boys in primary, secondary, and tertiary education.

- Ratio of literate women to men ages 15 to 24.
- Share of women in wage employment in the non-agricultural sector.
- Proportion of seats held by women in National Parliament.

The Government of India is committed to achieve the goal of gender equality, which is also enshrined in the Constitutional Rights. Furthermore it has committed itself to ensure gender equality in several international conventions and Beijing Platform for Action. Almost all the political parties have inserted gender equality, in some form or the other, as an objective in their party election manifestoes. Today no person or politician can oppose gender equality as an undesirable development goal.

Without going into international comparisons, I shall restrict myself to data over-time on India to measure the extent of inequality, measures taken to reduce it through policy measures and empowerment of women as well as the effects of the measures taken so far and the road ahead. We can find answers to most of these questions through gender analysis of data on economic, social and political position of women. Sex segregated data are not available for all sectors before 1971. However, importance of collection of such data was impressed by the report of the Committee on the Status of Women 1975 and since then special efforts have been made to collect sex-segregated data. The Report, "Towards Equality" (1975) was the first exhaustive document, which brought out extreme gender inequalities in the country and the need to change this status. The Report was presented to the Parliament, which recommended that the Prime Minister take specific administrative and legislative measures at removing economic and social injustices and disabilities and discrimination to which women continued to be subjected even after a quarter century of Independence. It is interesting to note that the Parliament's recommendations did not take note of gender inequalities as such but was concerned about injustices and discrimination and disabilities which when removed would no doubt create a more enabling environment but will not necessarily lead to equality because importance of measures for gender equlity were overlooked by the Parliament in its recommendations. It is only in subsequent research and pressure from the women's movement that these measures have been introduced in piecemeal.

Since then through gender specific Government policy measures, international gender focused development aid and collective actions by women's groups have led to reduction of inequalities somewhat. The results of this collective efforts of almost quarter of a century (1975-2001) has been captured through collection and analysis of exhaustive data in the report prepared by Dr. Sarla Gopalan (retired Secretary Department of Women and Child Development, Government of India) "Towards Equality—The Unfinished Agenda—Status of Women in India—2001" commissioned by the National Commission for Women. Gender analysis of the data on several sectors of Indian society reveals, however, that after 55 years of Independence gender equality is nowhere in view.

What has Changed Since 2001 Towards Equality?

Following policy measures are significant:

- A section included on Women's Development in the 6th Five Year Plan and each Plan subsequent to it. The focus in the Plan priorities in this section has progressively changed from women's welfare programmes to women's empowerment and gender equality issues including insisting on all sectoral Ministries to appoint a gender specialist and evaluate their budgetary allocations for gender budgeting.
- Setting up of National Human Rights Commission (NHRC) and National Commission for Women (NWC). While the NHRC takes up all human right violations, a slogan has been coined by women activists that human rights are women's rights. Thus NHRC thus is more often faced with cases of gender violence. NWC has been constituted especially on demand from women activists in order to have national level and states level commissions, which will recommend policies, which will bring about gender equality and women's empowerment. These are spoke-persons on women's issues. However, because of its limited powers and being under the Government the NWC has often been seen to be not doing enough to strengthen women's voices. At states level few Commissions are strong enough to mobilize public opinion and manage State level political bosses for effective pro-women policies and programmes.
- Amendments to existing laws and new laws passed by the Parliament with gender perspective.

The laws passed before and since CSWI report which focused on protection of women rights and safety were:

1. In the area of personal rights, besides the Hindu Women's Right to Property–1922 and 1937, and Hindu Succession Act, 1956, Shariat Law/Muslim Marriage Act, 1939 and the Hindu Code Bill, 1952, the additional laws passed to protect women's personal rights since 1971 are the Hindu Women's Succession Act, 1978–Hindu Succession Act (Andhra Pradesh), 1985. Hindu Succession Act, 2005. The 1956 Act did not give equal rights to married women in self-acquired property and even the unequal share was not given to them by the use of instrument of 'Will' and/or by coercing the daughters to give up their share in favour of their brothers. Also daughters had no right in the inherited property. Andhra Pradesh was the first to plug these loopholes in the Succession Act in 1986. Recently the Hindu Women's Succession Amendment Act, 2005 has given full equality to women in inherited property, which includes agricultural land.
2. In 1978 the Child Marriage Restraint Act of 1929 amended to raise the age of marriage for girls from 14 to 18 and that of the boys from 15 to 21.
3. The Dowry Prohibition Act (DPA) enacted in 1961 was amended in 1984 and again in 1986 to make it a non-bailable (cognizable) offence under Sections 304-B and 498-A which made strict punishment possible particularly in case of dowry deaths.
4. While Hindu women are protected in case of divorce which itself was made more difficult under the Hindu Code Bill, Muslim women are not under the purview of the act and when applied in case of Shahbano for maintenance there was protest from the Muslim community for interference in their personal laws and a special Muslim Women (Protection on Divorce) Act was passed in 1986. Because of uneven treatment of women from minority communities there has been a strong demand for Uniform Civil Code.
5. Adoption, Guardianship's Custody, Hindu Minority and Guardianship Act, 1956 gives unequal, exclusive right to the father. The Act was challenged in 1999 in the Supreme Court, which by its decision set precedence for mothers and father to have equal right.

6. Domestic Violence against women cases so far were registered under IPC 498-A under which few criminals have been punished and courts take ages to decide the cases. Under the new Act 'Prevention of Domestic Violence Act, 2005' in cases registered under the Act women have very broad rights and it is mandatory for the courts to decide the cases within a stipulate period. The Act has been termed as the Magna Carta for women's liberation.
7. Regulation & Prevention of Misuse of Sex Determination (RPMSD) Act, 1994: Unfavourable sex ratio, particularly in 0-5 age group through female foeticide has led the Government to pass the RPMSD Act. Under the Act both the parents and the doctors assisting in foeticide are liable for severe punishment including jail. However not a single conviction has been reported during the 11 years since the law was passed.
8. Rape is a major and a very severe form of violence against women. Rape is a cognizable offence. Section 376 of IPC deals with all cases of rape including that of minors. Section 354 of IPC deals with. As a result of Mathura (a minor) Rape case 114A amendment in Evidence Act was added, it was approved that the Courts shall not ask for evidence in rape cases. There are very strict rape laws with immediate imprisonment and punishment up to 7 years. There has been a discussion on extending death penalty in case of rape. However, it has been opposed by women's groups who insist that implementation of the existing laws and speedy verdicts by the Courts will be enough deterrent and death penalty will only further reduce conviction rate; currently conviction rate is very low and cases drag on for years.
9. Bill on Sexual Harassment at work place (with a very broad definition of 'work place') is expected to be passed in the winter session of the Parliament.
10. Immoral Traffic Prevention Act, 1986 prevents trafficking in women and children and child prostitution.

There has been rapid progress in amending laws and making new laws for protection of women and for stricter punishment for crimes against women. However, as laws have become stringent there is increase in reported crimes against women. However, large majority of cases are not registered and from among the ones registered a number of them are withdrawn and the ones that reach up to the courts few get completed and criminals punished.

But once the cases go up to the judiciary, judges have been liberal in interpretation of the laws in dealing with cases involving women's issues:

- *Policies for Gender Equality and Empowerment of Women.*
- *The National Policy on Education* (1992) brought focus on girl's education—Education being mother of all Institutions. The data on Literacy Rates by sex show much greater increase in female literacy than the males (Table 8.1) rapid increase in number of girls per 100 boys at Primary and Secondary Levels (Table 8.2) and at University and Higher Levels of Education (Table 8.3). Education is the main empowerment instrument for women.

Table 8.1: Literacy Rates—India 1951-2001

Year	*Persons*	*Male*	*Female*
1951	18.33	27.16	8.86
1961	28.31	40.40	15.34
1971	34.45	45.95	21.97
1981*	43.67	56.50	29.85
1991*	52.21	64.13	39.29
1997@	62.00	73.13	50.29
2001#	65.38	75.85	54.16

Note: Literacy Rate for 1951, 1961, 1971 and 1981 related to population aged five years and above.

* Census could not be held in Assam in 1981 and J&K in 1991.

@ An analysis based on NSSO Survey 1998, National Literacy Mission.

\# Census, 2001

Source: Registrar General, India.

Table 8.2: Number of Girls per 100 Boys in Primary and Secondary Schools 1950–2002

Year	*Primary Classes*	*Secondary Classes*
1950-51	39	16
1960-61	48	23
1970-71	60	35
1980-81	63	44
1991-92	71	50
2001-02	90	na

Table 8.3: Index of Growth in Enrolment of Girls in Higher Education 1970-71 and 1993-94

Category		*1993-94*
Graduation	Total	292
	Female	445
Post Graduation	T	331
	F	456
Research	T	470
	F	784
Diploma/Certificate	T	231
	F	300
1. Total for General Education	T	296
	F	447
Professional Education		
Graduation	T	302
	F	465
Post Graduation	T	425
	F	632
Research	T	467
	F	164
Diploma/Certificate	T	219
	F	222
2. Total for Professional	T	316
	F	458
Grand Total for 1 + 2	T	298
	F	448
Females as % age of Total 1970-71	22%	33%

Source: Based on Data of the University Grants Commission quoted in NCRT Report 'Fifty Years of Women's Education in India' (1947–97).

Educational Development

Literacy rate of females has gone up from 8.9 per cent in 1951 to 54.2 per cent in 2001. Both, rise in income and increase in the age at marriage are responsible for it as well as Government's implementation of compulsory universal Primary Education and non-formal education programmes like National Elementary Education Mission, Operation Black Board, and Sarva Shiksha Abhiyan and innumerable NGOs initiatives. The latest policy announcement of investment of Rs.80,000 crores on universal Secondary Education in the next five years is a major initiative of the Government of India. Contribution of NGOs in imparting literacy to adult women and drop-out girls is of significant.

Disparities between male and female education still exist. However, the importance of education has been brought out by the National Committee on Women's Education 1959 and National Plan of Action

1976. National Policy on Education 1986. Programme of Action on Education for Women's Equality brought out the importance of education for empowerment of women, development and gender equality. National Development Council Committee on Literacy-1992-93 reported on a World Bank study in 29 countries, which showed an inverse relationship between child mortality and women s education and direct correlation between immunization status of children and female education.

Education thus is not simply a matter of social justice but more a matter of economic growth, social well-being and social stability. The Committee also found that absence of facilities for higher levels of education leads to higher drop-out rate at primary levels of education.

The drop-out rates at primary school levels is high for both boys and girls and was almost equal in 1996-97. In fact in secondary class the drop out rates of boys is higher than that of girls (Gopalan: Table 5.10, p. 148).

One of major recommendations of the BPFA is allocation of sufficient resources on education. The GoI committed itself to double the expenditure on education from 3 per cent to 6 per cent of GDP. However, ten years later (in spite of levy of education cess of 2%) the expenditure on education as percentage of the GDP has gone up to 4 per cent.

The growth in the number of institutions between 1950-51 and 1998-99 for different levels of education is given in Table 8.4.

Table 8.4: Growth in Recognized Educational Institutions

	Primary	*UP*	*HS*	*College*	*Profes-sional*	*Univer-sity*
1950-51 =100	209671	13596	7416	370	208	27
1993-99	626737	190166	112438	7494	2113	237
Index	334	715	660	4937	9843	1139

In spite of increase in educational institutions, India will achieve universal primary education target 30-years after Korea and 20-years after Malaysia (UNDP Human Development Report, 1999).

National Policy on Empowerment of Women

The 73rd & 74th Amendment to the Constitution reserving 1/3 of seats for women in elected village and urban local bodies passed in 1972 and 1974 have brought millions of women across the country into local level decision-making bodies not only as members of the institutions but also as their heads and chairpersons (Tables 8.5 and 8.5-A).

Table 8.5: Women Representatives in Panchayati Raj Institutions 2002

(*Percentage*)

S. No.	India/States/Union Territories	*1997* GP	*1997* TP	*1997* ZP	*2002* GP	*2002* TP	*2002* ZP
1.	Andhra Pradesh	33.84	37.01	33.21	33.00	33.65	33.24
2.	Arunachal Pradesh	Arunachal Pradesh Panchayati Raj Act not yet passed					
3.	Assam**	18.01	26.09	—	33.45	34.73	30.00
4.	Bihar	Post 73rd Amendment Election not held in the State, not available					
5.	Chhattisgarh	—	—	—	33.74	34.33	34.67
6.	Delhi	NCT Government propose conduct of Panchayat Elections					
7.	Goa	36.53	—	—	31.76	—	34.00
8.	Gujarat	17.29	33.43	33.38	33.35	33.48	33.54
9.	Haryana	30.74	35.31	80.53	33.59	34.65	34.71
10.	Himachal Pradesh	32.93	33.59	33.33	36.78	33.90	34.66
11.	Jammu & Kashmir	State proposes adopting 73rd Amendment					
12.	Jharkhand	Current Figures not available					
13.	Karnataka	43.79	40.21	36.45	44.86	42.24	38.09
14.	Kerala	Current Figures not available			36.21	38.40	34.20
15.	Madhya Pradesh	32.93	34.84	2.99	33.82	33.44	33.79
16.	Maharashtra	33.33	33.31	33.31	33.33	36.06	33.73
17.	Manipur	35.67	—	36.07	35.48	—	36.07
18.	Meghalaya						
19.	Mizoram	Traditional Councils performs duties of local Government.					
20.	Nagaland						
21.	Orissa	33.35	33.35	33.26	35.88	35.14	34.66
22.	Punjab	35.69	Election due		35.69	32.78	31.90
23.	Rajasthan	29.73	31.67	3.21	34.52	36.29	36.11
24.	Sikkim	37.34	—	30.43	36.88	—	31.52
25.	Tamil Nadu	25.07	35.31	3.40	26.86	26.94	26.37
26.	Tripura	33.55	34.18	34.29	33.33	35.81	34.15
27.	Uttaranchal	—	—	—	—	—	—
28.	Uttar Pradesh	15.08	23.11	23.60	—	—	—
29.	West Bengal	36.33	35.18	33.94	22.46	22.42	21.58
	Union Territories						
30.	Andaman & Nicobar Islands	34.33	—	33.33	34.39	37.31	33.33
31.	Chandigarh	Currents Figures are not available			33.95	40.00	30.00
32.	Dadra & Nagar Haveli	34.07	—	25.00	36.29	—	33.33
33.	Daman & Diu	39.68	—	33.33	26.98	—	40.00
34.	Lakshadweep	37.97	—	36.36	37.97	—	36.36
35.	Pondicherry	Not available. Post 73rd Amendment Elections not held in the State					
	India	6.40	8.86	3.38	3.66	2.70	2.99

Note: Annual Report 2000-01, Ministry of Rural Development, Government of India.

Source: India, Ministry of Human Resource Development, Department of Women and Child Development. (2004). Government of India, II & IIIrd Periodic Report on the Convention on the Elimination of All Forms of Discrimination against Women: CEDAW Periodic Report, New Delhi, p. 86.

Table 8.5A: Women in Panchayati Raj Institutions (1995-2001)

(Figures in thousand)

Year	Women	Men	Total
1995*	318 (33.5%)	630	948
2001*	725 (26.6%)	1,997	2,722

Besides while the proportion of women voters voting has been lower than that of the men, the difference in the proportion has narrowed from 15 percentage points to just 8 (Table 8.6).

Table 8.6: Percentage of Female and Male Voters casting their Vote in each Election in India

(In per cent)

Election Year	Percentage of Female Voters	Percentage of Male Voters
1952	—	—
1962	29.07	59.68
1967	41.79	65.12
1972	44.37	66.03
1977	43.26	62.15
1980	39.28	58.48
1985	41.37	58.04
1993	49.72	68.22
1998	53.35	66.45

Women's representation in higher level decision-making bodies, i.e. in the Land Upper House of the Parliament (Table 8.7) and in State Legislatures (Table 8.8) show that women constitute a miniscule proportion of the total members. However, their proportion in Parliament (Lok Sabha and Rajya Sabha) has gone up from 6.1 per cent in 1989 to 8.4 per cent in 1999. In 2004 it is 9.1 per cent. However, in State Legislatures has either gone down or has remained stagnant. Efforts at passage of the Constitutional Amendment for 33 per cent reservation of seats for women in national and state level legislatures has been strongly opposed by all political parties in spite of commitment in their party election manifestoes gender equality and one-third representation of women in all decision making bodies.

Table 8.7: Women's Representation in Parliament 1952–1998

Year	Lok Sabha			Rajya Sabha*		
	Seats	Women MPs	% Women MPs	Seats	Women MPs	% Women MPs
1952	499	22	4.41	219	16	07.31
1957	500	27	5.40	237	18	07.59
1962	503	34	6.76	238	18	07.56
1967	523	31	5.93	240	20	08.33
1971	521	22	4.22	243	17	07.00
1977	544	19	3.49	244	25	10.25
1980	544	28	5.15	244	24	09.84
1984	544	44	8.09	244	28	11.48
1989	517	27	5.22	245	24	09.80
1991	544	39	7.17	245	38	15.51
1996	543	39	7.18	223	19	08.52
1998	543	43	7.92	245	15	06.12
Average	527	31	5.91	239	22	09.11

Their representation in Council of Ministers has remained steady at 10 per cent for decades (Table 8.9). However there number among Cabinet Ministers is never more than one member. Even among party executives the proportion of women ranges from 5-12 per cent; lowest being among leftist parties who have never have had a woman as member of the Politburo until very recently. However, there are a number of political parties headed by women—Congress I, Samta, BSP, AIADMK and AITC (Gopalan Table 9.7, p. 290).

Women in Higher Civil Services: The proportion of women in Government Services has gone up from 11 per cent to 15 per cent. In senior level services, women constitute less than 10 per cent both in IAS (around 9%) and Indian Customs and Central Excise Service (around 7%) and in Indian Police Service (2%). Their proportion in Indian Statistical Services and in the Central Secretariat Services is also (5-7%). In some of the other services their proportion ranges between 15-20 per cent. Their higher proportion in the Indian Foreign Service is noticeable. Their proportion in some of the other services like the Police is almost negligible, (Table 8.10). While women among State High Court Judges is low but has been increasing but among Supreme Court Judges there has never been more than one woman (Table 8.11).

Table 8.8: Declining Representation of Women in State Legislatures 1952-97 (% of Women MLAs)

States	1952	1957	1960-65	1967-69	1970-75	1977-78	1979-83	1984-88	1989-92	1993-97	1998-99	State Avg.
Andhra Pradesh	2.9	3.7	3.3	3.8	9.1	3.4	4.1	3.4	3.7	2.7	9.5	4.6
Arunachal Pradesh	—	—	—	—	—	0.0	3.3	6.7	3.3	3.3	1.7	3.0
Assam	0.5	4.6	3.8	4.0	7.0	0.8	0.8	4.0	4.0	4.8	—	3.2
Bihar	3.6	9.4	7.9	2.2	3.8	4.0	3.7	4.6	2.8	3.4	—	4.3
Goa	—	—	—	6.7	3.3	3.3	0.0	0.0	5.0	10.0	5.0	4.4
Gujarat	—	—	8.4	4.8	3.2	—	2.7	8.8	2.2	1.1	2.2	4.0
Haryana	—	—	—	7.4	6.2	4.4	7.8	5.6	6.7	4.4	—	6.2
Himachal Pradesh	0.0	—	—	0.0	5.9	1.5	4.4	4.4	5.9	4.4	8.8	4.2
Jammu & Kashmir	—	—	0.0	0.0	5.3	1.3	0.0	1.3	—	2.3	—	1.5
Karnataka	2.0	8.7	8.7	3.2	5.1	4.0	0.9	3.6	4.5	3.1	2.3	4.3
Kerala	0.0	4.8	3.9	0.8	1.5	0.7	3.2	5.7	5.7	9.3	—	3.6
Madhya Pradesh	2.1	10.8	4.9	3.4	5.4	3.1	5.6	9.7	3.4	3.8	8.1	5.4
Maharashtra	1.9	6.3	4.9	3.3	9.3	2.8	6.6	5.6	2.1	3.8	4.2	4.6
Manipur	—	—	—	—	0.0*	—	0.0	0.0	1.7	0.0	—	0.3
Meghalaya	—	—	—	—	1.7	1.7	0.0	3.3	—	1.7	5.0	2.2
Mizoram	—	—	—	—	0.0	3.3	3.3	2.5	0.0	0.0	0.0	1.2
Nagaland	—	—	—	0.0	—	—	0.0	1.7	—	0.0	—	0.5
Orissa	9.6	3.6	1.4	3.6	1.4*	4.8	3.4	6.1	4.8	5.4	—	4.0
Punjab	2.2	5.8	5.2	1.0	5.8	2.6	5.1	3.4	5.1	6.0	—	4.0
Rajasthan	0.0	5.1	4.5	3.3	7.1	4.0	5.0	8.0	5.5	4.5	7.0	5.0
Sikkim	—	—	—	—	—	—	0.0	0.0	6.3	3.1	3.1	2.5
Tamil Nadu	0.3	5.9	3.9	1.7	2.1	0.9	2.1	3.4	9.0	3.8	—	3.6
Tripura	—	—	—	0.0	3.3	1.7	6.7	3.3	—	1.7	—	3.0
Uttar Pradesh	1.2	5.8	4.4	2.8	5.9	2.6	5.6	7.3	3.3	4.0	—	4.1
West Bengal	0.8	3.6	4.8	2.9	1.6	1.4	2.4	4.4	7.1	6.8	—	3.4
Delhi	4.2	—	—	—	7.1	7.1	7.1	—	—	4.3	12.9	7.3
Pondicherry	—	—	6.7	3.3	0.0	0.0	3.3	3.3	1.7	3.3	—	2.6
Period Average	1.8	6.3	4.9	2.9	4.4	2.8	3.8	5.3	4.5	4.0	6.0	4.1

Table 8.9: Representation of Women in the Central Council of Ministers, 1985-2005

Year	*Females*	*Males*	*Total*
1985	4 (10.0%)	36	40
2001	8 (10.8%)	66	74
2005	7 (10.4%)	—	—

Note: Figures within parentheses indicate percentage to total.
Data refers to 9 States—Gujarat, Haryana, Karnataka, Kerala, Madhya Pradesh, Punjab, Rajasthan, Tripura and West Bengal.
For whole of India (as on 18.10.2001)

Source: National Informatics Centre, Parliament House, New Delhi.

Table 8.10: Number of Judges by Sex in Supreme Court/High Courts as on 2003

Name of the Courts	*Number of Judges*		
	Female	*Male*	*Total*
Supreme Court	1	24	25
High Courts			
Allahabad	1	69	70
Andhra Pradesh	2	32	34
Bombay	3	49	52
Calcutta	1	41	42
Chhattisgarh	—	4	4
Delhi	1	27	28
Guwahati	—	12	12
Gujarat	1	30	31
Himachal Pradesh	1	5	6
Jammu & Kashmir	—	11	11
Jharkhand	—	12	12
Karnataka	1	31	32
Kerala	1	25	26
Madhya Pradesh	1	23	24
Madras	1	30	31
Orissa	—	12	12
Patna	1	23	24
Punjab & Haryana	1	30	31
Rajasthan	1	25	26
Sikkim	—	2	2
Uttaranchal	—	4	4
Total	17	497	514

Note: Department of Justice, Ministry of Law, Justice and Company Affairs.

Source: India, Ministry of Statistics and Programme Implementation, Central Statistical Organization. (2003). Women and Men in India 2002, New Delhi. p. 98.

Table 8.11: Percentage of Women in Civil Services Personnel

Category	*Year*	*Women % of Total*
Indian Administrative Service	1993	9.3
	1996	9.9
Indian Audits & Accounts Service	1993	15.7
	1995	17.8
Indian Customs and Excise	1993	6.7
	1996	7.8
Indian Economics Service	1993	18.8
	1996	16.2
Indian Foreign Service	1993	10.1
	1995	13.7
Indian Information Service	1994	15.8
	1996	15.6
Indian Postal Service	1993	17.5
	1996	21.4

Economic Empowerment of Women

Women experience barriers in access to economic resources land, natural resources—wood, water, capital, credit technology etc. Also most of the economic activities they do are not counted as work because they are not paid for it. Besides they have to spend lot of their time in household chores to which no economic value has been put so far.

According to time use surveys the time spent by women and men (completed in six States during the year June 1998 to June 1999) 55 per cent of the work time was contributed by women and only 45 per cent by men. However, women are at the margin of economic activities in unorganized, low skill, low paid jobs, thus large majority of women are not counted as workers. Globalization increases vulnerability of those at the margins of economies. Many women have been forced to accept low pay and poor working conditions and thus have become the preferred workers.

The process of globalization involves economic reforms in terms of stabilization measures and structural adjustment. Stabilization includes reduction in budget deficits—involving expenditure control, devaluations, imports control etc. Structural adjustments involve deregulation and liberalization—liberalization of foreign trade delicensing of industries—control on inflow of foreign capital, slashing subsidies etc.

Protected economies with extensive periods of control feel more the pain and slow process of adjustments. Economic measures for liberalization of the economy in India have included sale of uneconomic public enterprises, reduction in size of the Government. After a period of devaluation, the rupee now has remained stabilized. In short-term, liberalization leads to increase in unemployment.

Work Participation Rates

- Exact change in work participation rate for women is rather difficult to gauge because of frequent changes in the definition of "worker" used in the Population Censuses—which give the most reliable and exhaustive country-wide data. The changes in definition effects the count of workers in agriculture and unorganized industrial sectors where large majority of the women work. Thus change in definition effects Female Work Participation Rate (FWPR) than Male Work Participation Rate (MWPR).
- However, the data show that FWPR has almost doubled between 1971–2001 from 13.9 to 25.7 per cent (Table 8.12); the WPR of the males, however, has remained almost unchanged around 52 per cent. The main reason for increase in FWPR has been more scientific methods of counting 'workers'. There has, however, been increase in WPR of both rural and urban women. The rapid increase in education and skill development of women and their capacity to work at lower wages has also contributed the increase in the FWPR.

Table 8.12: Work Participation Rates, 1971-1991, India

Workers	India			Rural			Urban		
	Total	*Female*	*Male*	*Total*	*Female*	*Male*	*Total*	*Female*	*Male*
1971 (Main & Secondary)	34.0	13.9	52.7	36.1	15.9	53.8	29.6	7.2	48.9
1981 (Main & Marginal)	36.8	19.8	52.6	38.9	23.2	53.8	30.0	8.3	49.1
1991 (Main & Marginal)	37.5	22.3	51.6	40.0	27.1	52.5	30.2	9.7	48.9
2001	39.3	25.7	51.9	42.6	31.0	52.4	32.2	11.5	50.8

Source: Census of India, 1971, 1981, 1991 and 2001.

Distribution by Sector

Proportion of women workers in agriculture has remained almost unchanged although there has been a one percentage point decline in agricultural labour and a one-percentage point increase as cultivators. There has been a one percentage point decline in women working in the household sector which has not been fully compensated by increase in 'Other than Household Industry' which recorded an increase of only one-quarter of one percentage point. The major addition has been in women working in the 'Other services' sector where 8.26 per cent of women work as against around 7 per cent in 1981, (Table 8.13).

Although the percentage of women among total workers in the organized sector has gone up steadily from 11 per cent in 1971 to 17.4 per cent in 1999. Table 8.14 shows, however, that they are concentrated in the public sector and in services and in industries where wage levels are low. The employment in the public sector has reached a saturation point. Women's employment has not increased in most of the sectors except in trade, hotels and restaurants and in casual labour and there has been feminization of poverty.

Poverty Eradication Programmes include almost 13 schemes of which DWACRA is specially targeted to women. In 1999-2000 several of the schemes were merged into one scheme SGSY and several others were restructured. Laws were passed for protection of 'Home Based Workers' in some industries, e.g. *Beedi* workers. In 1996 the Government of India signed the ILO Convention for protection of HBW and also prepared a National Policy on Home Based Workers. Several labour laws have been passed during the past 40-years, which give equal wages, maternity relief and unemployment insurance etc.

The Government has launched innumerable schemes for increase in employment and income of the poor rural population with some specifically aimed at women, e.g. DWACRA. The latest Rural Employment Guarantee Bill passed by the Parliament recently reserves 33 per cent of additional employment for women.

Other policies announced but no plan of action or implementation of the policy has been the fate of the followings:

- National Plan of Action for Women 1976.
- National Perspective Plan 1988-2000.
- National Plan of Action for the Girl Child 1991-2000.

Table 8.13: Occupational Shift of Workers through Decadal Census, 1971-1991

Category	Persons			Male			Female		
	1971	*1981*	*1991*	*1971*	*1981*	*1991*	*1971*	*1981*	*1991*
I. Cultivators	43	41.46	38.41	46	43.71	39.63	30	33.09	34.22
II. Agricultural Labourers	26	25.12	26.44	21	19.57	21.06	50	46.34	44.93
III Livestock, Forestry, Fishing, Hunting, Plantations, Orchards and Allied Activities	2	2.25	1.90	2	2.37	1.99	3	1.83	1.60
IV. Mining and Quarrying	1	0.58	0.62	1	C.63	0.70	—	0.35	0.34
V. Manufacturing, Processing, Servicing and Repairs:									
(a) Household Industry	4	3.45	2.42	4	3.18	2.09	4	4.57	3.53
(b) Other than Household Industry.	6	7.85	7.76	7	8.91	8.89	3	3.60	3.88
VI. Construction	1	1.66	1.95	1	1.87	2.32	1	0.87	0.66
VII. Trade and Commerce	6	6.33	7.46	6	7.41	8.98	2	2.04	2.26
VIII. Transport, Storage and Communications	2	2.75	2.81	3	3.36	3.54	—	0.37	0.32
IX. Other Services	9	8.55	10.23	9	8.99	10.80	7	6.94	8.26
	100	100.00	100.00	100	100.00	100.00	100	100.00	100.00

Source: Census of India, 1971 [Series 1-India Part II-B (i) General Economic Tables]. Women in India, A Statistical Profile—1997, Department of Women and Child Development.

Table 8.14: Proportion of Women among the Total Workers in the Organized Sector and Percentage Increase 1971–1999

Category	*1971*			*1999*			*Variation*
	Total	*Women*	*% of Women*	*Total*	*Women*	*% of Women*	*In women employees 1999/1971 (%)*
Public Sector							
Central Government	27.60	0.78	2.82	33.13	2.52	7.60	323 %
State Governments	40.50	3.53	8.50	74.58	13.79	18.50	390%
Quasi-Central and	19.23	0.86	4.50	63.86	5.89	9.22	684 %
State Local Bodies	18.78	3.44	18.30	22.59	5.91	26.16	171%
Total	107.10	8.61	8.03	194.15	28.11	14.50	327%
Private Sector							
Large Establishments	67.40	10.63	15.77	77.70	17.92	23.06	195%
Small Establishments				9.28	2.77	29.85	
Total	67.40	10.63	15.77	86.98	20.69	23.12	195%
Grand Total	174.50	19.24	11.02	281.13	48.30	17.18	251%

Source: Directorate General of Employment and Training.

Other Good Initiatives

- Setting up of Department of Women and Child Development in the Ministry of Human Resource Development has led to gender sensitization in budgetary allocations and in design of Government supported development programmes. The department functions like a watchdog in the Government.
- Central Social Welfare Board and Rashtriya Mahilakosh (which was set up as a result of the findings of the Shram Shakti Report) have welfare and micro-credit as their main functions towards empowerment of women.
- Parliamentary Committee on Empowerment of Women was set up in 2000, the year of Women's Empowerment in India. The Committee examines measures taken by the Union Government for Women's Empowerment and in Five Years Plans and on gender budgeting.

Women and Health

Table (8.15) on expectation of life at birth of men and women shows that while the female expected to live longer than the males during 1901-21, their life expectancy fell below that of the males from 1931 to 1961, it was about the same in 1971. However, since 1971, the female expectation of life has increased more than that of men and in 2001 women expected to live three years longer than the men (Table 8.15).

Table 8.15: Life Expectancy at Birth of Males and Females during the Past 100 Years 1901-2001

Year	*M*	*F*	*Both*	*F-M*
1901-11	22.6	23.3	22.9	0.7
1911-21	19.4	20.9	20.1	1.5
1921-31	26.9	26.6	26.8	-.3
1931-41	32.1	31.4	31.8	-.7
1941-51	32.4	31.7	32.1	-.7
1951-61	41.9	40.6	41.3	-1.3
1961-71	46.4	44.7	45.6	-1.7
1970-75	50.5	49.0	49.7	-1.5
1976-80	52.5	52.1	52.3	-.4
1981-85	55.4	55.7	55.4	.3
1986-90	57.7	58.1	57.7	.4
1987-91	58.1	58.6	58.3	.5
1991-95	59.7	60.9	60.3	1.2
1992-96	60.1	61.4	60.7	1.3
1996-01*	62.4	63.4	62.9	1.0
2001**	63.9	66.9	65.3	3.0

* Planning Commission July 2000.
** Statistical Outline of India 2003-04, Tata Services Ltd., p. 31.

Source: Country Statement, International Conference on Population Development, Cairo, 1994 and Registrar General of India.

Data on Infant mortality rates by sex over the years (1972-1998) given in Table 8.16 do not show any pattern. However, the data on sex ratio given in Table 8.17 show a continuous decline from 1901 to 1961 when it was 841. The ratio has improved some since then but is still low at, having gone up from 927 to 933 in 1991.

They are however higher in rural areas then in urban and higher in the southern States than in the northern States. The high expectation of life at birth of females and continuous low sex-ratio only indicate the oft-focused issue of female foeticide. The earlier explanation of female feticide is not borne by data of infant mortality rates given in Table 8.16, however, female children if allowed to be born, more or less live as long as the males. However, such low ratio of girl children to boy children born only reconfirms wide prevalence of female feticide. The sex ratio of children born has no relevance to poverty; it has been noticed in poor states as well as the richest. (Mapping Adverse Child Sex Ratio in India November 2003, UNPF, Ministry of Health and Office of the Census Commissioner of India).

Table 8.16: Estimated Infant Mortality Rates by Sex and Residence, India 1972-1998

Year	Total			Rural			Urban		
	Total	Males	Females	Total	Males	Females	Total	Males	Females
1972	140	132	148	151	141	161	85	85	85
1978	125	120	131	136	130	142	70	69	71
1983	105	105	105	114	114	115	66	69	63
1988	94	96	94	102	103	102	62	65	59
1993	74	73	75	82	81	84	45	47	44
1994	74	75	73	80	80	78	52	53	51
1998	72	70	74	77	76	79	45	42	49

Source: Women in India—A Statistical Profile, 1997. Sample Registration System, Fertility & Mortality Indicators—1982–1993, Registrar General of India; SRS Bulletins 1996, 1998, RGI.

Table 8.17: Sex Ratio of Population (F/M 1000) 1901-2001

Year	1901	1911	1921	1931	1941	1951	1961	1971	1981	1991	2001
F/M	972	964	955	950	945	946	841	930	933	927	933

Source: Registrar General of India 1991 series 1 and 2001- Provisional Population Totals.

The decline in sex ratio has occurred in states having large concentration of the population, is concentrated more in the northern states than in the south and includes some of the riches districts of Punjab, Haryana and Delhi. In a recent study supported by UNFPA on 'Infanticide in Rajasthan—Causes and Practices' conducted by VIHAAN, Society for Child Development, the sex ratio has been declining from 1961 to 1991. In 2001 it recorded a small increase over 1991. However, in 2001, 70 districts in 16 States and Union Territories have recorded more than 50-point decline in the child sex ratio. The sex ratio is the lowest in Kurukshetra (Haryana), Ahmedabad and in South Delhi. Some of the most prosperous district of Punjab, Haryana, Gujarat and Himachal Pradesh have the lowest sex ratios. Besides the decline between 1991-2001 has been recorded in all districts of Punjab and all district have a sex ratio of below 850 (2001). A similar trend has been recorded in Haryana, Gujarat and Delhi.

Besides, aborting female foetuses is common among upper-middle class educated couples.

In spite of the Passage of Regulation and Prevention of Misuse of PDI (Pre-natal Diagnostic Investigation Technology) in 1994, in 2001 the child sex ratio (0-6 age group) has declined from 945 to 927. Death rate by sex among 0-4 age group is 24.8 for girls and 22.7 for boys (Gopalan, Table 4.4, p. 74). However, over-all sex ratio has gone up somewhat between 1991-2001 primarily due to longer expectation of life for women than men. Better outreach of health facilities (the per capita expenditure on health has gone up from Rs. 0.61 in 1951 to Rs. 85.10 in 1994-1995 (Gopalan, Table 4.9, p. 125). Intensive healthcare awareness programmes run by the Government and the NGOs have contributed to better health so has use of soap and more availability of water.

However, increase in sex ratio between 1991-2001 is primarily due to higher life expectancy of women. Programme for awareness on family planning and late marriage have led to decline in fertility rates from 5.3 (1970) to 3.4 (1991).

Gender Equality, Women's Movement and Women's Empowerment

The Women's Movement in India today is a rich and vibrant movement, which has spread to various parts of the country. Women's movement consists of female-headed NGOs working for welfare of women who became activists and raised the issue of social, economic, legal and political discrimination suffered by Indian women who because of the patriarchal structures of the society were far from equal with men. During the last thirty years the movement has gone through a process of maturing from a demand and demonstration phase to mature interventions in policy, active participation in creating new institutional and organizational structures, professionalism and participation in discussions at local, national and international levels. The issues taken up by the movement are located within the sphere of the private, and are directly linked to the body and a woman's position within the family while understanding fully that what happens within the four walls of the house has direct correlation with what is happening outside of it and that public and private are connected and what happens inside the home is the result of patriarchy which pervades all institutions—private and public.

Since women are not equal to men, there is need to fight the current social attitudes and change institutional frameworks that stand in the way of equality. The efforts will have to be made at two levels: one to trigger the process of equity and second to empower women through awareness generation on their current situation and their rights, with particular focus on grassroots women. They held demonstration, and public campaign and used theatre as a median for awareness generation. .

The seventies consisted of various scattered movements that addressed issues ranging from the creation of the Self-Employed Women's Association (SEWA) and their struggle for equal wages to the *Chipko* movement's effort to raise the issue of a woman's right to land. Towards the end of the 1970s and the beginning of 1980s there was enormous increase in establishment of various women's groups, many registering themselves as NGOs in order to be able to collect funds for their activities.

In addition to earlier methods, the eighties were characterized by attempts to develop structures, coalition groups, support institutions as well as developing global sisterhood. Women's organizations were active on a whole range of issues, which they saw were inter-linked to control over women's bodies. They took up the issues of rape, domestic violence, dowry, personal law, *sati* and fought for the creation of new women-sensitive laws or amendments to existing laws.

The eighties began with the first National Conference on the Women's Movement that addressed the issue of rape and harassment of women. The *Mathura Rape Case* in 1980 ended the silence on violence against women, which led in 1983 to an amendment in laws pertaining to rape. Among the provisions was a minimum seven years imprisonment. However, these laws still remain anti-women and ineffective and sustained advocacy efforts to amend and implement these further continues.

When it came to domestic violence women's groups were successful in their struggle to introduce a legal provision which established this as a crime which the police were obliged to register and act upon—Section 498-A under the Indian Penal Code (IPC). Under this violence inflicted on a woman by her husband or his family was declared a non-bailable, non-cognizable offence. But as has been mentioned earlier there is a steep rise in the reported crimes against women among which domestic violence incidences constitute a major form of violence and that while women do find the courage to report the cases 80 per cent of them are dropped. Recently, under pressure from the movement and its spokesperson, the National Commission for Women, the Government of India has passed a 'Protection of Women from Domestic Violence Act' which has given the women wide rights in case of domestic violence.

In 1982 an amendment was passed to the Dowry Prohibition Act, which stated that an official inquiry would be conducted if the wife died within seven years of marriage. If found guilty of dowry harassment the husband (and co-accused) faced seven years in jail as well as fines. Some of the other important decisions taken were those surrounding *streedhan* (the gifts given to the bride by her parents, husband and in-laws).

Personal law was addressed in 1985 with the *Shahbano case*. It raised the issue of divorce (*talaaq*) and maintenance of Muslim women. Despite official roadblocks the Supreme Court gave a landmark ruling. It established the right of a Muslim woman to maintenance. However, the Government of India bowed down to religious sentiments against interference in the personal laws of Muslims.

In the early campaigns, groups learnt from day to day that targeting the State was not enough and that victims also needed support. So a further level of work was needed: awareness raising or conscientization so that violence against women could be prevented, rather than only dealt with after it had happened. Legal aid and counselling centres were set up, and attempts were made to establish women's shelters. It was only when groups began to feel sucked into the overwhelming volume of the day-to-day work of such centres that they began to feel that it was not enough to do what they now saw as 'reformist' and 'non-campaign' work. Knowledge was recognized as an important need. India is such a vast country; what did activists in Karnataka, a State in southern India, know of what was going on in Garhwal in north India? And yet, everywhere you looked, there was women's activity, activity that could not necessarily be defined as 'feminist', but that was, nonetheless, geared towards improving the conditions of women's lives.

In the nineties women took up the issue of communalism and globalisation and feminisation of poverty through wider networking both at the national and international level. As the movement moved from women's issues to issues of livelihood and sustainable development they began to address the issue of population policy demanding the exclusion, e.g. of Norplant in the Family Planning Programme because of its devastating side effects, taking forward the campaign against hazardous contraceptive technologies started in the eighties. The politics of population growth and control was further brought out at the international level in the 1994 UN International Conference on Population and Development.

The movement has realized that unless women have access to elected offices, even democracies will simply be masking another kind of dictatorship—the imposition of male agendas over female ones. In view of the current patriarchal structures, however, the only way women can enter the sphere of elected political offices is through reservation of seats for women. Earlier the 73rd and 74th Amendment to the Constitution mandated 33.3 per cent of seats for women in Panchayati Raj Institutions and urban local bodies. The women's groups are demanding same reservation in State legislatures and in the National Parliament.

The increasing feminization of poverty and violence against women emerge as the two main concerns at the Beijing Conference of delegates.

Strategies for Women's Empowerment used by the feminist groups is reformist and therefore, techniques used are advocacy, lobbying, litigation, and monitoring; they use any technique that has promise and they change tactics if necessary. The favoured techniques have been, *Protest, Litigation, Lobbying and Monitoring.*

Feminists have been able to rely on pre-existing structure of traditional women's groups (AICC, YWCA, NFIW, AIDWA, MDS, etc.) whose shared interests in many feminists issues has produced numbers and assistance that would otherwise have been costly and probably unattainable. Women's Reservation Bill, The Dowry Bill and the *Mathura Rape Case* served as a model of structure and tactics and provided considerable assistance to the incipient feminist movement as it developed. Thus despite the existence of numerous autonomous sub-groupings that compete for funding and influence, there is extensive coalition building and remarkable unity among feminists. Friendship among group activists as well as multiple group membership provide linkages and facilitate the sharing of information, mobilization of activity on a specific activity and for planning of joint strategy and *ad hoc* coalitions are also created as necessary within the movement and with allied groups. Thus women's movement has consciously engaged in network building and has created an effective mechanism for both specialization and cohesion. The movement is capable of energetic action through inter-group mobilization when necessary. (The amount of energy unleashed by the movement during the time of international conferences like the Fourth Women's Conference in Beijing effected partly by the huge funds invested by the united nations and several other donor agencies needs to be analyzed.)

Organization of grassroots women into groups around an economic activity or a social issue and conscientizing the group on their rights, the effect of patriarchy and the advantages of working as groups for enhancing their economic activity has greatly empowered grassroot women. Through group strength the women have been able to punish their tormentors. They have been successful in closing shops selling liquor and even in killing their tormenters within the premises of a courtroom in the presence of law enforcement authorities whom they found incapable of enforcing laws (*The Nagpur Aku Yadav case*).

In spite of limitation of resources feminists have been creative in the utilization of the resources they do posses. They have exercised flexibility in choosing strategies and approaches and have employed traditional pressure-group tactics with considerable skill. The simultaneous reliance on groups' specialization and coalition activity has maximized the effective use of existing resources. To change the society in which one lives it first necessary to modify the belief system of the people living in the society—one available route is education.

Selection of Issues and Success

When issues the movement has taken over are perceived as affecting role equity rather than role change, the opportunities for favourable impact on the political system has been maximized for feminist groups. Several factors are crucial to attainment of goals: the issue should have a broad based support, the issue is narrow enough not to challenge basic values or divide supporters; a policy network with access and capability of providing information to decision-makers, the willingness to compromise with constituents groups and in the political process and ability to define success in terms of increments of change rather than total victory. A major factor that influences success is the image of broad-based support. Focus on role of equity has found considerable support. When more dominant values are in question and role change issues are at stake the possibility of success becomes limited. Confrontation has not been the chosen strategy. The tactics of the women's groups has been centrist. Women also compromise on their demands—as are being pressurized to do in passage of the Women's Reservation Bill.

Successful group provide informational resources to bureaucrats and Members of Parliament as well as State Legislatures. To this day the most detailed and comprehensive public record of discrimination has been provided by the publication of *Equality Now* (1975).

Present Status of the Movement

The movement has become professionalized and bureaucratized. However, the early emphasis of the movement on "sisterhood" has remained a unifying force as a result, inclusivity remains as an operating principle. The commitment has remained movement oriented negating the role of the media stars or charismatic leaders preventing organizational demise with the departure of the key individual. Feminists possess little resources but high status.

The movement has gained widespread acceptance as a legitimate pressure group

The multiple activities and success related to equity question will keep the movement alive. In addition more laws affecting equity roles are enacted and the law protects more rights for women, sufficient momentum will be sustained to keep the movement as a viable force in politics. The movement, so far, made limited use of the courts, however if women's reservation bill fails the courts may be one avenue for redress that the movement may use to greater advantage. In the coming decades more emphasis on cementing gains already made, by continued monitoring of the commitment of administrative agencies to enforcement of existing laws. Effective impact on some issues will involve participation of women affected by it with the feminist activists.

If the women's movement continues to abide by the "political rules of the game" and if parameters are drawn around issues selected for advocacy, feminist will continue to be successful political participants in an age of declining parties and pluralist expansion of interest group politics. Today the women's movement in India is rich and vibrant, which has spread to various parts of the country.

In spite of limitation of resources feminists have been creative in the utilization of the resources they do possess. They have exercised flexibility in choosing strategies and approaches and have employed traditional pressure-group tactics with considerable skill. The simultaneous reliance on groups' specialization and coalition activity has maximized the effective use of existing resources. To change the society in which one lives it first necessary to modify the belief system of the people living in the society—one available route is education.

When issues the movement has taken over are perceived as affecting role equity rather than role change, the opportunities for favourable impact on the political system has been maximized for feminist groups. Several factors are crucial to attainment of goals; the issue should have a broad based support, it should be narrow enough not to challenge basic values or divide supporters; a policy network with access and capability of providing information to decision-makers, the willingness to compromise with constituents groups and in the political process and ability to define success in terms of increments of change rather than total victory.

A major factor that influences success is the image of broad-based support. Focus on role equity has found considerable support. When more dominant values are in question and role change issues are at stake the possibility of success becomes limited.

Successful groups provide informational resources to bureaucrats and members of Congress. To this day the most detailed and comprehensive public record of discrimination has been provided by the publication of *Equality Now* (1970) which has only recently been updated by the 'Unfinished Agenda'. Women also compromise on their demands—as are being pressurized to do so in passage of the Women's Reservation Bill. Confrontation is not the chosen strategy and the tactics of the women's groups have been centrist.

The movement has become professionalized and bureaucratized. However, the early emphasis of the movement on "sisterhood" has remained a unifying force as result inclusivity remains as an operating principle. The commitment has remained movement oriented negating the role of the media stars or charismatic leaders preventing organizational demise with the departure of the key individual. Feminists possess little resources but high status.

The movement has gained widespread acceptance as a legitimate pressure group

The multiple activities and success related to equity question will keep the movement alive. In addition more laws affecting equity roles are enacted and more rights for women are protected by the law sufficient momentum will be sustained to keep the movement on a viable force in politics. The movement, so far, made limited use of the courts, however if women's reservation bill fails the courts may be one avenue for redress that the movement may use to greater advantage. In the coming decades more emphasis on cementing gains already made, by continued monitoring of the commitment of administrative agencies to enforcement of existing laws. Effective impact on some issues will involve participation of women affected by it with the feminist activists.

If the women's movement continues to abide by the "political rules of the game" and if parameters are drawn around issues selected for advocacy, feminist will continue to be successful political participants in an age of declining parties and pluralist expansion of interest group politics.

REFERENCES

Canadian International Development Agency (CIDA): Setting the Stage for the Next Century: The Federal Plan for Gender Equality, Status of Women Canada, 1995.

National Commission for Women: Towards Equality—The Unfinished Agenda Status of Women in India—2001.

OECD-DAC (Development Assistance Committee) Guidelines for Gender Equality and Women's Empowerment in Development Co-operation, Development Co-operation Guidelines Series.

9

Reaching the Unreached: Public Services

K.V. RAGHAVULU

Introduction

In the milieu of the new banking culture fostered by reforms, lending to agriculture or priority sectors generally, became unfashionable. The relative magnitude of the flow of credit to agriculture shrank and the interest rate regime which was designed to pamper the private corporate sector openly discriminated against agriculture. Even Public Sector Banks (PSBs) merrily defaulted on the credit targets of 18 per cent of net bank credit to agriculture and 40 per cent of net bank credit to priority sectors. The Reserve Bank of India (RBI), which was obsessed with implementing the Basle norms as the main thrust of banking reforms, winked at the defaults. While the priority sector target has been attained during the last couple of years, the target for agriculture continues to remain unattained: lending to agriculture is around 15 per cent today, as compared to the target of 18 per cent of net bank credit. While highly rated corporate entities could raise money from banks or PSBs at interest rates as low as 6 or 7 per cent, the small farmers were required to pay a rate of 12 per cent. It was left to the Government of India to reduce the lending rate to small farmers to 8 per cent only in August 2003. The neglect of agriculture and the rural sector generally was total during the 1990s, there is a deterioration of the health of all institutions involved in rural credit: PSBs, Cooperative Credit Institutions and Regional Rural Banks (RRBs)—all of which showed high Non-Performing Assets (NPAs). In other words, the financial sector reforms which were transplanted from the so-called Basle norms were ill-concerned and unrelated to the specific socio-economic milieu of the Indian soil, and they succeeded in creating an adversarial environment for rural credit. In fact, the credit-deposit ratio of RRBs declined sharply over the 1990s, indicating a reverse flow of funds from the rural to the urban areas. Even RRBs were permitted to invest in the PSU bonds. No wonder, the annual average rate of agricultural growth decelerated sharply from 4.7 per cent in the Eighth Five Year Plan to only 2 per cent in the Ninth Plan. The rural credit delivery system had slowed down.

It is indeed gratifying to note that the Tenth Five Year Plan endorses this view-point: It is being evident, however, that the organised financial sector in India is either unable or unwilling to finance a range of activities that are of crucial importance both for growth and development. Agriculture, unorganised manufacturing and services, and various types of infrastructure are instances of such sectors. The recent financial sector reforms have naturally focused primarily on improving the viability and stability of financial institutions without adequately addressing this issue. It is, therefore, necessary to consider methods of encouraging the financial sector to finance such activities, without impinging on its viability or compromising on prudential concerns. One thing should be made quite clear: it was not "inability" of banks because the banking system was highly liquid during the 1990s, with the credit-deposit ratio hovering around 50 per cent.

Poverty

The process of economic development, based on nature and its resources, quite often, causes damages to them. Major factors responsible for such damages are poverty, growing population and poor industrial/farming techniques adopted (Jon Martin Trolldalen, 1989, *Sustainability and Dryland Management, Constraints and Challenges*). Poverty is considered as a major factor for indiscriminate use of natural resources at a pace higher than its regenerative capacity. Our population constitutes around 16 per cent of the world population, and with just 2 per cent of the world's land resources, the pressure on land is immense when compared to overall world situation.

It is often believed that there is a strong trade off between emphasis on sustainability and poverty eradication measures. It is disheartening to note that with all the efforts to control the poverty in the region by 2020, as many as two out of every five children in south Asia will be under-nourished [*as estimated by the International Food Policy Research Institute (IFPRI), Washington D.C.*]. It underscores the view that the measures for sustainability/environmental protection should not be at the cost of curtailing agricultural production which is basis for poverty eradication. There should not be any conflict between the conservation of production system and attaining food security.

Population Growth

The issue of sustainable development in agriculture is likely to deteriorate further in view of the increased foodgrain requirement, changing dietary pattern, etc. of the growing population as viewed in unison with stagnant/declining productivity and shrinking natural resources (R.S. Paroda, 2000, *Policy Support Essential, Survey of Indian Agriculture)*. It is estimated that the share of productivity is more than 70 per cent in the total increase in the production followed by area expansion (around 20 %) and the rest by factors like improvement in cropping intensity. The proportion suggests that by 2030 AD, more than 120 million ha. have to be additionally brought under cultivation to meet our food requirements at that time. Given the limitation of area that can be brought under crop cultivation and in view of ever increasing urban areas, land put to industrial use, etc., the only possibility is to increase productivity/intensity adding to the stress on land, and thereby aggravating environmental problems and endangering our biological diversity. It is, therefore, felt that in order to maintain the food security, increased food production alone is not sufficient, but the population growth also need to be controlled/slowed down.

Cost

Monetary cost of sustainability measures is also a matter of concern. Efforts to bring waste/dryland under cultivation to ease pressure on agricultural land are often constrained by high cost of investment compared to the net benefits derived from such investments. Similar problems are also confronted in the case of treating of degraded lands and water bodies. Public investment, by estimating social costs and benefits of such projects, is required to make such projects feasible. It is commonly accepted that the economic benefits (short-term and long-term) of sustainability/conservation practices should be equal or more than cost involved.

Equity Concerns

The concept of social justice was in-built in the concept of sustainable agricultural development. It has been believed that overall growth in agriculture would lead to equitable distribution of the gains through generation of additional income and employment. However, we have seen that agricultural growth on its own could not substantially reduce the incidence of poverty, even in regions that have experienced remarkable growth in agriculture. This was partly because the poor were out of the purview of costly seed-fertiliser-irrigation technology and partly due to weak trickle down effect of the technology itself. Similarly, the emphasis on maximisation of yields led to the input intensification in the developed areas, leaving a large part of the country particularly the dryland areas, devoid of the new technology which consequently gave rise to inter-regional disparities in income and incidence of poverty. Hence, while considering the sustainability issues, social justice consideration should also be given due importance.

Table 9.1: CD Ratio of Rural Financial Institutions

(In per cent)

Agency-wise	*March 1999*	*March 2000*	*March 2001*	*March 2002*	*March 2003*
Sch. Com. Banks@	37.95	37.12	35.67	37.52	38.87
SCBs	85	87	91.5	88	88
DCCBs	92.1	82	85	86	87
RRBs	41.96	40.94	41.33	41.83	44.23

@ In respect of rural and semi-urban branches.

Source: Basic Statistical Returns—RBI.

Credit Flow During Plan Period and Likely Shortfall During X Plan Period

Credit disbursement during the IX Plan period at Rs.229956 crore exceeded the target of Rs. 229750 crore. Estimated credit flow during X Plan period—

Table 9.2: Progress in Credit Flow during 2004-05 as on 30 September 2004

(Rs. in crore)

Agency	*Disb. upto Sept. 2003*	*Target by GoI for 2004-05*	*Disb. upto Sept. 2004*	*Absolute increase over Sept. 2003*	*% Growth*
Comm. Banks	18,930	57,000	30,304 (53%)	11,374	60
Co-op. Banks	14,356	39,000	16,825 (43%)	2,469	17
RRBs	3,506	8,500	6,462 (76%)	2,956	84
Total	36,792	105,000	53,591 (51%)	16,799	46

Rs.736570 crore. First two years of X Plan period witnessed a shortfall of Rs. 31490 crore. Even taking into account GoI announcement of doubling of credit in next three years, a shortfall of Rs.161079 crore *vis-à-vis* plan target is likely at the end of X Plan period.

Highlights of Performance on GoI Announcement of 18 June, 2004 (As on 30 September, 2004)

Disbursement by all Banks stood at Rs. 53,590.65 crore forming about 51 per cent of projected credit flow of Rs. 1,05,000 crore for 2003-04. Commercial Banks, Co-operatives and RRBs achieved 53 per cent, 43 per cent and 76 per cent of their respective targets. There has been a growth of 46 per cent during the first half of the current year compared with the corresponding period of last year.

Maximum Growth rate achieved by RRBs (84%) followed by Commercial Banks (60%) and Co-operatives (17%). Under 'Financing new Farmers', 60.5 per cent of the target was achieved. Commercial Banks have extended loans to 5503 farmers indebted to informal sources like money-lenders to redeem their debt with them and provided Rs. 13.85 crore up to 30 September 2004.

Progress in Implementation of KCC Scheme

Up to 31 March 2004 all Banks issued 414 lakh cards. Cooperative Banks issued 242.58 lakh cards (59%), RRBs issued 38.99 lakh cards (9%) and Commercial Banks issued 132.43 lakh cards (32%).

Strategy for 2004-05

To bring all farmers including oral lessees, defaulters (by motivating them to clear their dues) under the KCC fold. The scheme has been modified recently to include term loans for agriculture and allied activities and a reasonable component to meet the consumption needs, besides the existing facility of providing crop loan limit.

Rural Financial Sector in India

Over 139,000 rural retail outlets of formal banking system are working in the country, viz. Commercial Banks—33,000, RRBs—14,500, Co-operatives -92,000. The population per rural outlet is 4,700. Historically, rural credit has been targeted through Co-operatives.

RBI had fixed certain targets in agriculture, priority sector, etc., viz. minimum priority sector lending (40%), minimum agricultural lending (18%), minimum weaker section lending (10%) and regulated interest rates on deposits and loans. Special poor-targeted programmes are also implemented.

Poverty Alleviation Programmes

IRDP

Number of families assisted were 54.5 million Bank Loans disbursed were Rs. 253.77 billion Subsidies disbursed were Rs.145.00 billion.

Programmes for SC/ST/OBC/Minorities were also implemented. These loans almost equal to those under IRDP. Annually 40 per cent rural loans are "small loans" of ≤ Rs. 25,000. Average loan is Rs. 5,250.

Due to fixed margins and high transaction costs, banks cut down on appraisal and monitoring costs; which resulted in high defaults. Leakages of subsidized resources were also observed in various studies conducted. Tremendous outreach was achieved but 36 per cent of cash dues of rural households still come from informal sources (1992) and 33 per cent (27% ?) of population continued to be "poor". The poorer of the poor were still outside the fold of banking system. A common view was that very poor are not bankable.

The normal savings and credit products did not suit the needs of the poor and procedures were complicated and cumbersome and it was amounted to high transaction costs for the poor. Even the special programmes did not recognise savings as important and resources handled were often larger than the poor's capacity to handle. Others decided everything for the unreached people. The poor need *HASSLE-FREE* Credit Mechanism to keep safe thrift and tiny surpluses and credit to meet emergencies and for micro Enterprises. It is necessary to design a system wherein savings and credit products are taken care. The system should also improve the access of the poor to financial services which reduce transaction costs of the poor, reduce transaction costs of the financing agencies and enhance participation by poor in decision-making. It should build up handling capacities of the agencies who can provide these services. The system has to identify new partners, workout roles for different partners and integrate formal banking system to ensure sustainability.

SHG Programme

What are SHGs ?

The SHG is a homogeneous group with common problems, affinity and neighbourhood. It generally consists of 10 to 20 participative members, they meet frequently, they have mutual trust and appreciation of others' view and take transparent decisions. It provides for simple accounting, pooled thrift is bankable, pooled thrift can be used to meet emergent needs. The pooled credit needs are bankable.

They build up capacities, larger needs are met and needs of more people are met. The SHG-Bank linkage programme increases outreach to poor, it is cost effective for both banks and the poor. It reduces dependence on informal source. It is a door step, hassle-free, self-managed financial services. It helps access of formal banks to savings and credit of the poor. It helps graduation from consumption to production loans and graduation to higher income levels with no subsidy and initiates the process of empowerment. About 0.263 million SHGs were financed by banks which accounts for 4.5 million poor households and about 22.5 million poor people. 90 per cent SHGs are of women groups and the average loan size was Rs. 1,072 per member. Member level on-time repayment rates were near 100 per cent and about 8,000 bank branches of over 314 banks and 1030 NGOs are involved in the SHG programme. It is proposed to enhance outreach to 20 million families or 100 million rural poor through 1 million SHGs by 2007-08 and the estimated demand for micro-Credit is Rs. 150 billion to Rs. 500 billion per year.

The Strategy

The proposed strategy is to focus on States where more poor live, thrust on promotion of large number of SHGs, ensuring rating/grading of SHGs and Training and Sensitisation. It is necessary to invest in human capital development and sharing best practices to develop the SHG programme.

More emphasis on Group Formation and Nurturing, Group Composition—thrust on affinity and homogeneity need to be given. The members own stake in the group in the form of savings and members learn to maintain financial discipline. The collective wisdom in credit decisions help good recoveries. The peer support minimises the risks of failures. Savings and credit is a continuing process and freedom of selecting loan purpose, with benefit of peer counselling. Poor gets a FACE-empowerment.

Issues and Challenges

It is necessary to expand the programme in backward areas. Banks to ensure post linkage management and federations of SHGs. NGOs as financial intermediaries have to develop MIS and both Banks and NGOs to help graduation from micro-finance to micro-enterprises and contain supply side anxieties.

Role Played by NABARD

Programme is conceptualised by NABARD. Grants are given to NGOs, Banks, Individual Rural Volunteers (IRVs), Farmers' Clubs etc. for promoting and nurturing of SHGs. NABARD designs training and capacity-building programmes for stakeholders. Cumulatively, 21,744 programmes were conducted in which 7.95 lakh participants were trained. NABARD evolved region-specific micro-Finance alternatives, viz. Grain Bank in Orissa, Smart Card in Andhra Pradesh, Computer Munshi in Orissa and Jharkhand, Barefoot Book Writers in Uttar Pradesh and Bihar, Joint Liability Groups and collaboration with post offices in Tamil Nadu.

Documentation and dissemination of Best Practices of micro-Finance through its periodicals "Saving Grace" in English and "Akshaya Patra" in Hindi is carried out.

Future Strategies

- Target for 2005-06 is 2.5 lakh SHGs linkage (Credit linkage of 3.47 lakh SHGs during 2004-05).
- To pay focused attention on 13 priority States having 70 per cent of the rural poor of the country (in Non-southern States, coverage increased—37% to 41%).
- Promotion of micro-Enterprises among matured SHGs, Pilot Projects in 10 Districts of 10 States in the country and support marketing strategies for SHG products.
- Credit rating of m-FIs with 75 per cent subsidy, capacity-building of m-FIs and to encourage bulk lending to m-FIs at cheap rate of interest, 7.5 per cent per annum presently to evolve new innovations.

Recent Policy Initiatives in Rural Credit by RBI and NABARD

- Special Agricultural Credit Plan (SACP).

- **Opening of specialised agricultural finance branches.**
- **Rashtriya Krishi Bima Yojana (RKBY).**
- **Kisan Credit Card Scheme (KCC).**
- **Scheme for purchase of land for agriculture purposes.**
- **Relief for drought affected farmers.**
- **Rural Infrastructure Development Fund (RIDF).**

PART II

PARTICIPATION AND EMPOWERMENT

10

Governance for Development at Grassroots

G. PALANITHURAI and V. RAGUPATHY

Introduction

In the recent decades, federal governments in the world have started decentralising political, financial and administrative responsibilities to lower level institutions and even to the private sector.[1] Decentralization is more specific and intensive in developing countries as they face a number of problems in governance.[2] But it is ultimately due to the extreme political pressure on the central system of governance with an expectation of proper delivery of services, effective mobilization of resources and allocations, and stability in the macro-economic system along with equity.

A variety of experiences have been gained by nations in the whole process of decentralization of power.[3] There are ups and downs in the experiments and a stage has come not to argue or debate on whether decentralization is "good" or "bad" or "unproductive" since decentralization is a political reality throughout the world. Against this background India made a historical and significant attempt through the 73rd Amendment to Constitution to decentralize powers to local bodies. This write up provides a descriptive narration of the whole process of decentralization of powers from the initiative of introducing the 73rd Constitution Amendment Act to the functioning of grassroots level institutions against the backdrop of the conceptual framework.

Rural India has undergone a metamorphosis after the introduction of the 73rd Amendment to the Constitution of India and the positioning of the Panchayats in place for governance, administration and management of development.[4] The development envisaged at the grassroots is equity based, social justice-based. Developments to be achieved at the grassroots are designed with the active participation of the people through a process of planning. The supply driven rural development activities are turned into demand driven ones. The hitherto excluded social segments and the women are brought into the fold of governance at the grassroots through reservation of seats especially for Women and *Dalits*. The hitherto managers of rural development, namely the bureaucracy, has been pushed to background, projecting the elected representatives to the front for taking up the development initiative. To deliver goods and services to the people, but these potentials and initiatives have to be tapped properly by all stakeholders. How these new initiatives are working in India can be seen in this paper. Before moving on to the real issue, one has to fully understand the context in which the new dispensation has been introduced. First the global context has to be seen.

Multiplicity of Concepts

For "decentralization of power" a variety of concepts have been used by academics and administrators and the meanings of the concepts are not the same. Concepts such as Local Self-Government, Local Government, Grassroots Democracy, Grassroots Institutions, Panchayat Raj, Panchayati Raj, Panchayati Raj System, New Panchayati Raj System, Local Body Institutions, Empowering People, Small Government Approach and Decentralized Governance are used interchangeably to connote the same meaning of devolving powers and responsibilities from higher level governing institutions to lower-level governing institutions. Of course, all the concepts are not the same and there are semantic differences among the concepts.

Some of the concepts are western in origin and many others are indigenous in nature. The concepts have contextual meanings also. Academics and Administrators have been using those concepts interchangeably to connote a broader meaning, namely that the problems of the people at the community-level are being tackled through service facilities rendered by legitimately created institutions. At present "Decentralization of Powers" is an accepted concept all over the world and hence it is being used everywhere. But, at the same time, some other concepts are used for rhetoric, such as empowering people and grassroots democracy.

Background of the New Initiative

It is essential to look at the emergence of new institutional mechanism to perform certain duties and responsibilities in this context. All over the world countries have introspected on their development path and initiated the corrective mechanism in the light of the reforms introduced in the domain of the economy of the developed countries. The new developments in India are the outcome of an introspection which resulted in the introduction of new policy initiatives which are quite opposite to the path-ways of prosperity which India has followed so far.

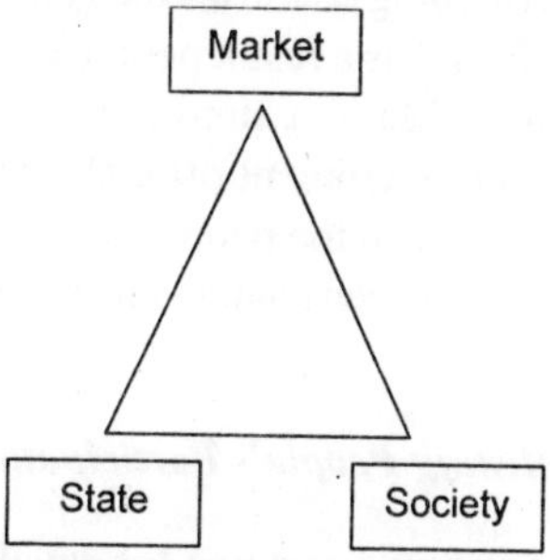

At present the three forces —the State, the market and the society are at loggerheads, because of the increased market oriented economic activities and the resultant activities of controlling the power of the State and the society. The State cannot abdicate its responsibilities as the people's participation in the development process has increased and the influence of the democratic institutions has been expanded. States are not able to control the influence of the market and hence a parallel organisational set-up is needed to contain the activities of the market when they are exploitative in character. To maintain equilibrium in the system it was thought of empowering the people through an organisational mechanism.

Basic Objectives of Decentralization of Power

Objectives vary from country to country depending on the needs and the conditions of the respective country. In India the objectives are broader and deeper as they have to create far reaching results in the attitude, behaviour, performance and the life-style of the people who live in the rural areas. The objectives are:

(a) Deepening of Democracy

The confined democracy within the framework of the Parliament and State Legislatures has to be broadened to percolate into the society to the level of the last rung of the society. Collectivities who have been hitherto marginalised and deprived have to be brought into the fold of democracy. More numbers from women and *Dalits* have to be brought into the system of governance. This should be an affirmative action by which the marginalised and depressed are given an exclusive space through reservation of seats in the local bodies. It is a process by which democracy is being widened and deepened.

In many of the developing countries the gap between the people and the Government is wide and as a result people are not being treated with self-respect and dignity. Many a time the poor are illtreated in the Government Offices by Government officials. The local bodies will be able to bridge the gap between the people and the Government. Further development activities have been changed from supply driven to demand driven mode.

(b) Ensuring and Enlisting People's Participation

People's participation is a *sine qua non* for development. Development initiatives are to come from the people. People are the initiators and the Government is the facilitator. To make the Government a facilitator as different from the role of a provider, people have to take the initiative and accept new responsibilities. Their participation in developmental activities will bring about accountability on the part of the Government. People have to assess their requirement and chalk out schemes and programmes according to their own needs. They should be people's programmes and the Government has to participate in the programmes by enabling the people to carry out their programmes on their own.

(c) More Responsibilities to the People

In the past fifty years all the responsibilities of the people and the community have been taken away from them and vested with the Government Departments. Communities had enormous responsibilities to manage their affairs on their own and they had their own management system. In the name of nation-state, Governments have expanded their departments. The expansion of the Government has increased only the management cost not the efficiency in delivery. To deliver services worth one rupee the Government takes six rupees as delivery charges. Moreover the quality of the service is also poor. In this structural expansion of the Government, people lost their orientation towards the responsibilities that have to be discharged. Ultimately what we find is that people have voters' orientation and not citizens' orientation. In order to make the people responsible, more responsibilities have to be given to them. Hence development works should be allowed to be done with the involvement of the people.

(d) Reduction of Government Structure

Many of the countries in the world are facing crisis as a result of the expansion of the departments of Government. The bigness of Government is a burden on the society. It will pave the way for the disintegration of the nation. The unwieldy expansion of Government departments becomes a threat to the elected representatives. Moreover people are terrorized by the Government through its procedures, rules and regulations. The larger the size of the Government the lesser the contact with the people. Hence there is a gap between the people and the government. In order to avoid this complication of the bigness of the Government, it was felt that the structure of the government should be reduced to the minimum level and this could be done only by giving more responsibilities to the people through institutions created legitimately.

(e) Harnessing the Human Energy

It is a truism that development depends upon the way in which the human energy is being utilized. Human energy has different forms and different levels. The State should find a way to raise the level of the energy of the people and more particularly the State should work out a detailed mechanism through which the energy of the people can be utilised. The developed nations have demonstrated to the world how best the energy level of the people could be enhanced and how best the enhanced energy could be utilised for activities to accelerate economic growth. To utilise the energy of the people, space should be created appropriately—preferably an institutional arrangement to utilise the energy of the people. An institutional mechanism can be created for this purpose. Local body institutions are one such institutional device or mechanism to make use of human energy for productive functions.

(f) Administering Development and Social Justice

It is a well known fact that countries which are in the process of decentralizing powers have different objectives. In India, it was a conscious decision of the Central Government that Panchayati Raj Institutions have to work for administering development and social justice by evolving their own plan of action. Traditionally Panchayati Raj has been looked at as an institution meant for providing services and facilities. But in the new context it is meant for not only services and facilities but also growth with equity.

(g) Planning from the Below

So far the planning exercise was done at the central level which resulted in large scale demands of the people being unmet. In fifty two years of governance, centralised planning has achieved tremendous results in the macro-economics of this country and yet the rural realities have not been changed drastically towards achieving development. Still people are living in sub-human living conditions in the rural and urban areas. In order to provide basic facilities, to make use of the facilities to grow and to bring economic activities to the rural areas planning exercise has to be done at the micro-level. This exercise was thought of many times in independent India—whether planning institutions could be created at the bottom either at the block level or at the district level. Now this concept has gained currency and the planning exercise has to be initiated at different levels starting from Gram Panchayat.

(h) Building the Capacity of the People

It is imperative in decentralization of power that people should take more responsibilities and discharge the same effectively and efficiently to deliver services of quality to the people. To perform the new responsibilities the capacity of the people has to be increased. Till date people have developed an attitude that the Government would provide everything and people have to receive the same. Now the role has been reserved. People have to manage their affairs on their own. Therefore, the new system envisages a new task of building the capacity of the leaders and the people. In a market driven economy for every act people require skill and efficiency. People's skills have to be updated and upgraded to match the requirement of the market. Against this background capacity-building assumes significance.

(i) Establishing the Principle of Subsidiarity

The whole exercise has to revolve around the principle of subsudiarity. Activities that can be carried out at the lower level can be allowed to be carried out at that level and things that cannot be carried out at that lowest level can be allowed to be carried out at the next higher level. In this manner activities are to be carried out at different appropriate levels depending on the capacity of the people. Based on the above principle powers are to be devolved.

Putting the Last First

Fifty years of development in India have demonstrated the postulate that development cannot be achieved without the involvement of the people in the development process as stakeholders. In all the years development programmes have been the programmes and schemes of the Government and not of the people. In the whole development process one would find the people as the last beneficiaries. What is more, development was not achieved as planned. Now programmes are not to be generated by the Government but evolved by the people. Programmes and schemes are to be owned by the people. Governments have to participate in the programmes and schemes of the people. Thus the people come first and the Government comes next.

Establishing Good Governance

The administration has demonstrated that it can only alienate the people. Proliferation of administrative procedures, norms and regulations have made the system totally untransparent. As long as administration is being run in a non-transparent way corruption will continue to be entrenched in every nook and corner of the society. To avoid such a malaise, the administration should simplify procedures, norms and regulations. The whole process has to be initiated from below. Administration should be transparent. Further it should accommodate every segment of the society, especially women and the marginalised.[5]

It Should be Pro-Nature

The alarming environmental problems and global warming caution our policy-makers and administrators that unless serious action is taken our life will become problematic. Hence the global scenario needs local action to protect the environment. The activities of the people should be oriented towards protecting the environment and maintaining the natural order. While taking decisions at the micro-level and implementing schemes and programmes for rural development enough care should be taken to protect the flora and the fauna. All the ponds, wells, tanks, channels and lanes are to be protected and maintained properly. All the commons are to be well secured and managed. There is a close link between environmental problems and poverty. The green is the lifeline of the poor. By protecting the green the poor can be given life.

It Should be Pro-Poor

All the activities are to be decided upon on the basis of equity. While taking decisions and fixing priorities for rural development activities, the Panchayat has to decide whether the benefits will reach the poor. Normally, in the absence of basic facilities and amenities, the rich do not bother about it, because they can afford to purchase the same. But in the case of the poor, their life will be miserable in the absence of public facilities and basic amenities. Hence, priority has to be given to the poor. Unto this last has to be the policy of the Panchayat. The poorest of the poor have to be given priority in deciding upon benefits.

It Should be Pro-Women

There is yet another important sector which needs care and attention. Panchayat activities have to be carefully designed so as to provide enough scope for women both in terms of participation and benefits. Hitherto the neglected section of the society were women, who should be integrated into the mainstream of economic and political activities. Further, of the problems faced by the rural society, 70 to 80 per cent are being faced directly by women in the absence of basic amenities, facilities and services. Hence the decentralized political institutions nearer to the people should have resilience to accommodate women folk in their activities.

With the above posited objectives an amendment was introduced in Parliament to amend the Constitution of India so as to create a systematic structure at the lowest level of the socio-political system.

A Framework for Decentralization of Powers in India

The process by which the 73rd and the 74th Amendments to the Constitution of India were enacted, the subsequent enactment of conformity legislation in the States, the conduct of elections to grassroots institutions, the constituting of committees to suggest ways and means to devolve powers to the three tier Panchayats, the constituting of State Finance Commissions to make recommendations to augment resources, the subsequent steps taken to implement the recommendations of the committees and commissions, the perception of the newly elected leaders of the new dispensation, the understanding of the new system by the officials in the administrative system, the outlook of the people towards the new institutions, the capacity of the leaders of the grassroots institutions and the activities carried out at the grassroots under the new dispensation have brought to light the conflicting tendencies in the whole range of activities of the institutions, procedures and processes, which, incidentally, indicate the confusion in the minds of the players and the stakeholders in the institutions about the new institutions. For any meaningful exercise perspective is more important than any other aspect. Obviously variations are bound to emerge. But conflicting tendencies need not and should not figure in the process as they will affect the whole system. While looking at the scenario of the process of devolution and decentralization taking place in India one notices the stagnation and the conflict which are the result of poor understanding of the package of devolution. Even academics, while writing articles and books on the Panchayati Raj System, have not set the perspective as the backdrop for their analysis. To look at the issues of grassroots institutions, one has to have the right perspective. The prime objective of this write up is to develop a perspective for devolution of powers in the Indian context against the backdrop of the initiatives which have been taken in the political dispensation.

Backdrop

No doubt the new dispensation has come through a political resolution and Constitutional provision. The political resolution and Constitutional provision is not a sudden development in Indian polity. It has its own historical background. The new dispensation is not an isolated event. It is contextual. When a perspective is developed against the backdrop of history one cannot wish away the context while looking at the historicity. Democratization, Decentralization, Devolution and Developmentalization are contemporary issues and movements all over the world. Though the process of creating democratic institutions in India has a history of its own, the very interface between national and global activity and global and local activity cannot be ignored. Both the Indian tradition of building up grassroots democratic institutions and the contemporary globalization and localization of economic and political activities form the core for building a perspective for devolution of powers in India.

There are ample historical evidences to argue that democratic practices had been developed even in ancient India. But, in modern India, it started from Mahatma Gandhi. British administrators, excepting Rippon, had looked upon institution building at the grassroots not as a part of democratizing the society but as an instrument to ease the frustration of the people by delivering some services. Rippon's resolution reiterated an important fact that the people of India should be given their first lessons in popular and political education through local boards. The same British had dismantled the existing system, despite knowing the fact that natural resources were being controlled and managed by the communities. The community structure was strong and hence they were able to manage the natural resources. By making the land to be owned by individuals, the British weakened the community structure and practices. It was only during the freedom struggle that the democratic process at the grassroots was initiated. Gandhiji's framework of *Gram Swaraj* was the basic foundation for subsequent discourses on empowering people. Jawaharlal Nehru initiated the process of looking for institutional framework at the grassroots within the established democratic politico administrative system despite staunch opposition.[6] Next to Mahatma Gandhi, Jayaprakash Narayan made a fervent plea through his framework for developing a grassroots democratic system. In fact, he developed his thesis on the basic propositions of Mahatma Gandhi. Finally, Rajiv Gandhi started this exercise not as a dream but to find a solution to the unsolved problems of the people over three decades. His framework of devolution was not instantaneous. It was a framework evolved through a process of consulting the millions of stakeholders who were in the process of administering development and social justice. Rajiv Gandhi did not advocate an instrument to negate the State system as Mahatma Gandhi and Jayaprakash Narayan did. In the process of decentralizing governance and administering development and social justice Rajiv Gandhi wanted to rejuvenate and reinvent the Government machinery both at the Centre and in the States.

Rajiv Gandhi's Framework

Devolution of powers for revitalizing Panchayat institutions is the outcome of a series of serious discussions and deliberations held among the Chief Ministers, Secretaries to Government, Ministers of Rural Development, officials and representatives of the people in local bodies. This process was started in the year 1985 itself. As a young Prime Minister, Rajiv Gandhi wanted to look at administration from a modern perspective of new public management. He identified the crux of the issue, namely, the poor delivery of services and the reason for the poor delivery of services to the people. He lamented that our administration is unresponsive because it is not representative in character. He organised a series of workshops and conferences and analysed the reasons through the host of stakeholders.[7] The framework he developed was the distilled essence of the views of thousands of elected local body representatives, hundreds of District Magistrates, scores of senior Government servants and dozens of Ministers and Chief Ministers. His perception of development was of two types. One was macro and the other was micro. Governments had paid adequate attention to macro-development but that much of attention has not been given to micro-development. When very big dams were constructed, we totally forgot to construct small checkdams at the community level. When very big industries were established we failed to pay attention to small and tiny industries, including cottage industries. When universities and research institutions were developed, elementary schools and primary schools were not given that much of importance. When higher level medical facilities were developed, primary healthcare was not given that much of importance.

Rajiv Gandhi pointed out the neglect of the micro-activities. Why this stagnation? Why this neglect? He tried his best to find answers through Gandhiji and hence he quoted Gandhiji "True democracy cannot be worked by twenty men sitting at the Centre; it has to be worked from below by the people of every village". Further he quoted Gandhiji: "Swaraj is to be obtained by educating the masses to a sense of their capacity to regulate and control authority". To underline the importance of people's participation, he quoted Nehru: "It is not good enough to work for the people, the only way is to work with the people, and give them a sense of working for themselves; development work can be executed only if the people take the responsibility in their own hands."

Rajiv Gandhi first listed the ills in the present system. He started arguing that the democratic structure at the top and the middle are strong because these two structures were positioned only through the conduct of elections promptly whereas the third tier is weak because elections are erratic. When the third level governance is weak, it will certainly affect the other two tiers. Without a strong structure at the grassroots, democracy cannot flourish at the Central and State levels. It was promised during the freedom struggle that democracy and development would be ensured down to the lowest level. But those promises were not fulfilled after Independence. People at the top were busy strengthening their own positions in politics as well as in administration. Democracy and development were part of our freedom struggle. When they have not been achieved, one can presume that the freedom struggle is not over and it will continue. The form and mode may vary.

Development programmes are designed in the capitals by intellectuals regardless of the needs at the grassroots and, as a result, the intended targets and results are not achieved. The centralized development planning initiatives improved the conditions of people sitting at the top but did not remove the difficulties of the people at the grassroots. In the absence of elected representatives at the grassroots, power brokers emerged in our country. For any benefit the people had to approach the power brokers. The administrative system was being managed by the best brains in the country and yet our administration was weak, unresponsive, inefficient, unsympathetic, often callous, sometimes even cruel to those whom they were meant to serve.

It is the considered opinion of talented civil servants that despite all the goodwill in the world, all their sincerity and all their dedication, a bureaucratic administration cannot be responsive until it is rooted in a democratic polity. A responsive administration needs must be a representative administration, an administration which is responsible to the people. The bureaucracy is responsible only to its masters at the next higher level.

No one denies the importance of the bureaucracy. They are essential as they are experts in their art. But people should not be depend solely on the bureaucracy. Excessive dependence on the bureaucracy by the people will slow down progress. Normally people see the administration, not as servants of their representatives, but as a form of bureaucratic dictatorship. Yet another problem Rajiv Gandhi identified was that our administration was plagued by enormous wastage. The cost of administration was eating up a disportionately high share of the total outlay for schemes and programmes. Further there were leakages as power brokers were in the pipeline.

There is yet another malaise: people have increased their expectations and demands from the Government. This is also the creation of our polity. We have reduced our people to recipients of development benefits instead of being participants in the development process. There is a vast gap between what the people demand and what is possible to achieve. This is a dangerous trend in our polity. Yet another issue in polity is the neglect of women in the progress of governance. Women are under-represented. At the grassroots they are totally neglected and marginalized. It is women who undertake much more than half the economic activities of rural India. It is the women of rural India who are the main repository of India's great cultural tradition of moral values, which are fundamental to the survival of our civilization. But, many of the women issues do not surface in the decision-making bodies. *Dalits* and tribals are another marginalized segment and they have to be brought under the new dispensation. By pushing the people to the periphery the civic culture of the people has been dismantled.

Framework for Action

Rajiv Gandhi emphasized two important concepts, namely democracy and development, which are cardinal principles. The two aspects are very important contextually. The rising aspirations of the people in respect of democracy and development are very high at present. How development is going to be achieved through a democratic process at the grassroots is now the main concern. First a system has to be created at the bottom through a Constitutional Amendment. Regular conduct of election to these institutions will stabilize the system as in the case of the National Parliament and the State Legislatures. Devolution of powers, functionaries and finance will ensure delivery of services to the people on a sustainable basis. Provision has to be made to induct women and *Dalits* into the portals of governance. To ensure the regular conduct of elections and regular flow of finance for independent activities, a State Election Commission and a State Finance Commission have to be created. The power given to the State Finance Commission is not the same as the power of the Central Finance Commission. A list of 29 subjects have been incorporated in the 11th Schedule for decision-making. The Eleventh Schedule does not confer any legislative power on local bodies. To ensure people's direct participation in governance the Gram Sabha has been created. To involve the people in the development process, and to alter the trend of supply driven activities to demand driven activities, a District Planning Committee is created at the district level to develop a plan from the people. The planning process involves the people in the exercise and integrates the social justice component in the plan.

The above system can be created through an amendment to the Constitution. This will be a charter for ending bureaucratic oppression, technocratic tyranny, crass inefficiency, bribery, red tapism, nepotism, corruption and the million other instances of malfeasance that afflict the poor of the villages, towns and cities.

The Essence of the Framework is: "we trust the people, we have faith in the people, it is the people who must determine their own destinies and the destiny of the nation. To the people of India, let us ensure maximum democracy and maximum devolution. Let there be an end to the power-brokers. Let us give power to the people".[8] In this framework one can notice a few limitations: the 29 subjects mentioned in the 11th Schedule do not give power to legislate, but only to take decisions. The State Finance Commission's recommendations are not mandatory in nature. The whole responsibility of giving power to the Panchayat system is given to the State Government. The Centre enacts the law and State has to operationalise the law which has been enacted by the Centre.[9]

Characteristics of the 73rd Constitutional Amendment Act

The 73rd Amendment to the Constitution of India is a guideline and a basic framework evolved to guide the State Governments to take the initiative of organising grassroots democracy as a system of governance to administer development and social justice. It opens up a new chapter in the modern political history of India. It has certain key characteristics which cannot be set aside by the State Governments. They are:

(a) Constitutional Entity

The New Panchayati Raj System came into being as an institution of governance with the approval of the Constitution and it is placed in the third stratum of the hierarchy of the institutions of governance. It enjoys the status enjoyed by the Parliament and the State Legislature in terms of its existence and conduct of elections. It is a permanent institution with earmarked role and responsibilities to be discharged by the newly created institutions.

(b) A New Phenomenon in Federalism

In a federal structure there should be two constituent units maintaining their relationship in administering development and peace through the institutional mechanism. There shall be division of powers between the Central Government and the provincial governments. Likewise there shall be sharing of resources between the Central and State Governments. But, now there is a third institution which has come into being as a institution of self-governance with distinct and categorical political, financial and administrative powers to administer development and social justice. Now the concept of federalism takes on a new meaning which explains the relationship among the three institutions, namely, federal, regional and local. Hence the 73rd Constitutional Amendment brought forth a new phenomenon in Indian federalism.

Diagram of Framework

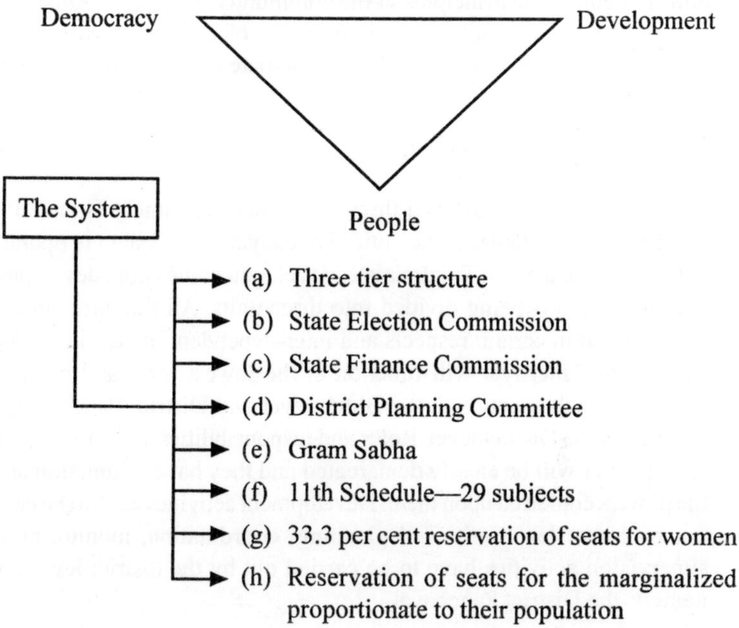

(c) From Representative Democracy to People's Democracy

Indian democracy has the character of a representative from Government and the decision-making has been based on the representations made by the representatives of the people both in the National Parliament and State Legislatures. It is an irony that one hundred crore people are being represented by around four thousand and seven hundred representatives of the people through State Legislatures and the National Parliament. It is natural that a representative cannot represent millions of people. Hence, even after fifty years of Independence, basic facilities and amenities have not been provided to the people. Now the concept of representation has been changed to direct participation of the people in the governing process and thereby an institutional mechanism, namely, the Gram Sabha, was created to enable the people to participate in the development process. The Gram Sabha is the basic institution created for the benefit of the people to participate in and administer development. It has been created to make the people associate themselves with the community to discuss the problems of the community and to find out solutions to the problems.

It should be out and out an organisation of the people. This body will nurture democratic principles in the community. The Gram Sabha gives meaning to the concept of "Empowerment of People". Cutting across caste, creed and colour people will participate in the meetings on equal footing.

(d) Three-Tier Structure

The new system envisages a three-tier structure, namely Gram, Block and District Panchayats. The entire Panchayati Raj System is operating in the area of a district. To administer development the entire development administration is being divided into three units. All the three units are independent in certain respects and inter-dependent in certain respects. The Gram Panchayat will function at the lowest level and the Block Panchayat will function at the Block level and District Panchayat will function at the District level. Roles and responsibilities, powers and duties for each tier will be clearly demarcated and they have to function as per the powers conferred upon them. Development activities have to be carried out by lower level units and planning, coordination, monitoring and supervision activities have to be carried out by the district level body, namely, the District Panchayat.

(e) 11th Schedule Earmarked 29 Subjects

The New Panchayati Raj System is grounded perfectly in the domain of governance authoritatively only through the 11th Schedule conferring 29 subjects on the rural local bodies. As indicated in the Constitution 29 items of work are reserved for the Panchayati Raj System. Of the 29 subjects, which are to be allocated to Gram Panchayat and the other tiers could be spelt out by the State Government through legislation. The role of the State Government is to allocate appropriate duties and responsibilities to each tier of the Panchayati Raj System. The 73rd Constitutional Amendment has indicated and suggested the items. It is not an exhaustive list. If the State Government wants to give more powers to the local bodies there is no bar. It is more appropriate if the 11th Schedule is given in the fourth list as Central list, State list, concurrent list and Panchayat list.

(f) Reservation of Seats for Women

Reservation of seats for women in all the tiers of local bodies is a new phenomenon in governance. When Rajiv Gandhi brought the 64th Amendment Bill only 25 per cent of the seats were earmarked for women. The one-third reservation of seats for women was the contribution of P.V. Narasimha Rao. The one-third reservation of seats for women has propelled the nation to moot another demand, namely, proportionate reservation of seats for women in the State Legislatures and the National Parliament. It is a moot question as to why reservation of seats to women is available in only local bodies? It is unequivocal from the reply given by many in the authoritative bodies that in the absence of basic facilities the worst sufferers are women. The sufferings of women are not felt by the policy-makers since they are mostly men. By giving women the opportunity of taking decisions they will reduce the burden of the women. There is yet another argument that the women have a legitimate space in governance which should be ensured to them. All along they have been marginalised and now it is felt that it should be given to them. It is their apace which is being occupied by men. The next explanation is that, while deepening democracy, deprived sections have to be included in the fold of the governing bodies and hence they are included by making exclusive reservation of seats for women.

(g) Reservation of Seats for Dalits

Reservation of seats for *Dalits* in the State Legislatures and in the National Parliament is not so difficult as these institutions are far away from the communities. Hurdles are not so great for them to function as members of legislative institutions. They are always looked at from the party perspective and not from the caste perspective. But it is very difficult for the *Dalits* to emerge in local bodies as they are seen by the society from the caste perspective. The whole lot of human rights violations, discrimination, social injustice are frequent at the grassroots level. Voices are raised vociferously from the marginalised. Both Central and State Governments have not fulfilled the objective of administering social justice. Hence it is felt that the responsibility of administrating social justice could be handed over to the lower level institution of governance and thereby it has been handed over to local bodies. This process starts from providing reservation of seats to the *Dalits* in the local bodies as it has been done for women, but proportionate to their population. By creating this clause in the Constitutional Amendment, the marginalised have been given adequate representation. This is yet another revolutionary step taken by the Government knowing fully well that the caste ridden Indian society has to pass through social conflict for some time.

(h) Creation of Finance Commission

It is a well known fact that the Panchayats had been crippled in the past by reducing the financial allocation to Panchayats as that power was vested with the State Government. In order to overcome this eventuality the Amendment has incorporated a provision by which a Finance Commission has to be constituted in every State to suggest the ways and means to share the resources between the local bodies and the State Government and suggest how local body finance could be strengthened. By creating this provision in the Amendment Act, financial flow to local bodies has been ensured regularly and this provision enables the local body institutions to keep themselves on a permanent footing.

(i) Creation of Election Commission

To elect near about 4700 representatives for State Legislatures and the National Parliament, an independent Election Commission has been created and now functions in all the States. As per the new provision, representatives to the tune of 30 millions have to be elected for which a massive machinery is needed. Against this backdrop a permanent Election Commission has been created. Conducting the election to the local bodies is the main earmarked function of the State Election Commission.

(j) District Planning Committee

Though planning is a much loaded concept and publicised word among the people since the days of Jawaharlal Nehru, the Planning Commission at the Centre and the State are not Constitutional entities. But the District Planning Committee is a constitutional creation. This District Planning Committee has come to stay with the specific purpose of altering the planning process. As per the new dispensation the planning exercise has to be initiated from the below. For the development of the whole district the District Planning Committee has to prepare a perspective development plan. Thus the Central concept of planning has been recognized at all the three levels as institutions of governance functioning at all the three levels to administer development.

Ensuring Regular Election to Panchayati Raj System

Election to local bodies was the will and pleasure of the State Governments. That is why election was not conducted for local bodies for about fifteen years in the past. Any number of reasons could be projected for postponing the elections. But through the new dispensation election has been ensured as it has been ensured to the Parliament and State Legislatures. The provision which is applicable to the election of Lok Sabha and State Legislature is applicable to the New Panchayati Raj system. Hence now the Panchayati Raj System becames a permanent entity. No State Government can withhold elections beyond six months. The above stated features are the key characteristics of the 73rd Constitutional Amendment Act and the spirit of the above Act has to be carried out by enacting a suitable conformity Act at the provincial level.[10]

To realize the potential of the 73rd Amendment, local bodies have to cross several stages and obstacles as it gives power to many by destabilizing the existing power holders. To make the Panchayats function and deliver goods and services as they are designed by the 73rd Amendment the following are imperative:

Political Will

Since it is a political and bureaucratic exercise, at different levels strong coordination and direction are essential. This direction has to be given by the Chief Minister of the respective State. The Chief Minister should have the political will and determination of devolve powers, functionaries and finance. The 29 subjects mentioned in the 11th Schedule are not in the hands of the Minister of Panchayati Raj and Rural Development. Several Ministries have to be coordinated. It could be done only by the Chief Minister. The Chief Minister should have political commitment.

Conformity Act

The basic principles of the 73rd Amendment have to be confirmed through a state legislation for the establishment of the New Panchayati Raj System. The new conformity legislation of the State Government should reflect the spirit of the 73rd Amendment to the Constitution of India. The new institutional arrangement is for governance not for implementing the schemes and programmes of the State and Central Government.

System Creation

In order to keep the Panchayati Institutions as institutions of governance, the Act has to be followed meticulously by evolving suitable rules and the rules have to be notified. All the details should be compiled and make into a compendium or handbook. This will be in the regional languae and that will be in all offices from Gram Panchayat to District Panchayat. The Panchayat administration and management have to be carried out and guided by a set of well regulated rules, norms and procedures.

Capacity-Building

To carry out the task the officials involved at all levels from Gram Panchayat to State Government have to be motivated, oriented and trained for a new task. Equally all the Panchayat leaders from Ward Members of Village Panchayat to District Panchayat President—the elected representatives—have to be trained properly to carry out their role and responsibilities. The training should be rigorous and serious. The skill, the capacity and the capability of all the individuals who are involved in grassroots governance have to be enhanced.

Awareness Building

Since the participation of people in governance is the main focus of the new dispensation, people have to be conscientised and they are to be mobilized. Against this background, with the help of the media, awareness has to be generated among the masses. The full potential of the new dispensation and the roles have to be explained.

Support Structure

When Panchayat leaders are in action that too women and *Dalits*, they have to face a lot of problems in the field as ours is an unequal society divided on caste and class and political lines. To overcome the problems the Panchayat leaders need a support structure which will provide needed information and resource persons.

Follow-up Machinery

The devolution process has to be tracked and then the information has to be given to different stakeholders to correct it. Monitoring of devolution is imperative: otherwise it would move in a reverse direction. By evolving a devolution index, the devolution has to be evaluated.[11]

Evaluation

Initially, to enact conformity legislation to confirm the spirit of the 73rd

Amendment to the Constitution of India, pressure had to be exerted on many States by the Centre. After the enactment of conformity legislation pressure had to be exercised on the States to conduct the elections to the local bodies. Much has not been done on devolving powers, functionaries and finance to the local bodies. There are six States, namely, Maharashtra, Gujarat, West Bengal, Kerala, Karnataka and Madhya Pradesh in the forefront in setting up models in devolving powers to the local bodies. These States are repeatedly referred to for decentralized governance in India by the scholars who are analyzing local governance. Having got positioned, the local body institutions have started their functions and, as a result, the demand to the representatives by the people and the demand to the State Governments for powers, functions, functionaries, finance and freedom by the elected representatives of the local bodies are increasing gradually.[12] The ills affecting the society have started influencing the Panchayats also. The traditional control over natural resources has been questioned and the service delivery mechanism has been thoroughly examined by the elected representatives of the local bodies.[13] The entry of women and *Dalits* has its own dimension of conflict and recognition depends on the maturity of the local political culture.[14] The presence of elected local body institutions has made a difference in grievance redresssal mechanism. Barring a few States many of the States have not shown commitment in devolving powers, functionaries and finance to local bodies.[15] Panchayats have started asserting themselves as institutions of governance. By their intervention in the rural areas, the petitions submitted to the Government have been drastically reduced. Many of the Panchayat leaders have started assuming responsibilities and carrying out activities. Women and *Dalit* leaders have started asserting themselves as factors to reckon with.

Dalits have brought to the fore the problems which they have been facing over a period of time.[16] Women ieaders have formed their own organisations to articulate their problems in the Panchayati Raj.[17] Women leaders, especially *Dalit* women leaders, face an array of personal, familial, social, economic and political hurdles to their access to reserved quotas and empowered political participation in governance which negate their ability to produce gender-responsive policy outcomes.[18] Basic services are being ensured by the local body leaders. Through local governance Panchayats have established a responsive administration. Good governance is being achieved at the grassroots.[19] Never have Government departments responded positively to the poor as is being done by the local bodies. Hitherto neglected social development issues are being discussed in the Gram Sabha, the Panchayat Council, the District Planning Committee and other Panchayat fora. Of course it is all being driven from the State Government. In many places it alters the paradigm of development from supply driven to demand driven mode. Panchayat leaders have evolved their own schemes to help the ultra poor. Dalit and women leaders have embarked on the issues of natural resource management.[20]

Many of the women leaders have changed the governance landscape from rulership to leadership.[21] For women leaders sanitation, water supply, street lights, livelihood are the major issues. They develop a new culture of democracy by listening to the people. They allow people to discuss, interact and debate. Some of the leaders are real models for the State Government to introduce new schemes. Panchayat leaders have formed alliance with varied stakeholders to work in the Panchayat together. Leaders having needed skill, capacity and capability manage the Panchayats effectively and deliver goods and services efficiently. The above narration is based on case studies conducted in the whole country by various organisations. No systematic study has yet been conducted in the whole country on the performance of the Panchayati Raj Institutions. In many of the Panchayats, with the support of the community based organisation and non-governmental organisations, planning was initiated at micro-level as there is no initiative from the State Government for micro level planning.[22] Panchayats have mobilised the community for social development activities.[23] Now Panchayats are spear heading another movement for creating knowledge revolution in the villages with the support of developmental organisations.[24] But these are all stories of experiences from different parts of the country at the initiative of some of the development organisations and leaders, but not throughout the country.

The Panchayat leaders are swimming against the current. In many of the States, conditions are not favourable for devolution of powers. The hurdles are many. Officials are not cooperative. There is no clear cut activity mapping with regard to devolution of powers. The orders passed by the Government do not reach the Panchayat leaders. The negative mindset of lower level officials, the corrupt administrative system, the caste ridden society dividing the community on caste lines, political parties' pressure on the system, the poor capacity of the leaders, the poor resources of the Panchayats are working against the Panchayat. The Panchayat leaders have to overcome all these problems and then only they can perform. Women and *Dalit* leaders have to face a battery of problems, as they are women and *Dalits*. There are training institutions and they do conduct training programmes for the local body leaders. There are State Institutes of Rural Development all over the country working for capacity-building. Even the capacity-building is not being done very rigorously. Many development organisations have come to the rescue of the Panchayat leaders when they are in trouble while in the field. Barring Kerala, micro-plan exists on paper not in action. But many of the Panchayat leaders on their own have taken steps to evolve plans for their villages with the active participation of the people. The leaders are working on different issues in the Panchayats. One would find variations in the functioning of the Panchayats among the States. Of the three governments more number of representatives are in the third tier of governance but they are weak as the Panchayat is a State subject. In order to overcome the existing impasse in devolving powers to local bodies and associated problems, the present UPA Government at the centre created a full-fledged Ministry for Panchayati Raj. In consequence of the creation of the Ministry, seven round table discussions have been held with State Government ministries and the higher level officials and 150 resolutions have been passed unanimously on 150 issues for furthering the devolution process. Action points have been fixed and a monitoring mechanism has been evolved to monitor the process. Thus second phase of decentralization of powers has been initiated.[25]

Conclusion

The 73rd Amendment to the Constitution of India provides an opportunity to change the paradigm of development from the supply driven to the demand driven mode, from exclusive to inclusive strategy, from top down planning to bottom up planning and from bureaucratic activities to people centric activities. It alters the character of our democracy from the representative to the participatory. All the above are only opportunities provided by the Constitutional Amendment. But the outcome of these opportunities depends on how these opportunities are utilized by the people. Limited experience from case studies shows that Panchayats have potential to bring about transformation and needed change in the society. But there are an array of hurdles in the process. It is natural that people who hold power want to retain the same with them. To change the scenario the State Governments have political will. The elected representatives have to be sensitized on administration management and development. They have to be trained. The needed support structure has to be created.

Officials at all levels have to be trained. A linkage between Self-Help Groups and Panchayats has to be established. Now models are emerging. By using these models, Panchayats have to be activated throughout the country. When more activities are carried out by the Panchayat, it can win over the community. Though the decentralization process is centrally driven, to reach the target it should be made a people's programme. The potential has to be tapped by making it a mass movement. Civil society organisations, development organisations and institutions, media, academic and research institutions have to play a vital role in the transformation process.

REFERENCES

1. James Manor, *The Political Economy of Democratic Decentralization*, Washington DC: The World Bank, 1999.
2. The argument started from the proposition that public institutions have not delivered the goods and services to the level of the demands of the public. For more details see. Devesh Kapur and Pratap Bhanumehta (Ed.) *Public Institutions in India: Performance and Design*, New Delhi: Oxford University Press, 2005.
3. Gert W. Kueck, S.C. Mathur and Klaus Schindler, *Federalism and Decentralization,* New Delhi: Konrad Adenauer Foundation, 1998.
4. D.N. Gupta, *Decentralization: Need for Reforms*, New Delhi: Concept Publishing Company, 2004.
5. G. Palanithurai, *Towards Decentralization of Powers,* Gandhigram: Rajiv Gandhi Chair for Panchayati Raj Studies, 2001.
6. Ashok Mehta, *Report of the Committee on Panchayati Raj Institutions,* Government of India: Ministry of Agriculture and Irrigation, 1978.
7. (a) Rajiv Gandhi's Inaugural Address to the 13th All India Panchayat Parishad, New Delhi on 22nd September 1986.
   (b) Rajiv Gandhi's Inaugural Address at the Workshop of District Collectors and Magistrates on Responsive Administration held at Hyderabad on 13th February 1988.
   (c) Rajiv Gandhi's Inaugural Address at the Workshop of District Collectors and Magistrates on Responsive Administration held at Imphal on 2nd April 1988.
   (d) Rajiv Gandhi's Inaugural Address at the Workshop of District Collectors and Magistrates on Responsive Administration held at Jaipur on 30th April 1988.
   (e) Rajiv Gandhi's Inaugural Address at the Workshop of District Collectors and Magistrates on Responsive Administration held at Coimbatore on 18th June 1988.

(f) Rajiv Gandhi's Inaugural Address in the Panchayati Raj Sammelan of Northern States New Delhi on 27th January 1989.
(g) Rajiv Gandhi's speech at the Conference of the Chief Ministers on Panchayati Raj, New Delhi on 5th May 1989.
(h) Rajiv Gandhi's Address at a Seminar of Municipal Officers in New Delhi on 5th June 1989.
(i) Rajiv Gandhi's Address at a Conference of Chief Minister's New Delhi 7th July 1989.

8. (a) Rajiv Gandhi's Statement while introducing the Constitution (64th Amendment) Bill on Panchayati Raj in Lok Sabha 15th May 1989.
(b) Rajiv Gandhi's Statement in the Lok Sabha while introducing the Constitution (65th Amendment) Bill on 7th August 1989.
(c) Rajiv Gandhi's Text of reply to the debate on the Panchayati Raj and Nagarpalika Bills Rajya Sabha on 13th October 1989.

9. (a) Government of India, *Powers and Functions of Panchayati Raj Institutions: A Framework* (Occasional Paper 5), New Delhi: Ministry of Rural Development, 1994.
(b) Government of India, *Renewing Local Self Government in Rural India,* New Delhi: Ministry of Rural Development, 1994.
(c) D. Bandyopadhyay and Amitava Mukerjee (eds), *New Issues in Panchayati Raj,* New Delhi: Concept Publishing Company, 2004.

10. Government of India, *73rd Constitutional Amendment Act,* New Delhi: Government of India, 1993.
11. Devolution Index has been prepared in the Round Table Discussion and it was approved by all the State Governments for Implementation. Ministry of Panchayati Raj, *Proceedings of the Round Table and Resolutions,* New Delhi: Government of India, 2005.
12. George Mathew (Ed.), *Status of Panchayati Raj in India,* New Delhi: Concept Publishing Company, 2002.
13. G. Palanithurai, *Dimensions of New Panchayati Raj System,* New Delhi: Concept Publishing Company, 2004.
14. G. Palanithurai, *Process and Performance of Dalit and Women Gram Panchayat Leaders,* New Delhi: Concept Publishing Company, 2005.
15. G. Palanithurai, *Dynamics of New Panchayati Raj System in India,* Vol. I, II, III, New Delhi: Concept Publishing Company, 2003.
16. G. Palanithurai, *Dalit Leadership in Panchayats,* Gandhigram: Rajiv Gandhi Chair for Panchayati Raj Studies, 2003.
17. G. Palanithurai, *Building Alliances and Capabilities among the Elected Women Representatives of Grassroots Institutions,* Gandhigram: Rajiv Gandhi Chair for Panchayati Raj Studies, 2003.
18. Jayshree P. Mangubhai, *To what extent does affirmative action in democratic decentralized governance in India enable women to assert their rights to political participation and socio-economic development?,* M.A. Dissertation, University of London, 2005; G. Palanithurai, "The

Empowerment Process: Women Representatives in Grassroots Institutions in Tamil Nadu", *Indian Social Science Review,* Vol.6, No.1 (Jan-June 2004).

19. G. Palanithurai, "Good Governance at Grassroots", *The Indian Journal of Political Science,* Vol. LXVI: No. 2, (April-June 2005).
20. G. Palanithurai, *Leadership Matters in Grassroots Governance: Pathfinders Series I to XVII,* Gandhigram: Rajiv Gandhi Chair for Panchayati Raj Studies, 2005.
21. G. Palanithurai, "Role of Elected Women Representatives in Evolving Alternative: Governance at Grassroots", *Social Change and Development,* Vol. 3, (July-2005), pp. 118-139.
22. Experiences of other States in micro-planning are captured in the V volume of the multi-volume set. G. Palanithurai, *Dynamics of New Panchayati Raj System in India,* Vol. V, New Delhi: Concept Publishing Company, 2005.
23. G. Palanithurai, *Panchayat, Communities and Social Development,* Gandhigram: Rajiv Gandhi Chair for Panchayati Raj Studies, 2003.
24. M.S. Swaminathan Research Foundation, *Role of Panchayati Raj Institutions in Spearheading a Rural Knowledge Revolution,* Chennai: MSSRF, 2005.
25. G. Palanithurai, *Second Generation Reform in Decentralization,* Gandhigram: Rajiv Gandhi Chair for Panchayati Raj Studies, 2005; Ministry for Panchayati Raj, *Proceedings of the Round Table Resolutions,* New Delhi: Government of India, 2005.

11

Rural Governance and Panchayati Raj Institutions

MAHI PAL

Introduction

The essential aim of our freedom struggle was to remove poverty, hunger, illiteracy, unemployment and all sort of diseases from which India had been suffering for two centuries under the British regime. Thus, after Independence, the Government of India had launched various developmental and welfare-oriented programmes/schemes through Five Year Plans to enable the people to lead a life with dignity. The Community Development Programme and the National Extension Scheme were introduced for the all-round development of rural society. In order to bring about people's participation in rural development, the Government of India constituted the Balwantray Mehta Team in 1957. This team suggested democratic decentralization for enlisting people's participation in developmental activities and, towards this end, suggested the creation of Panchayati Raj Institutions (PRIs) at district and sub-district levels. But after an initial phase of ascendancy and growth of Panchayats, these institutions became a victim of the lack of political will, hostility of bureaucracy and apathy of people. The Report of the Asoka Mehta Committee (1977) pointed out that the PRIs had failed in their objectives on account of unfavourable political environment. Later on, the G.V.K. Rao Committee (1985) and the Singhvi Committee (1986) were constituted to suggest ways and means to strengthen the PRIs. But despite all this, the viability, sustainability and empowerment of these institutions remained at the mercy of the State Governments.

The enactment of the 73rd Amendment to the Constitution (hereafter referred to as the Central Act) was a watershed in the field of decentralization as it, *inter alia*, provided certainty, continuity and strength to the Panchayati Raj System (PRS) in the country. Later on, PRS was also extended to the Scheduled Areas by the Provisions of the Panchayats (Extension to Scheduled Areas) Act, 1996 (hereafter referred to as the Extension Act). Panchayats have been in existence in India in one form or other even before independence and more than one decade has elapsed after the enactment of the 73rd Amendment to the Constitution and more than 5 years have elapsed after the enactment of the Provisions of Panchayats (Extension to Scheduled Areas) Act, 1996.

In this backdrop, it appears appropriate to review the functioning of the Panchayats from the point of view of ascertaining as to what extent have these institutions enabled the people to participate in delivering good governance at local levels and in the light of experiences, suggest ways and means for further strengthening the PRS in the country. Hence, the present paper endeavours to assess the empowerment and entitlement of the PRIs, effectiveness of these institutions in delivering rural good governance with people's active and functional participation, including of those who hail from the excluded and marginalised groups, and the policy prescriptions for effective local governance.

The paper has been divided into seven Parts. Part I of the paper deals with the rural governance in the context of PRIs. The salient features of the Central Act and the State Act have been outlined in Part II. The issues of empowerment and entitlement has been discussed in Part III. Part IV deals with the marginalised Groups and Women. The role of civil society organizations in promoting rural local self-government has been discussed in Part V. Access to the marginalised groups of common property resources has been discussed in Part VI. This is followed by some suggestions and the policy prescriptions for empowering the Panchayats adequately to enable them in delivering good governance in rural areas. In the last Part, some concluding remarks and observations have been presented.

PART I
Rural Governance

Governance may be defined as the management of all the processes which, *inter alia*, promote individual's capacity and capability as well as create environment of their fulfilment. In a broader sense, overall thrust of governance is to create a social order where human amourpropre is secured. There are several opinions on the issue of governance. In the context of governance, the National Human Development Report 2001 of the Planning Commission gave several possible approaches pertaining to the Governance issue. Two of these are being given here to facilitate our discussion in the context of governance and Panchayati Raj Institutions. The World Bank defined governance "...as the manner in which power is exercised in the management of a country's economic and social resources. The World Bank has identified three distinct aspects of governance (1) the form of political regime; (2) the process by which authority is exercised in the management of a country's economic and social resources for development; and (3) the capacity of Governments to design, formulate and implement policies and discharge functions" (Human Development Report, 2002:115). The United Nations Development Programme viewed governance "...as the exercise of political, economic and administrative authority in the management of a country's affairs at all levels. It comprises mechanisms, processes and institutions through which citizens and groups articulate their interests, exercise their legal rights, meet their obligations and mediate their difference" (Human Development Report, 2002: 115).

This could only be achieved if proper socio-economic environment is created for holistic development of each and every individual and thereby the entire society. This requires not only liberty and deliverance of the oppressed, depressed and suppressed, but also establishment of such institutions which are not only democratic but also close to masses as well as have adequate representation of women and marginalised sections at various levels of governance. Panchayats are the such institutions which have been represented by more than 3 millions elected representatives in the form of presidents and members at the three levels of it across the country. This is perhaps the broadest democratically elected representative base which exists in any country of the world.

We will discuss in the following part of the paper that the Panchayats have been made as institutions of self-government. Although, the term 'Institution of Self-Government' (ISG) has not been defined in the Act, the percept and the connotation are well understood. In order to function as ISG, the Panchayats have to fulfil three basic conditions, namely (a) institutional existence in the sense that the decisions are taken by the people's representatives; (b) institutional capacity in the form of empowerment of institutions to make their rules independently; and

(c) financial viability in terms of sufficiently empowering Panchayats in raising resources to meet their responsibilities. In other words, the Panchayats should enjoy functional, administrative and financial autonomy (Mahi Pal, 2004:138). In this context, the issue of governance has been addressed across the various sections of the paper.

PART II
Salient Features of the Central and the Extension Act

The main features of the Central Act and brief comments on these are being given in this section. The Central Act has both mandatory and enabling provisions for Panchayats. Following are the salient features of this Act: (i) Formation of Gram Sabha at the village level; (ii) Uniform three-tier system at the village, block and district levels with exemption for intermediate level in States with population of less than two million; (iii) Direct elections to all seats for all members at all levels; (iv) Indirect elections of the chairpersons at the intermediate and apex levels, the mode of election for the lowest level has been left to the State; (v) Reservation on rotational basis for the Scheduled Castes (SCs) and Scheduled Tribes (STs) in proportion to their population both for the membership as well as chairpersonships of the Panchayats at all levels; (vi) Reservation of not less than one-third of the seats and offices for women; (vii) Five year term; (viii) Fresh elections within six months in case of suppressions or dissolution; (ix) Devolution of powers and responsibilities by the State in the matters preparation and implementation of plans for economic development and social justice including implementation of schemes relating to the subjects listed in the Eleventh Schedule of the Act; (x) Setting up the State Finance Commission after every five years to review the financial position of these institutions and to make suitable recommendations to the State on the distribution of funds among the Panchayat; (xi) Setting up the State Election Commission for holding Panchayat elections under its superintendence, direction and control; (xii) The 74th Amendment to the Constitution provides for constitution of District Planning Committee to prepare a draft plan for the whole district comprising Panchayats and Municipalities. Although, this provision is a part of the 74th Amendment Act, the States have agreed to extend it to the 73rd Amendment Act. Hence, for all practical purposes, it is also assumed to be a part of the Central Act.

The Central Act gave birth to the third generation of the Panchayati Raj in India. The main lacuna of this Act, however, is that, instead of clearly specifying the functions and powers of Panchayats, it has left it to the discretion of the State Governments. The Article 243G of the Constitution clearly says: the legislature of a State may, by law, endow the Panchayats with such powers and authority as may be necessary to enable them to function as institutions of self-government and such law may contain provisions for the devolution of powers and responsibilities upon Panchayats at the appropriate level subject to such conditions as may be specified therein, with respect to (a) the preparation of plans for economic development and social justice, and (b) the implementation of schemes for economic development and social justice as may be entrusted to them including those in relation to the matters listed in the Eleventh Schedule.

This Article is the corner-stone of the entire thrust on democratic decentralisation as visualised in the Act. It is, however, not mandatory on the part of the State Government to implement this because the word "may" is used five times in this Article. Thus, it is at the discretion of the State Government to "give" or "not give" powers to the Panchayats. It implies that the power sharing with the Panchayats depends on the political leadership at the State level. We have the example of West Bengal where Panchayats elections were being held since the seventies, i.e. even before the Amendment because the political party in power has political will for strengthening the grassroots democracy.

The salient features of the Extension Act are as follows: (i) Every village shall have Gram Sabha consisting of persons whose names are included in the electoral rolls for the Panchayats at the village level; (ii) Reservation for STs shall not be less than half of the total number of seats at all tiers of Panchayats. Reservation of tribal communities shall be on the basis of proportion to their population. The chairpersonship shall be reserved for STs at all levels of Panchayats. In case some ST communities have no representation at intermediate or district level Panchayats, the State Government shall nominate such under-represented STs. But such nomination should not exceed one-tenth of the total number of elected members of the Panchayats; (iii) Every legislation on the Panchayats in the Fifth Scheduled Area shall be in conformity with the customary law, social and religious practices and traditional management practices of the community resources; (iv) Gram Sabha shall be competent to safeguard and preserve the traditions and customs of the people, their cultural identity, community resources and customary mode of dispute resolution; (v) Gram Sabha is empowered to approve plans, programmes and projects for social and economic development, to identify persons as beneficiaries under the poverty alleviation and other programmes and to give certificate of utilisation of funds for various plans and programmes; (vi) Gram Sabha or Panchayat at the appropriate level would be consulted before making the acquisition of land for development projects and before resettling or rehabilitating persons affected by such projects. However, actual planning and implementation of the projects shall be coordinated at the state level; (vii) Recommendation of the Gram Sabha or the Panchayats at the appropriate level is mandatory for the grant of prospecting licence or mining lease for minor minerals, and grant for the exploitation of minor minerals by auction; (viii) Gram Sabha and Panchayats are endowed with powers to enforce prohibition or to regulate or restrict the sale and consumption of any intoxicant; ownership of minor forest produce; to prevent alienation of land; to manage village markets; to exercise control over moneylending; to exercise control over institutions and functionaries in all social sectors; to exercise control over local plans and resources for such plans, including tribal sub-plans; (ix) Planning and management of minor water bodies shall be entrusted to the Panchayats at the appropriate level; and (x) State legislatures may endow Panchayats with such powers and authorities, as may be necessary to enable them to function as institutions of self-government. The statutes shall contain safeguards to ensure that Panchayats at the higher level do not assume the powers and authority of any Panchayat at the lower level or of the Gram Sabha. Hence, the cardinal principle of allocation of functions has been provided to a greater extent in the Act itself; (xi) State legislatures shall endeavour to follow the pattern of the sixth Schedule to the Constitution while designing administrative arrangements in the Panchayats at district level.

Thus, the Extension Act is an important step in enabling State legislatures to make laws for the tribals that ensure their control and rights over natural resources and conserve and preserve their identity and culture in a participatory manner through the institution of the Gram Sabha.

PART III

Empowerment and Entitlement of the Panchayati Raj Institutions

The State Governments were supposed to transfer the functions, finance and functionaries pertaining to 29 subjects listed in the Eleventh Schedule to the Constitution to the PRIs to enable these bodies as institutions of

Self-government. The connotation ISG has been defined in terms of enjoying functional, financial and administrative autonomy in their functioning at different levels. Let us discuss the trends and the present position with regard to these three aspects of the decentralisation as in November 2001, May 2003 and April 2004 in order to know the extent of empowerment of the Panchayats.

Transfer of Functions

The basic principle (that is, what can be done at a particular level should be done at that level only) has not been adhered to by the States while allocating functions among the three tiers of Panchayats. As mentioned earlier, the State Governments were supposed to transfer to Panchayats functions pertaining to 29 subjects listed in the Eleventh Schedule of the Constitution.

As per the information on the transfer of functions to the Panchayats, the Report of the Working Group on Decentralized Planning and Panchayati Raj Institutions for the Tenth Five Year Plan, out of all the States and Union Territories, only Karnataka, Kerala, Rajasthan, Sikkim, Tamil Nadu, West Bengal, and Daman & Diu have devolved 29 subjects to the Panchayat. The States of Arunanchal Pradesh, Assam, Bihar, Jharkhand, Goa, Gujarat, and Union Territories of A&N Islands, Chandigarh and Pondicherry have not transferred any functions while others have transferred only some functions ranging from 3 in Dadra & Nagar Haveli to 25 functions in Orissa (Table 11.1). As is evident from this Table, although several States have given a varying number of responsibilities to the Panchayats, these have not been put into practice. This has in turn adversely affected the efficiency and effectiveness of the Panchayats. This is the position of devolution in November 2001. In this context, the Working Group on Decentralised Planning rightly observed, "Items listed as responsibilities in the States are couched in vague terms. A glance at the variety of these items reveals that they are a shopping list of sectors and sub-sectors, broad sub-sector and activities in a sub-activities/specific responsibilities under a broad activity, with no role clarity ... in some States, the line departments still exercise the powers of supervision and control over the schemes of subjects transferred to the Panchayats" (Working Group on Decentralised Planning, 2001:13).

On the issue of functional devolution on Panchayats, the Task Force on PRIs observed: "In State like Andhra Pradesh, the State Government is implementing even a decentralized programme like Janmabhoomi

Table 11.1: State-wise Status of Devolution of Departments/Subject with Funds, Functions and Functionaries to Panchayati Raj Institutions (in November 2001)

Sl. No.	States/UTs	No. of Departments/Subjects Transferred to Panchayats with			No. of Departments/Subjects Yet to be Transferred to Panchayats with		
		Funds	Functions	Functionaries	Funds	Functions	Functionaries
1.	Andhra Pradesh	05	13	02	24	16	27
2.	Arunachal Pradesh	—	—	—	29	29	29
3.	Assam	—	—	—	29	29	29
4.	Bihar	—	—	—	29	29	29
5.	Jharkhand	—	—	—	29	29	29
6.	Goa	—	—	—	29	29	29
7.	Gujarat	—	—	—	29	29	29
8.	Haryana	—	16	—	29	13	29
9.	Himachal Pradesh	02	23	07	27	06	22
10.	Karnataka	29	29	29	—	—	—
11.	Kerala	15	29	15	14	—	14
12.	Madhya Pradesh	10	23	09	19	06	20
13.	Chhattisgarh	10	23	09	19	06	20
14.	Maharashtra	18	18	18	11	11	11
15.	Manipur	—	22	04	29	07	25
16.	Orissa	05	25	03	24	04	26
17.	Punjab	—	07	—	29	22	29
18.	Rajasthan	—	29	—	29	—	29
19.	Sikkim	29	29	29	—	—	—
20.	Tamil Nadu	—	29	—	29	—	29
21.	Tripura	—	12	—	29	17	29
22.	Uttar Pradesh	12	13	09	17	16	20
23.	Uttaranchal	12	13	09	17	16	20
24.	West Bengal	12	29	12	17	—	17
25.	A&N Islands	—	—	—	29	29	29
26.	Chandigarh	—	—	—	29	29	29
27.	D&N Haveli	—	03	03	29	26	26
28.	Daman & Diu	—	29	—	29	—	29
29.	NCT Delhi	Panchayati Raj System is yet to be Revived					
30.	Lakshadweep	—	06	—	29	23	29
31.	Pondicherry	—	—	—	29	29	29

Source: *Report of Working Group on Decentralised Planning and Panchayati Raj Institutions for the Tenth Five Year Plan* (2002-2007), Government of India, Planning Commission, November 2001.

programme mainly through bureaucratic apparatus, which is against the spirit of 73rd Constitutional Amendment Act. It has a content of people's participation and social mobilization, but it bypasses the Panchayati Raj bodies and in fact this programme poses a serious challenge to the system of Panchayati Raj due to its poor involvement in this programme. The State Government of Haryana had set up Village Development Committees which negates the provision of 73rd Constitutional Amendment Act regarding reservation of SC, women and seems to replace the elected Gram Pancha243yat" (Task Force on PRIs, 2001:9).

Table 11.2 gives the status of functions transferred to the Panchayats as on May 29, 2003 and Table 11.3 gives the same as on April 1, 2004. A comparison between the status of powers devolved to Panchayats in November 2001, on May 29, 2003 and on April 1, 2004 indicates the following:

(i) In the State of Assam, functions relating to 29 subjects have been transferred to the Panchayats only in 2003. This is a positive development towards Panchayats' empowerment. In Bihar, Gujarat, Himachal Pradesh, Goa, Kerala, Chhattisgarh, and A&N Islands, some more functions have been devolved on the PRIs during the last one and half years. But these States have not transferred all 29 subjects to the Panchayats. Bihar is the only State which has devolved 5 more functions to the Panchayats between May 2003 and April 1, 2004 as Table 11.3 indicates. On the other hand, Uttar Pradesh is the only State which, instead of devolving more functions to these bodies, had in fact, withdrawn one function from the Panchayats as the Table indicates. But none of the States except Assam, Karnataka and Chhattisgarh has devolved all the 29 subjects which were expected to be devolved on them immediately after the coming into force of the Central Act. The case of Haryana is interesting because here it is indicated that 'almost all 29 subjects' have been transferred to the Panchayats. These words conceal more than they reveal as to the status of functional devolution to the Panchayats in the State. But as per the latest information available on this issue, powers pertaining to 16 departments have been transferred to the Panchayats.

(ii) In Goa, Punjab, A&N Islands and D&N Haveli, very little devolution of functions has taken place as column 4 of Table 11.2 indicates and no progress has been made from May 2003 to April 2004 in this regard as the Table indicates.

(iii) Surprisingly, in Arunachal Pradesh, Jharkhand, Pondichery and Chandigarh, none of the functions pertaining to the 29 subjects have been transferred to the Panchayats during the last decade.

(iv) Instead of decentralisation in terms of devolving powers to the Panchayats, it has been the reverse, i.e. the centralisation, that took place in Sikkim, Uttaranchal and Daman & Diu, because some of already devolved functions to Panchayats have been taken back as the minus sign given in column 4 of Table 11.2 indicates. From May 2003 to April 2004, further centralisation took place in Uttar Pradesh where one function has been withdrawn by the State from the Panchayats as indicated by the minus sign in column 4 of Table 11.3.

Table 11.2: Status of Devolution of Departments/Subjects with Funds, Functions and Functionaries to Panchayati Raj Institutions (as on 29-5-2003)

Sl. No.	*States/UTs*	*No. of Departments/Subjects Transferred to Panchayats with*			*No. of Departments/Subjects Yet to be Transferred to Panchayats with*		
		Funds	*Functions*	*Functionaries*	*Funds*	*Functions*	*Functionaries*
(1)	(2)	(3)	(4)	(5)	(6)	(7)	(8)
1.	Andhra Pradesh	05	17(+4)	02	24	12	27
2.	Arunachal Pradesh	—	—	—	29	29	29
3.	Assam	—	29(+29)	—	29	—	29
4.	Bihar	—	20(+20)	—	29	9	29
5.	Jharkhand	—	—	—	29	29	29
6.	Goa	6(+6)	6(+6)	—	23	23	29
7.	Gujarat	15(+15)	15(+15)	15(+15)	14	14	14
8.	Haryana	—	Almost all 29 subjects	—	29	—	29
9.	Himachal Pradesh	02	26 subjects (+3)	*	27	3	22
10.	Karnataka	29	29	29	—	—	—
11.	Kerala	15	26(+3)	15	14	3	14
12.	Madhya Pradesh	10	23	09	19	06	20
13.	Chhattisgarh	10	29(+6)	09	19	—	20
14.	Maharashtra	18	18	18	11	11	11
15.	Manipur	—	22	04	29	07	25
16.	Orissa	05	25	21(+18)	24	04	8
17.	Punjab	—	07	—	29	22	29
18.	Rajasthan	—	29	—	29	13	29
19.	Sikkim	24(-5)	24(-5)	24(-5)	5	5	5
20.	Tamil Nadu	—	29	—	29	—	29
21.	Tripura	—	12	—	29	17	29

(Contd.)

Table 11.2 (*Contd.*)

(1)	(2)	(3)	(4)	(5)	(6)	(7)	(8)
22.	Uttar Pradesh	12	13	09	17	16	20
23.	Uttaranchal	(-12)	11(-2)	11(+2)	29	18	18
24.	West Bengal	12	29	12	17	—	17
25.	A&N Islands	6(+6)	6(+6)	6(+6)	23	23	23
26.	Chandigarh	—	—	—	29	29	29
27.	D&N Haveli	—	03	03	29	26	26
28.	Daman & Diu	5(+5)	9(–20)	3(+3)	24	20	26
29.	NCT Delhi	Panchayati Raj System is yet to be revived					
30.	Lakshadweep	—	06	—	29	23	29
31.	Pondicherry	—	—	—	29	29	29

* No functionary has been transferred. However, GPs and IPs have been made the appointing authorities with regard to 9 functions on contract basis.

Note: The provisions of the Constitution (73rd Amendment) Act, 1992 are not applicable to the States of J&K, Mizoram and Nagaland.

Source: Agenda Paper: Ist Meeting of the Empowered Sub-Committee on Financial and Administrative Empowerment of Panchayati Raj Institutions held on 26-6-2003, Ministry of Rural Development, Department of Rural Development.

Table 11.3: Status of Devolution of Departments/Subjects with Funds, Functions and Functionaries to Panchayati Raj Institutions (as on 1st April 2004)

Sl. No.	*States/UTs*	*No. of Departments/Subjects Transferred to Panchayats with*			*Status of Constitution of DPC's*
		Funds	*Functions*	*Functionaries*	
(1)	(2)	(3)	(4)	(5)	(6)
1.	Andhra Pradesh	05	17	02	Not yet constituted. However, an Ordinance has been issued by the Government of AP in December 2003 for constitution of DPCs.
2.	Arunachal Pradesh	—	—	—	Not constituted.
3.	Assam	—	29	—	Not constituted.
4.	Bihar	(+8)	25	—	37 districts out of 38 districts constituted on ad hoc basis. Chairman ZP is the Chairman of DPCs

(*Contd.*)

Table 11.3 (*Contd.*)

(1)	(2)	(3)	(4)	(5)	(6)
5.	Jharkhand	—	—	—	Panchayat elections yet to be held. Constituted. President of ZP is the Chair-person of DPC.
6.	Goa	6	6	—	—
7.	Gujarat	15	15	15	Not constituted.
8.	Haryana	—	16	—	16 districts out of 19 districts. Rest under consideration.
9.	Himachal Pradesh	02	26 sub-jects	11	Only in 6 districts out of 12. Minister is Chairperson of DPC.
10.	Karnataka	29	29	29	Yes, in all Districts. President ZP is Chair-man of DPC.
11.	Kerala	26 (+11)	26	15	Yes, Chairman of District Panchayat (DP) is Chairman of DPC.
12.	Madhya Pradesh	10	23	09	Yes. District incharge Ministers are Chair-persons.
13.	Chhattisgarh	10	29	09	Constituted. Minister is Chairperson of DPC.
14.	Maharashtra	18	18	18	Not constituted.
15.	Manipur	—	22	04	Yes in 2 districts out of 4, Adhyaksha, DP is Chairperson.
16.	Orissa	09 (+4)	25	21	26 Districts. Minister is Chairperson of DPC.
17.	Punjab	—	07	—	Not constituted
18.	Rajasthan	18 (+18)	29	18 (+18)	Yes, Chairman of DP is Chairman of DPC.
19.	Sikkim	24	24	24	Yes
20.	Tamil Nadu	—	29	—	Yes, Chairperson of DP is Chairperson.
21.	Tripura	—	12	—	Not Constituted.
22.	Uttar Pradesh	04 (1–8)	12 (–1)	06 (–3)	DPCs are not functional.
23.	Uttaranchal	—	11	11	Yes. Minister is Chair-man of DPC.
24.	West Bengal	12	29	12	Yes, Chairperson of DP is Chairperson of DPC.

(*Contd.*)

Table 11.3 (*Contd.*)

(1)	(2)	(3)	(4)	(5)	(6)
25.	A&N Islands	06	06	06	Yes, Chairman of DP is Chairman of DPC.
26.	Chandigarh	—	—	—	Not constituted.
27.	D&N Haveli	—	03	03	Yes, Chairman of DP is Chairman of DPC.
28.	Daman & Diu	05	09	03	Yes, Chairman of DP is Chairman of DPC.
29.	Lakshadweep	—	06	—	Yes. Collector-cum-Development Commissioner is Chairperson.
30.	NCT of Delhi	—	—	—	Panchayati Raj System is yet to be revived.
31.	Pondicherry	—	—	—	Panchayat Elections yet to be held.

Note 1: The provisions of the Constitution (73rd Amendment) Act, 1992 are not applicable to the States of J&K, Meghalaya, Mizoram and Nagaland.

Note 2: This Table has been taken from the Mid-Term Appraisal of the Tenth Plan, Planning Commission.

Source: Ministry of Rural Development, Government of India.

Transfer of Finance

Functional devolution to Panchayats without matching funds is a delusion. This is in fact what has happened with the State Panchayat Acts. Before listing the functions to be performed by the Panchayats, a number of states have introduced certain qualifying clauses. Thus, in Andhra Pradesh, Haryana and Tamil Nadu, it is 'within the limit of its funds'. In Punjab, it is 'to the extent its funds allow to perform'. In Himachal Pradesh and Madhya Pradesh, the clause is 'as far as the Gram Panchayat funds at its disposal'. This makes the picture about finances dimension of the Panchayats empowerment hazy because it is the resources at their disposal that determine their role in performing developmental activities in their respective areas of operation. Gujarat and Karnataka do, however, have Panchayat-friendly provisions in this regard. In Gujarat, as per Section 204 of the Panchayat Act, if the income of a Gram Panchayat is not commensurate with its functions, the Taluka Panchayat shall take necessary steps to provide resources to enable it to perform its functions effectively. The Karnataka Panchayat Act provides for grant arrangements for the Gram Panchayat under which a grant of Rupees one lakh is to be given to it by the State Government (Section 208).

On the issue of the role of the State Finance Commissions (SFCs) is financially empowering the Panchayats, the Mid-Term Appraisal of the 9th Plan observed that: "More buoyant taxes like sales tax and excise are kept out of the purview of PRIs. All SFCs have put great emphasis on internal revenue mobilization but none has suggested any effective mechanism for PRIs to generate their own revenue... The SFCs reports have paid less attention to the issue of autonomy, financial management and auditing procedures. The main deficiency of the reports lies in the fact that the recommendations are not based on clear statement of the spending responsibilities of local bodies. Indeed, the absence of attention to the elementary principle, that the expenditure assignment must precede any tax or revenue assignment, has made most of the SFCs recommendations suspect." (Mid-Term Appraisal, 2000:175-176).

The Working Group on the State Resources also commented in the same manner: "The suggestions of the first generation SFCs if fully implemented, will go a long way in improving the revenue generation capacity of the local bodies. However, the SFCs have failed to take a comprehensive view and the resource sharing between the State and local bodies still remains limited and unsettled. The State Governments have also been slow and hesitant in accepting the recommendations of the SFCs" (Report of the Working Group, 2002: 93)

The extent of transfer of financial powers to the Panchayats may be seen in Table 11.1, Table 11.2 and Table 11.3 of the paper. Table 11.1 presents the financial status at the fag end of the year 2001, Table 11.2 depicts the same as on May 29, 2003 and Table 11.3 shows the same just one year after as on 1st April, 2004. From a perusal of these Tables, following conclusions emerge:

(i) During the period of one and half years, some positive changes towards financial devolution have taken place because four more States have devolved financial powers to the Panchayats in 2003 as comparison of column 3 of Table 11.1 and Table 11.2 indicates; but from May 2003 to April 2004, no positive changes have taken place in this regard as no further functions have been transferred to these bodies. Conversely, centralisation in the State of Uttar Pradesh took place where finance pertaining to one subject from the Panchayats has been withdrawn as Table 11.3 indicates.

(ii) Karnataka is the only State which has devolved funds to Panchayats pertaining to 29 subjects of the 11th Schedule to the Constitution.

(iii) There are two States, namely Sikkim and Uttaranchal, where, instead of financial decentralisation, financial centralization took place because some of the already devolved financial powers have been withdrawn as minus signs against these states indicate.

The insult to injury was added on this account by the finding of the Report of the Task Force of Devolution of Powers and Functions to the PRIs brought out by the Ministry of Rural Development which pointed out that the mandatory provisions of the Central Act have not been implemented in letter and spirit by most of the States/Union Territories even about a decade after the coming into force of the said Act.

This Report also worked out the total annual plan outlay of the State Plans which partly or wholly pertained to 29 subjects indicated in the 11th Schedule to the Constitution. An amount of Rs. 40,818.65 crore was calculated as plan outlay under the annual plans for 2000-01 for the schemes/programmes of the State Governments under the subjects listed in the Eleventh Schedule to the Constitution (Table 11.4). It may also be seen from this Table that scheme/programmewise outlay ranges from Rs. 4.64 crores for Plantation to Rs. 8128.44 crores for Roads and Bridges. In between, there are eleven schemes/programmes under which the annual outlay was more than Rs. 100 crore. Similarly, Rs. 31,575.87 crore outlay has been earmarked for the annual plan 2000-01 for the schemes/programmes of the Central Ministries under the subjects listed in the Eleventh Schedule to the Constitution (Table 11.5). It may be seen from this Table that outlay under different schemes/programmes ranges from Rs. 1.20 crore for the Nutrition scheme to Rs. 5548.15 crore for the Roads and Bridges scheme. There are 9 schemes/programmes under which annual outlay was more than Rs. 1,000 crore. A huge amount of Rs. 72,394.52 crore, which legitimately belongs to the Panchayats, has actually been under the domain of either the Central Government or the State Government.

Another important issue relates to the finance of the Panchayats pertaining to their own contribution through tax and non-tax measures. The report of the Working Group on State Resources has thrown light on this issue. Table 11.5 shows that own revenue (both tax and non-tax) of the PRIs as ratio of GDP has been declining. The decline is sharper in the case of revenue from tax in comparison with that from non-tax.

Table 11.4: Plan Outlay for Annual Plan 2000-01 for the Schemes/ Programmes of the State Governments under the Subject Mentioned in the Eleventh Schedule to the Constitution

(Rs. in crores)

Sl. No.	*Name of the Scheme/Programme*	*Outlay 2000-01*
1.	Crop Husbandry	1268.77
2.	Soil and Water Conservation	643.50
3.	Animal Husbandry	354.19
4.	Dairy Development	49.87
5.	Fisheries	205.95
6.	Forestry and Wildlife	1393.03
7.	Plantations	4.63
8.	Agricultural Research & Education	292.21
9.	Agricultural Fin. Institutions	61.63
10.	Other Agricultural Programmes	98.18
11.	IRDP and Allied Programme	516.16
12.	NREP/JRY	1161.22
13.	Other Programmes like EAS etc.	1062.46
14.	Land Reform	294.08
15.	Minor Irrigation	1907.93
16.	Non-Conventional Sources of Energy	119.43
17.	Village and Small Industries	800.87
18.	Roads and Bridges	8128.44
19.	Road Transport	1265.30
20.	Inland Water Transport	16.74
21.	Other Transport Services	614.50
22.	General Education	4972.07
23.	Technical Education	422.49
24.	Art and Culture	171.07
25.	Medical and Public Health	3306.64
26.	Water Supply and Sanitation	6031.74
27.	Housing	1936.57
28.	Welfare of SCs, STs & OBCs	1972.99
29.	Social Welfare	904.26
30.	Nutrition	721.79
31.	Other Social Services	119.86
	Total	**40818.65**

Note: The outlay mentioned against each scheme/programme is the sum of the outlay for all the State Governments allocated against each of their scheme.

Source: *Report of the Task Force on Devolution of Powers and Functions upon Panchayati Raj Institutions*, Government of India, Ministry of Rural Development, August 2001.

Table 11.5: Plan Outlay for Annual Plan 2000-01 for the Schemes/ Programmes of the Central Ministries under the Subject Mentioned in the Eleventh Schedule to the Constitution

(Rs. in crores)

Sl. No	*Name of the Scheme/Programme*	*Outlay 2000-2001*
1.	Crop Husbandry	1380.22
2.	Soil and Water Conservation	63.42
3.	Animal Husbandry	77.07
4.	Dairy Development	36.45
5.	Fisheries	116.61
6.	Forestry and Wildlife	281.65
7.	Plantations	169.35
8.	Agricultural Research and Education	550.00
9.	Agricultural Programmes	28.76
10.	Other Agricultural Programmes	6.67
11.	Land Reforms	67.50
12.	Minor Irrigation	88.67
13.	Non-Conventional Sources of Energy	858.46
14.	Village and Small Industries	1125.35
15.	Road and Bridges	5548.15
16.	General Education	4433.06
17.	Technical Education	496.18
18.	Family Welfare	3200.00
19.	Water Supply and Sanitation	2071.03
20.	Housing	3869.73
21.	Welfare of SC, ST & OBCs	997.74
22.	Social Security and Welfare	2231.85
23.	Nutrition	2.90
24.	Other Social Services	1.20
25.	Special Programme for Rural Development	1075.45
26.	Rural Employment	2798.40
	Total	**31575.87**

Source: *Report of the Task Force on Devolution of Powers and Functions upon Panchayati Raj Institutions*, Government of India, Ministry of Rural Development, August 2001.

Hence, besides transferring resources from the Central and State Governments through various schemes and other constitutional mechanisms such as Central Finance Commission, State Finance Commission, the Panchayats must mobilise their own resources for acquiring fiscal autonomy which is an essential ingredient to enable them to function as ISGs. In this way, the principle of development from within will be applied. This may be defined "...as an endogenous development process by which people themselves assume the responsibility to manage their own resources, define their needs, goals and aspirations and make decisions affecting their well-being and sustainable potentials of the local community as well as meet their inner urges and preferences for development through self-reliant approaches." (Sundaram, 1997, p.134)

Table 11.6: PRIs Own Revenue-GDP Ratios

Year	*PRIs Own Revenue-GDP Ratio*	
	Tax	Non-Tax
1997-98	0.025	0.020
2002-03	0.023	0.019
2006-07	0.021	0.018

Source: Report of the Working Group on States' Resources for the Tenth Five-Year Plan (2002-2007), Volume II, Planning Commission, January 2002.

The Working Group on State Resources commented that "Simple reluctance to mobilize, lack of role clarity as between tiers, lack of staff support, unviable administrative area, creation of parallel channels of flow of funds like MP funds, MLA funds, state-sponsored schemes independent of PRIs, like Janmabhoomi in Andhra Pradesh, are among the major factors explaining low revenue levels of local bodies. There are States like Bihar and Tripura which have not collected any money from PRIs" (Report of the Working Group, 2002:94).

It is clear that a large amount of funds pertaining to the 29 subjects listed in the 11th Schedule to the Constitution is still in the control of the Central and State Governments. But the other side of the picture is that the Panchayats have not used the provisions contained in the State Panchayat Acts for mobilising either the tax or the non-tax revenue at their levels.

Transfer of Functionaries

Control over the personnel of the Panchayats pertaining to the 29 subjects of the 11th Schedule by Panchayats is no less important than transferring the functions and finances to them. In this context also, the current status is no better than that in the case of financial devolution as Tables 11.1, 11.2 and 11.3 show. The Panchayat Acts provide powers for inspection, inquiring into their affairs, suspension of Panchayat' Resolutions, issuing directions, etc., to the State Governments. Besides, the key functionaries, namely, Secretaries and Executive Officers, at all levels of Panchayats are State Government employees in most of the states. For ensuring administrative autonomy, the Panchayats should have their own service cadre. In Tripura, for recalling a State Government employee, a resolution to that effect has to be passed by a majority of the members of the Panchayat.

The status of transfer of the functionaries to the Panchayats during November 2001 and April 2004 was as follows:

(i) In the entire country, Karnataka is the only State which has transferred personnel to the Panchayats pertaining to 29 subjects given in the 11th Schedule.

(ii) During the period from November 2001 to May 2003, Gujarat, A&N Islands and Daman & Diu have transferred some functionaries under the control of the Panchayats. Orissa is the only State which has transferred the maximum number of the functionaries to the Panchayats as column 5 of the Table 11.2 shows. Rajasthan is the only State in the country which has transferred 18 functionaries under the control of the Panchayats from May 2003 to April 2004.

(iii) As many as 13 States/Union Territories have not transferred any of their functionaries to the Panchayats by April 2004 and unlikely by so far also.

(iv) In the State of Sikkim, in place of administrative decentralization, administrative centralization took place because instead of transferring more functionaries to the control of Panchayats, already transferred functionaries have been withdrawn during the period under reference as column 5 of Table 11.2 indicates. From May 2003 and onward as the Table indicates Uttar Pradesh is the only State where centralisation in this context took place as the functionary pertaining to one department has been withdrawn in the State.

It is also relevant to mention here that many officials of the State Governments posted at the district and sub-district levels do not want to work under the administrative control of the elected Panchayats. This situation may be described as democracy at national and state levels and bureaucracy at the district, block and village levels.

The foregoing discussion on the scenario regarding the functions, finances and functionaries (Triple Fs) of Panchayats gives adequate evidence that, with some exceptions (like States of Karnataka and Kerala), status of the Panchayati Raj Institutions in terms of making them autonomous in the areas of their operation remains more or less the same as it was before the enactment of the Amendment to the Constitution. Evidently, assuming these institutions as autonomous institutions is at best a delusion. Political participation of the representatives in these bodies perhaps has not gone beyond casting their votes.

For corroborating the gloomy picture on devolution of powers from the Central and State Governments to the Panchayats that has emerged from the above discussion, it would be worthwhile to quote here what the Task Force on PRIs had to observe in this regard. "Though some steps have been taken by the MORD to goad the State Governments to strengthen and deepen the process of democratic decentralization, it has not yielded the desired results. Most Central Ministries have not yet internalized the PRIs role in the delivery of services handled by them. What is true of the Central Government is also true of the State Governments. Leaving aside one or two states, such as Kerala, which have taken active steps for substantially empowering PRIs in planning and development, there is little by way of operationalising models of such empowerment. Even states like Kerala feel handicapped in regard to central schemes since the guidelines for implementation are drawn by the concerned agency of Central Government" (Task Force of PRIs: 2001).

While introducing the Constitution (Seventy-Third) Amendment Bill in Parliament on 1 December, 1992, the then Rural Development Minister G.Venkatswamy said that: "This casts a duty on the Centre as well as the States to establish and nourish the Village Panchayats so as to make them effective, self-governing institutions". And by introducing this Act the Government was fulfilling Mahatma Gandhi's dream of *Gram Swaraj*, he added. But the foregoing assessment to the extent of their autonomy shows that the Centre and the States have not performed their duties as expected.

The Extension Act has not only made the Gram Sabha a strong body, but also put '*jal, jungle* and *jamin*' (water, forest and land) under its control. All the States, barring Jharkhand, have enacted State Legislation in conformity with the Extension Act (Working Group Report, 2001:29). The State Legislature should have kept in mind the following aspects while amending their respective Panchayat Acts: (i) Gram Sabha should be made the corner-stone of the entire system of PRIs by way of according freedom to them in managing, protecting and preserving natural resources and through formulation of socio-economic development programmes, projects and schemes for tribal people; (ii) Relationship between the Gram Sabha and panchayats should be similar to the one between legislatures and the government. In other words, Panchayats should take the approval of the gram sabha in all matters affecting rural economy and society; (iii) The underlying spirit of the Extension Act is that of devolution of powers and authority (rather than delegation of powers) to the Gram Sabha which represents participatory democracy rather than empowering representatives of decentralised democracy in the form of Gram Panchayats and higher tiers of the PRS. And while devolving powers and authority on Panchayat or Gram Sabha, the state legislature should ensure that Panchayats at the higher level do not assume the powers and authority of any Panchayat at the lower level or of the Gram Sabha (Mahi Pal, 2000:1603).

But most states have not kept these principles in mind while enacting their conformity legislation. States have not gone into the spirit of this legislation and have tried to manipulate the provisions in a narrow way. It has been rightly commented on by the Mid-Term Appraisal Report of the Ninth Plan: "Implementation of the law has been severely hampered by the reluctance of most State Governments to make laws and rules that conform to the spirit of the law. What is perceived as weak-kneed political will has led to bureaucratic creativity in minimalistic interpretations of the law" (MTA, 2000: 180).

However, keeping in view the provisions of the concerned State Acts, it may be said that Conformity Acts of Maharashtra, Madhya Pradesh and Himachal Pradesh are comparatively more in conformity with the letter and spirit of the Extension Act. In order to implement the Extension Act in the right perspective, amendments in its related legislations such as Land Acquisition Act, Excise Act, Irrigation Act, Land Revenue Code/ Act, SC/ST Land Alienation Act, Moneylenders Act and Regulated Market Act are required to be addressed now.

The reference of the Approach to Mid-Term Appraisal of the 10th Plan is very contextual in order to know the status and the seriousness of the Government on strengthening the PRIs. It says that the "Governance forms the key element of the Tenth Plan, but there has been practically no movement on this front. The NDC had set up three empowered sub-committees on (a) Governance, including e-Governance, (b) Creating an investor-friendly environment, and (c) Empowerment of PRIs, which were expected to draw up blue-prints for governance reforms. There is nothing available from these sub-committees at present, except for a national plan for e-Governance, which has the in-principle approval of the previous Prime Minister" (Approach to MTA 2005:13). However, in the same breath, it says that: "The Tenth Plan has laid great stress on the role of

Panchayati Raj Institutions (PRIs), not only as the cutting edge of democratic decentralization but also for improving the efficiency and accountability of the delivery systems for a number of publicly provided services"(Approach to MTA, 2005:13). Does not it indicate that the masses of the country are being fooled?

Constitution of the District Planning Committees

Article 243G of the Constitution, which contains the powers, authority and responsibilities of Panchayats says, "Subject to the provisions of this Constitution, the Legislature of a State may, by law, endow the Panchayats with such powers and authority as may be necessary to enable them to function as institutions of self-government and such law may contain provisions for the devolution of powers and responsibilities upon Panchayats at the appropriate level, subject to such conditions as may be specified therein, with respect to (a) the preparation of plans for economic development and social justice, and (b) the implementation of schemes for economic development and social justice as may be entrusted to them including those in relation to the matter listed in the Eleventh Schedule."

Article 243W of the Constitution contains similar provisions for urban local bodies.

Article 343ZD of the Constitution says that: (1) There shall be constituted in every State at the district level a District Planning Committee to consolidate the plans prepared by the Panchayats and the Municipalities in the district and to prepare a draft development plan for the district as a whole.

(2) The Legislature of a State may, by law, make provision with respect to:

(a) the composition of the District Planning Committees;
(b) the manner in which the seats in such Committees shall be filled:

provided that not less than four-fifths of the total number of members of such Committee shall be elected by, and from amongst, the elected members of the Panchayat at the district level and of the Municipalities in the district in proportion to the ratio between the population of the rural areas and of the urban areas in the district;

(c) the functions relating to district planning which may be assigned to such Committees;

(d) the manner in which the Chairpersons of such Committee shall be chosen.

(3) Every District Planning Committee shall, in preparing the draft development plan—

(a) have regard to:

(i) matters of common interest between the Panchayats and the Municipalities including spatial planning, sharing of water and other physical and natural resources, the integrated development of infrastructure and environmental conservation;

(ii) the extent and type of available resource whether financial or otherwise.

(4) The Chairperson of every District Planning Committee shall forward the development plan, as recommended by such Committee, to the Government of the State.

These Constitutional provisions enabled the State Governments to make Panchayats viable and pulsating institutions so that they can prepare economic and social planning with people's participation at various levels emerging from Gram Sabha.

Keeping the above in view, it is expected that for the purpose of formulation of the decentralized plan, the first step is that it should be discussed in the Gram Sabha (GS) for onward consideration in the Gram Panchayat. The plans prepared by all GPs falling under the jurisdiction of the Panchayat Samiti (intermediate tier) are to be assembled at its level with the purpose to have detailed discussion on the plan proposals received from the various GPs and if there is any gaps or lacunae in the plans prepared by various GPs, they should be sorted out.

The plans so prepared by all the Panchayat Samitis of the district will be sent to the Zilla Parishad with the purpose of discussion at its level and if there are some gaps these should be sorted out. In this way, plan for the entire rural areas of the district will be prepared by the ZP. In a similar fashion plans of the urban area will be prepared by the Urban Local Bodies of the District. The plans so prepared by the urban local bodies will be integrated by the District Planning Committees. These Committees are expected to prepare the plan for the entire district and to send a draft plan to the State Government. This is the basic constitutional provision for the decentralized planning.

But even after more than a decade of the conformity legislation by the States, as many as 10 states/union territories are yet to constitute these Committees. Table 11.3 depicts the progress made towards constituting DPCs across the States and the Union Territories. It is clear from this Table that only twenty States/Union Territories, namely Bihar, Chhattisgarh, Haryana (only in 16 out of 19 districts), Himachal Pradesh (only in 6 out of 12 districts), Manipur (only in 2 out of 4 districts), Karnataka, Kerala, Madhya Pradesh, Orissa, Rajasthan, Sikkim, Tamil Nadu, Uttar Pradesh, Uttranchal and West Bengal, and four Union territories, namely, Andaman & Nicobar Islands, Daman & Diu, D&N Haveli and Laksadweep, have taken action in this regard. These bodies are not functional in the State of Uttar Pradesh as revealed by the Table. Moreover, out of 20 States/Union Territories, where these Committees have been constituted, only in 11 States/UTs, the chairperson of the Zilla Parishad has been made the Chairperson of this Committee. In the rest, either a Minister or a bureaucrat has been made the Chairperson of these bodies.

Issues for Discussion

As per the Constitution, Panchayats are the institutions of Self-Government. It implies that these institutions should be autonomous at their level functionally, financially and administratively. Keeping this in view, some issues have been identified for comprehensive and in-depth discussion.

1. It is clear from above discussion that except in one or two States, powers in terms of the (Triple Fs) to the Panchayats to the extent to enable them to function as institutions of self-government has not been devolved even after more than a decade of coming of the Central Act into force. Is there any inherent deficiency in the Central Act which impedes the process of making these bodies autonomous and for ushering in the era of good governance in rural areas?
2. There are two constitutional bodies, namely the State Finance Commission (SFC) and the State Election Commission (SEC), established in compliance with the Central Act in the States. The role of these bodies is very important in bringing continuity, certainly and strength to these bodies. The experiences of the functioning of these institutions across the States/UTs may be highlighted and deliberated upon.
3. The Extension Act is in many ways an improved version the Central Act as far as the functional jurisprudence of the

Panchayats is concerned. The experiences of the implementation of this Act in the Fifth Scheduled Areas may be discussed. The important concern in this connection is as to what extent have the conformity legislations of the States, like the Forest Act, the Excise Act, the Mineral Act, etc., been amended and the Tribal sub-Plans been dovetailed with the plans prepared by the Panchayats to enable the tribals to decide about the various facets of their society and economy with people's participation.

4. The DPCs have not been constituted in all the States even after more than a decade of the coming into force of the Central Act. What are the constraints in the process to constitute these committees before those States which have not constituted these so far may be discussed ? Composition and chairpersonship of these Committees are also relevant issues to be debated in the Conference.
5. The 73rd Amendment to the Constitution has given an enabling legal framework for promoting self-governance at district and sub-district levels. Although the 11th Schedule of the Amendment contains a list of 29 subjects (ranging from Agriculture to Maintenance of Community Assets) which the State may devolve to the Panchayats, it is silent on the important issue of speedy delivery of dispute resolution and inexpensive justice to the villagers. How could, without judicial functions, the Panchayat becomes an institution of self-government, as envisaged in the Article 243G of the Constitution? In view of the experiences gained so far in the functioning of the Panchayats, delegates may deliberate upon the feasibility of adding this responsibility to the Panchayats. It becomes more important to discuss this issue in the light of the recommendations of the 114th Report of the Law Commission of India.
6. The Gram Sabha is the head and heart of the entire scheme of local governance because it provides opportunity to each and every villager to partake in the decision-making process of the village development. There is direct relationship between its proper functioning and strengthening of the Panchyati Raj System in the country. The village plan which will be dovetailed with the district plan through intermediate and apex tiers of the Panchayati Raj System, is also to emerge from this institution.

The experiences about the functioning of this body across the States may also be discussed and debated?

PART IV
Marginalised Groups and Women

As indicated elsewhere in the paper that reservation for SCs, STs and women in the Panchayats for membership and chairpersonship has been provided by the Constitution. As a results of this provision, more than 6 lakh SCs/STs and more than 10 lakh women have been holding the offices of the members and chairpersons at different tiers of the Panchayats. It is evident from the various studies that the political space which has been provided by the Constitution to these groups in Panchayats and their subsequent exposure to decentralised governance and planning for development has both positive and negative outcomes. We shall discuss these in this part of the paper. The findings of a study of the working of Panchayats in six states conducted by PRIA reveals that: "...25 per cent women notice and remark on the visible change in their status within their family after they have been elected... about 60 per cent of women said that they would encourage women to stand for election. The same percentage (60 per cent) is contemplating to contest PRI election again" (PRIA, 2000).

On this issue, the Working Group on Decentralised Planning and PRIs observed that "One of the significant achievements of the provisions of the 73rd Amendment Act concerning reservation of seats and political offices in favour of women and the disadvantaged sections of the rural community is that it had improved their awareness and perception levels and has created an urge in them to assert their rightful share in the decision-making process at the local level" (Working Group, 2001: 47).

Thus, the affirmative action for these groups in local governance has resulted in social identities and political awareness among them and created an urge to become part of the mainstream political, economic and social life. With initial clashes between the 'higher' castes and the 'lower' castes, there are indications of social cohesion at local levels. The political space given to marginalised sections has, to some extent, dealt a blow to the asymmetrical social structure at the local level and given greater space for their participation and involvement in decision-making (Mahi Pal, 2001:1783).

But at the same time, the experiences of the functioning of Panchayats revealed that the caste prejudices emerged as a major stumbling block in the way of the functioning of the Panchayats particularly for those who belong to the marginalized sections of rural society. This, in fact, is the outcome of the unwillingness and grouse the dominant castes hold for having themselves become ineligible to share the powers and control they have long been used to in the PRIs, due to the constitutional provisions for the *Dalits*. Due to the prevailing strangle-hold of the caste structure in rural society in the country, neither the respect for the office of elected representatives nor the simple social values of giving equal regard to fellow human-beings, impel the fellow villagers and the fellow elected representatives to treat *Dalit* elected representatives as equal during the course of their functioning under the PRS. This has resulted the paradoxical situation, where, on the one hand, Panchayati Raj Act provides *de jure* powers to the office of the Sarpanches and, on the other, *de facto*, they remain bereft of these powers. The decentralized bureaucracy, which is expected to work under the guidance and supervision of the elected representatives of the Panchayats, is either generally away from the scene or succumbs to the pressure of the village politics and power game (Mahi Pal, 2004:3582).

But all has not been lost, and as they say, every cloud has a silver lining. The PRS has been instrumental, to a significant, in igniting the process of releasing the depressed, oppressed and suppressed energy of the these groups who got the opportunity to come forward as elected representatives. It was found that wherever the *Dalit* elected representatives were oppressed and obstructed by the dominant castes, they came out openly to resist and to struggle against the oppressors. Importantly, it was also found that whenever the women Panchayat leaders were literate, they were found to be more assertive than the others. The other side of the phenomenon is that the elected representatives of these Groups, especially the educated ones among them, had become quite visible, assertive and vocal whenever the circumstances allowed. It may be treated as the beginning of the end of the invisibility of these sections in the local governance scenario. This process of their becoming visible, assertive and empowered can be hastened by enlarging the scope of their functions and building their capabilities to function effectively through appropriate training and exposure opportunities (Mahi Pal, 2004: 3583).

Besides social assertion by the established forces to check the process of empowerment of these sections people have also approached the courts to scuttle the participation process and, interestingly, the States' attitude remained quite lukewarm in this regard. All the States and UTs covered under the Central Act have constituted Panchayats except Arunachal

Pradesh, Jharkhand and Pondicherry. But elections were not held in time in a number of States. It has, however, been seen that, in most of the cases, the delay in holding elections was due to court cases, natural calamities or law and order problems. The working group on decentralised planning and Panchayati Raj for the Tenth Plan observed that: "... most of the court cases, grievances of aggrieved parties related to reservation of seats for SCs/STs/OBCs/women, delimitation of constituencies, percentage of reservation of chairpersons of lower tier to higher tier, etc. In most of the cases, State Governments have not pursued these pending court cases vigorously for their early hearing so that a final decision of the courts become available as early as possible" (Working Group on Decentralised Planning and PRIs, 2001:9). It reflects the apathy of the States in ensuring early participation of women and marginalised groups in local decision-making bodies at the local level.

Issues for Discussion

Marginalised groups and women have been given space through the PRIs to have share in the functioning of rural local governance. On the basis of the experiences gained so far, following issues may be addressed in the discussion:

1. It is evident from the experiences gained so far that most of the court cases pertained to the implications of various Article and Clauses of the Constitutions and other relevant State Acts of different States/UTs. The merits of the issues on which the various layers of court have been approached by the respondents may be discussed.
2. What are the obstacles coming in the way of proper conduct of the business of Panchayats at the village, block and district levels may be discussed to arrive at a viable package for capacity building of these categories of the elected representatives.
3. It is also important to know whether these people are aware of the schemes of the development of marginalised groups and women and the extent of their implementation across the States.
4. In the context of the LPG which is based on the principle of 'survival of the fittest'; seeing the socio-economic profile of the women and the marginalised groups, they cannot be designated as the fittest for the market forces. The issue is as to what are the economic factors and forces that constrain these groups from participating in the democratic processes and mode of local governance.

5. It may also be discussed as to whether, to what extent, the reservation for these groups has lessened the caste and gender inequality across the States/UTs.

PART V
Civil Society Organisations and People's Participation in PRIs

The Article 243G of the Constitution envisages Panchayats to be autonomous in governance at their level and also to prepare plans for economic development and social justice including 29 subjects listed in the 11th Schedule. This has become more important in the light of the fact that most of the elected representatives are not only illiterate but also hail from the marginalized sections of rural society. Not only this, most of the women elected representatives of the Panchayats have been represented by their husbands and other relatives. The problem scaled up so much that the then Ministry of Rural Areas and Employment had to issue a letter to the Secretaries, Panchayati Raj and Rural Development of the States on May 20, 1997 stating that: "…it has been brought to our notice by a number of organizations and individuals including Hon'ble Members of Parliament in various forums, that quite often women representatives are not able to carry out their functions effectively and independently, if they are sought to be influenced by members of their family such as their husbands/brothers, etc., who may be accompanying them to attend meetings of Panchayats. An Hon'ble Member of Parliament has suggested during the meeting of Consultative Committee of Parliament attached to the Ministry of Human Resource Development, that women representatives and that their husbands by attending the meetings of Gram Panchayats diminish their role and status… the State Governments are requested to take necessary steps including this one and issue the guidelines to the Panchayats at all levels to ensure that the women members of PRIs are not reduced to figure heads or proxies." Indeed, there has not been much change in the participation of women in local governance.

Keeping in view the above background of the participation of the people in the PRIs, the role of the Civil Society Organisations (CSOs), comprising both registered and un-registered (in other words, NGOs and CBOs), assumes a critical role in making people aware about the rationale, responsibilities and rights of the elected Panchayat leaders in the functioning of local institutions. During the early phase of the implementation of the 73rd Amendment to the Constitution, not much emphasis was laid on the involvement and participation of the civil society in strengthening the PRIs. The M.L. Sighavi Committee, on whose recommendations the 73rd Amendment to the Constitution could see the light of the day, had this to say about the role of the civil society: "The Committee is of the view that the proposed National Institute of Local Self-Government should, in respect of its research and evaluation functions, utilize universities, research bodies and voluntary organizations, such as All-India Panchayat Parishad, which should have a consultative nexus with it and which would be an invaluable resource for dissemination of information as well as for proper motivation" (Committee for the Concept Paper:13). But the All-India Panchayat Parishad has not done much constructive work in this regard because it had itself been unable to sort out its internal problems, what to think about solving the problems of the Panchayats by way of facilitating the process of devolution and capacity-building.

However, some of the CSOs have played effective role in strengthening the grassroots democracy. The role of the Kerala Shastra Sahitya Parishad in strengthening the PRIs in Kerala is a folklore in this context. Besides organizing training camps on various issues of decentralized governance, planning and development, the Parishad has published a dozen hand-books like Guides to Panchayat Development, collected secondary data, written local history, developed an innovative approach to agricultural development, soil and water management, environmental cleanliness, etc., for mass consumption. "It were these activities, both human resources and the study materials, that acted as the spring board to the people's planning campaign" (Parameswaran, 2000:13). Besides, there has been the formation of the Statewide Voluntary Technical Corps (VTC) comprising 8,000-10,000 technically experienced persons recruited to examine and vet the projects prepared by the Panchayats. These VTCs have examined between 1 and 1.5 lakh projects prepared by the Panchayats for implementation during the 9th Plan period (Vargease, 1997). Similarly, various registered organizations have played their role in capacity-building of the Panchayati Raj functionaries through creating awareness about the Panchayats, imparting training about their rights and responsibilities, sharing information with them, helping them in formulation of village, block and district plans and activating the institution of Gram Sabha.

The role of the CSOs in empowering the Panchayats is undoubtedly very important. But, at the same time, these organizations have also to put their houses in order to be able positively to influence the PRIs by way of inducting participatory planning for development of villages and minimizing local interest based conflicts of the contending parties and persons.

Another important issue pertains to the respectability of the PRIs themselves in the eyes of the villagers. In this connection, the Mid-Term Appraisal of the 9th Plan has this to say: "...the Panchayat is not seen as a very valuable organisation by villagers; the main reason is perceived lack of individual benefits, also lack of transparency and lack of influence and a high presence of corruption. People place much higher value on other development activities which bring a larger number of individual benefits and are more transparent" (MTA, 2000:187). The Appraisal further says that: "The savings group has a purpose. All women meet to deposit money and to decide on who can take loans. We will get benefits. There is no work for the individual in the Panchayat, and especially not for women. What will we gain from participating there? Whether we go or not does not make a difference" (MTA, 2000:187).

In the context of people's participation in Panchayats' elections, a brief comment on the process of elections to the Panchayats in Haryana which were held in April this year appears relevant. The voter turnout was lower as compared to the 2000 Panchayat elections. The decline in the voters' participation can be attributed to two main factors: (i) voters being busy in harvesting, and (ii) their indifference towards the election process. The scene before the nominations was very interesting and deserves description for readers. The elections were notified on March 17 and the period for making nominations was two days, i.e., March 18 and 19. There was very less time for the people, particularly those who hail from the marginalised groups, even to toy with the idea of contesting Panchayat elections. So everything happened in a hurry. It has been gathered from the people that had more time been given for nominations, more serious candidates might have filed the nominations. Further, there are large number of discrepancies in the delimitations of the wards of the Panchayats. It was surprising to note that members of a family had been distributed among various wards of the Gram Panchayat as voters which was not as per the existing rules. When villagers were asked why did they not raise objections, their answer summarily was: "what to do". Is it not the symptom of growing apathy and amnesia among people about the administration and Panchayati Raj System itself? Another interesting feature of the Panchayat elections in the State was that a large number of wards of GP were lying vacant as nobody had cared to file his/her candidature for these seats. For example, as many as 66 wards in Gurgoan district and 43 wards in Sirsa district were lying vacant as no one came forward to file his/her nomination. This has also happened in other districts. Seeing this state of affairs, the SEC has directed the Returning Officers for allowing the filing of nominations for these wards on March 31, 2005 and April 1, 2005 and elections originally scheduled for April 3, 2005 were be held on April 9, 2005 along with second phase. Does it not it show the apathy of the people for the post of Panch, arising in all probability from their experiences of their negligible role in local governance previously. The reservation for women and SCs has also been mostly nominal rather than real. The social forces in terms of dominant castes and patriarchy in terms of manhood hold the balance in the rural setting. The representatives of SCs and women were rarely contesting elections on account of an urge from them and from their community but largely on account of the constitutional compulsion for this. The dominant sections of the villages wanted to conduit power through them as they themselves has been made ineligible to contest elections by the Constitution (Mahi Pal, 2005: 6).

In such a situation, there is a big challenge before the civil society to create an environment where everybody in the village sees a positive future in participating in the decentralized democratic channels opened by the 73rd Amendment Act through the PRIs. For this purpose, the CBOs' involvement and participation is very important.

There are umpteen associations and organizations in the forms of Mahila Mandals, Youth Clubs, Parent Teachers Associations, Self-Help Groups, Farmers Clubs, Health Committees, Education Committees, etc., which have been created by the State and constitute the components of civil society. But these are mostly lying inactive. Activating these institutions for strengthening the PRIs is strategically most critical because there is a positive relationship between the former and the latter as the various studies and appraisals have revealed. The experiences of the Empowerment and Sustainable Development Initiative (ESDI) an NGO working with Panchayats, is worth recalling here. The experiences revealed that neither the Sarpanches and Panches nor the electorate were aware about the meaning, significance and role of the Gram Sabha in the development of the village economy and society. The ESDI has been making efforts to strengthen grassroots democracy through the participation of the members of the Self-Help Groups in local institutions. The process of organisation of the meetings of Gram Sabhas, show that women who were members of the Self-Help Groups did participate in the meetings of Gram Sabha and were instrumental in starting social activities such as cutting and tailoring training centres for girls in the villages. Table 11.7 given below shows the participation of women who were members of SHGs, Mahila Mandal and others in the meetings of the Gram Sabha.

Table 11.7: Extent of Participation of Women in the Meetings of the Gram Sabhas

Sl. No.	*Name of Gram Panchayat*	*Date of Meeting*	*Participation*			
			Male	*Female*	*Total*	*%age of Women Participation to Total*
1.	Raipur Roran	2nd May 2004	35	30	65	46
2.	Poojam	2nd May 2004	15	25	40	63
3.	Butana	4th May 2004	20	25	45	56
4.	Arjaheri	7th May 2004	15	25	40	63
5.	Bairsal	14th June 2004	5	30	35	86
6.	Sandhir	15th June 2004	30	35	65	54

Source: Gram Sabhas put Social Development Agenda before Gram Panchayats, ESDI, 2004.

It may be seen from the above Table that women participation in the meetings of the Gram Sabhas was more than 50 per cent in one village. It was even more than 80 per cent in one village. This could have happened due to the initiative taken by the SHGs and, in turn these SHG members could take with them other women of the village because women felt comfortable in groups to attend the general meetings.

Here, it is important to note that when the SHGs have been formed through a proper social mobilization process, then only do these groups have the potential to make the PRIs vibrant and pulsating institutions. Secondly, there is need to promote social capital formation by forming and nurturing the SHGs and, networking of these for effective functioning of the PRIs. Our experiences show that with some exceptions (for example, Kerala) the stock of social capital in India is very low and whatever is available is lying latent. In fact, social capital is an asset that remains dormant until agencies activate this stock and use it to generate a flow of benefits to the people. Hence, the more capable NGOs should work towards activating various components of the social capital for not only strengthening the PRIs but also for activating the idle physical and financial capitals lying dormant in rural areas. Seeing the significance of social capital, the 10th Plan Document, in its Chapter on Poverty Alleviation in Rural India-Strategy and Programmes observed that: "The promotion of a movement which enhances social capital and forges linkages with other formal and informal stakeholders engaged in developmental activities would be a major thrust during the Plan". It may be mentioned here that, consequent upon the implementation of the 73rd Amendment Act, there is at least one elected representative in every street of the village across the country. In this perspective, it is easy to see that the NGO sector has a big challenge before it because if this sector does not have good rapport with the Panchayats or if the latter do not appreciate the positive worth of the work of voluntary sector, the elected head of the Panchayat may take a view that there is yet another head of the Panchayat meddling his/her work. As the work of the Panchayats is expected to increase in the years to come and the role of the voluntary sector is also expected to increase keeping in view the circumstances which are either already in vogue or are definitely emerging. In this situation, it is imperative that the Panchayats must be motivated to consider the CSOs as facilitators and not as interferers in their sphere. In fact, the sector should be both motivated and matured sufficiently to harness the full benefits of the environment which has been created by the government at the local level for the local people.

Hence, in order to enable the civil society to meet the emerging challenges in the wake of the added responsibilities and powers of the Panchayats, this sector has to enhance its capacity and equip itself suitably to have an in-depth understanding of the main issues of conflict among the Panchayats and resolve these effectively. In view of this, a long-term strategy and investment in their capacity-building is needed. Indeed, expertise of a different kind is required, strategies of a different nature need to be designed and personnel with different attitudes and orientation need to be mobilised to tackle the problems of the Panchayats. In the context of the role of CSOs in the near future, what the then Prime Minister had said about the national development while addressing a conference on this sector in 2002 is worth recalling here: "I would like to liken national building to a chariot that is driven by five horses. These are: the Central Government; the State Governments; Panchayati Raj Institutions; the private sector; and, last not the least, voluntary orgainisations and community-based groups. The chariot will run fast and in right direction only when all the five horses run in tandem" (All India Conference on Voluntary Sector, 2002: 24). What the Task Force on PRIs said while outlining the possible role of the NGOs and PRIs is worth quoting, too, in this context: "The NGO-PRI relationship should not be adversarial, but complementary. NGO-PRIs relationship would have to be based upon one basic premise that neither NGOs nor the PRIs should consider the other as contenders to their respective position. Primacy of the Panchayats, as a distinct tier in the representative system of governance has to be accepted by the NGOs. Similarly, the superiority of the NGOs in terms of access to knowledge, latest technological advancements as well as their capacity to deliver has to be appreciated by the PRIs" (Task Force of PRIs, 2001:77). In short, there is an imperative need for adopting collective and a collaborative approach by the civil society organizations for strengthening the PRIs.

The PRIs and the CSOs can help each other in providing good governance in the areas of: (a) capacity-building of elected representatives of the Panchayats through imparting training which includes skill, knowledge and attitude to them; (b) information-sharing with the Panchayat functionaries; (c) facilitating efforts of the Panchayats in the preparation and implementation of decentralised planning; and (d) facilitating the process of holding meetings of the Gram Sabha which is the head and heart of the entire PRS.

In the above context, the Participatory Research in Asia (PRIA) has been working with the PRIs in Haryana. This organisation has been mainly performing two activities:

(a) Spearheading the facilitating process of holding the meetings of the Gram Sabhas with the cooperation and networking of various Civil Society Organisations (CSOs) across the State. Before holdings the meetings of the Gram Sabhas, the PRIA organizes exposure meetings with the CSOs in which they are sensitised about the provisions and implications of the role and function of the Gram Sabha. Posters and pamphlets indicating the functions and powers of the Gram Sabha and role of the electorate are distributed among the participant organisations.

(b) PRIA is also intensively working in two Blocks, one in Fetahabad district and another in Mahendra Grah district for building the capacity of elected representatives, particularly the women, and the Civil Society Organisations in order to strengthen their functioning as institutions of proactive public participation in effective local governance. The formation of Nari (Women) Network comprising the members of erstwhile Sanjivani (the village level worker under the Government-sponsored project for women empowerment), the elected women Panchayat Member, Aaganwari worker and other active women from the rural and urban areas has been an innovative way of using the CBOs for the empowerment of women economically by taking up gainful activities and politically through participating in the various forums of the PRIs.

Indeed, there is, a great role of the CSOs in facilitating the efforts of the PRIs in preparing plans for economic development and social justice at their level while, at the same time the PRIs provide enabling environment for the fuller functioning of the CSOs. It has been found that where the NGOs intervened for facilitating the proper functioning of the Panchayats, better results were seen. In cases where the NGOs have formed the Self-Help Groups, the effectiveness of the Gram Sabha and the entire Panchayati Raj System has been enhanced through the more active and enlighted participation of the SHG-based members of the Gram Sabha and Gram Panchayat.

Issues for Discussions

In the foregoing paras, we have tried to spell out some aspects of the relationship between the CSOs, the people's participation and the strengthening of PRIs. This discussion has unfolded a number of issues for debate. Some of these may be formulated as follows:

1. The experiences of the relationship between CSOs (including also the NGOs and the CBOs) and the Panchayats across the States may be discussed.
2. It has been noticed that the Panchayati Raj System has functioned better where the CSOs have worked shoulder to shoulder with the Panchayats. But the other side of the phenomenon is that if the CSOs are not capable enough, they were also instrumental in creating problems *vis-a-vis* the Panchayats. Hence, capacity of the CSOs has to be built through a proper, well modulated and intensive training. What should be the strategy of capacity-building of these organisations, this needs adequately to be deliberated?
3. There are a number of Voluntary Organisations working at the micro, meso and macro-levels and performing virtually same sort of activities. Hence, the subsidiary principle of allocation of activities which was applied and debated a lot in the context of the PRIs, needs also to be equally applied on the case of the NGOs.

4. There are various forums and committees which are to be constituted by the Panchayats in accordance with the Panchayat Acts. These committees are very useful instruments to involve members of the Panchayats as well as other persons and organisations including voluntary organisations. Therefore, the experiences of the functioning of these Committees and associations of voluntary organistions may be discussed in depth.
5. The role of the SHGs is very important in promoting effective local governance as the experiences in a multiplicity of cases provided evidenced in this regard. The SHGs movement has been more widely and intensively spread in the southern parts of the country. Specific experiences of these states may be discussed in the conference and relevant lessons be derived therefrom.

PART VI
Access to Common Property Resources

As mentioned elsewhere in this paper, the Panchayats shall prepare plans for economic development and social justice at their respective levels. For this purpose, they need resources. No doubt, the Panchayats have been getting funds as a part of the various Central and State sponsored schemes. But these alone cannot enable them to become autonomous institutions of self-governance as they are quite often tied and conditions-laden and subservient to the granting Government's mandates. Hence, they must mobilise their own resources so that they can have the freedom to chalk out plans as desired by the Gram Sabha. Besides, the access of the common property resources (CPRs) to the marginalized groups for meeting their variegated basic needs, much as those for fuel, fodder and food or their livelihoods and sustenance is very important. Therefore, proper management of the CPRs determines significantly the welfare and development benefits of the marginalised groups and women and critically affects the participation of the weaker sections in the functioning of the Panchayats.

In the context of the CPRs, we shall now discuss in brief the scenario, status and experiences in the Haryana. As per the latest information available, out of the total cultivable common land, about 9 per cent has been encroached upon, mostly by the upper castes who have both muscle and money powers, and as many as 6961 cases were pending in the courts as at the end of the financial year 2002. It is interesting to note that, instead of removing the encroachment, the Government of Haryana has regularised the same. "The Government of Haryana has now regularized the encroachments upto March 2000 under a new policy. The encroachers were allowed to purchase such land which do not fall on the path-ways at the market rate. Gram Panchayats were empowered to pass resolutions to allow such purchases. The encroachers had to apply to the concerned Gram Panchayat. But very few of them came forward to do so. Since the traditional Panchayats have taken a back seat, elected (statutory) Panchayats are unable even to execute the decision made by the courts regarding common lands in favour of Panchayats. Though a full time Gram *Sachiv* has been deployed in the Gram Panchayat, even then they are unable to maintain the records of Common Property. This is what helps the encroachers. The Development & Panchayats Department has full time legal officers at the district level to defend the cases of the Panchayats in the Civil Courts but they have not been able to save the Village Common Lands from encroachers. Needless to say that their role requires to be reviewed" (Mor, 2004:371).

The Haryana Institute of Rural Development (HIRD), in collaboration with the National Commission for Women, organised a Conference of *Dalit* Women presidents and members of the PRIs in March 2004 in which 188 participants participated. One of the themes of the discussion was the management of CPRs. From the discussion, it emerged that CPRs, particularly the common land, are not being properly managed. Most of the *Sarpanches* said that some portion of the common land has been encroached upon by dominant sections of the villages. They pointed out that they had been trying their best to remove the encroachment and had succeeded in some cases, but, by and large, the encroachment persist and they have been still struggling for removing these encroachment. In this context, one *Sarpanch* from Mahendragarh district said that the district administration was not taking action against the encroachers in her village and a *phirni* (circular road) of the village could not be completed because persons of the dominant caste have encroached upon a piece of land falling in its way. When asked why she did not approach the Deputy Commissioner, she replied: "met the Deputy Commissioner along with Panchayat members on many occasions, but he did not take any action to remove the encroachment." Similar stories were also narrated by *Sarpanches* from other Districts of the State. The broad conclusions emerging from the interactions in this group were: encroachment on the land both outside and in the Abadi Deh was prevalent; women representatives were not aware about the nature and extent of CPRs and their management and legal provisions to deal with various categories of common land; there was lack of proper planning of CPRs; elected women representatives have made some efforts to develop and to conserve CPRs; and, in most of the cases, *Dalit* elected representatives failed to remove the encroachment due to the unfriendly attitude of the higher castes and the bureaucracy.

Issues for Discussion

The issue of CPRs is very important not only from the point of view of the livelihoods of the weaker sections but also for enabling these people to play their role effectively in the delivery of local governance. Possible points for discussion on this issue may as follow:

1. Here in this part of the paper, the experiences pertaining to the management of CPRs in Haryana have been discussed. The discussion should be compared and supplemented by a review of the happening in the other States on this score.
2. As indicated elsewhere in the paper, the *jal* (water), *jangal* (forests) and *jamin* (land) have been given to the villagers of the Fifth Schedule areas of the country as per the Extension Act. Specific concerns in these States may be brought to the fore in the discussions.
3. Is there any deficiency in the formulation of the various Acts and regulations meant to manage the CPRs or is the implementation machinery defective and, therefore, unable to watch over and defend the interest of the Panchayats and their elected representatives especially those belonging to the marginalised groups?
4. There might have been a number of successful experiments on the proper management of the CPRs across the States. The same may be shared with the participants for discussing the applicability and advisability of these experiments in different socio-cultural-*cum*-political settings.

PART VII
Suggestions and Policy Prescriptions

In the foregoing discussion a great deal of the analysis has been devoted to assessing the extent to which the Panchayats have emerged as institutions of self-government at their respective levels; the extent of people's participation, particularly that of the marginalised groups and women, in the functioning of the Panchayats; the role of the civil society organisations in facilitating the process of strengthening the Panchayati Raj Institutions; and the extent of access of the poor to the CPRs. The centrifugal concern and focus of this analysis was the degree of effectiveness of local governance. The analysis of these variables gave sufficient evidence that the Panchayats have not been playing their role as effectively they were expected to do. The question, therefore arises: what should be the future line of action to enable Panchayats to function as autonomous institutions to entail and ensure good governance in rural areas. In this part of the paper, we shall dwell on this issue.

1. Article 243E of the Constitution does not clearly mention the duration of the term of the office-bearers of Panchayats as five years, whereas, for Panchayats, the provision explicitly is "Every Panchayat, unless sooner dissolved under any law for the time being in force, shall continue for five years from the date appointed for its first meeting". For harmonious and meaningful functioning of the Panchayats, it may be suggested that the normal term of the Adhyakshas should be co-terminus with the life of the Panchayats. While most of the States have made the terms of office-bearers co-terminus with the term of Panchayats (i.e., five years), States such as Maharashtra and Karnataka and Union Territory of Andaman & Nicobar Islands have prescribed lesser period for the offices of Chairpersons of Panchayats. A tenure which is less than five years for the office-bearers of a Panchayat is particularly disadvantageous for the marginalized sections to build-up their capacity and experiences as local leaders. Hence, it is suggested that a provision may be added to Article 243K that the term of the Members and the Chairperson shall be co-terminus with the five year term of the Panchayats for harmonious and meaningful functioning of the Panchayati Raj Institutions.
2. As per Article 243E(1), "Every Panchayat, unless sooner dissolved under any law for the time being in force, shall continue for five years from the date appointed for its first meeting and no longer." In contrast, Article 243U(1) of the 74th Constitution Amendment for the Urban Local Bodies makes a specific provision to the effect that the elected body shall be given an opportunity to be heard before such dissolution. The Article reads, "Every Municipality, unless

sooner dissolved under any law for the time being in force, shall continue for five years from the date appointed for its first meeting and no longer:

Provided that a Municipality shall be given a reasonable opportunity of being heard before its dissolution."

It is clearly discriminatory in favour of the urban bodies, fairness demand that a similar provision should also be made in the Article 243E of the Constitution relating to Panchayati Raj Institutions.

3. Section 243D spells out the provisions of reservation for the SCs, STs, Women and Backward Classes. In the case of the SCs, STs and Women, explicit provisions have been made in the Central Act itself whereas for the Backward Classes an enabling provision has been which says, "Nothing in this part shall prevent the Legislature of a State from making any provision for reservation of seats in any Panchayat or offices of Chairpersons in the Panchayats at any level in favour of backward class of citizens." Due to this provision, different States have adopted different mode of providing reservation to Backward Classes. For example, in Haryana, one seat is reserved at each level of Panchayati Raj for backward classes at all levels where as there is no reservation for Chairpersonship at any level of the Panchayati Raj System. In the State like Karnataka, in comparative contrast, reservation for the post of Chairperson is provided for in the Panchayat Act for these classes.

   In view of above, it is suggested that the provisions of reservations for the Backward Classes should also be made mandatory as has been done for the SCs, STs and Women.

4. Article 243C of the Constitution made various provisions pertaining to the composition of the Panchayats and mode of elections of the Chairpersons at different levels. Various petitions were filed in the Courts in the States particularly in Kerala and Karnataka. Article 243(4) states, "The Chairperson of a Panchayat and other members of a Panchayat whether or not chosen by direct election from territorial constituencies in the Panchayat area shall have the right to vote in the meetings of the Panchayats." Article 243(5)(b) says, "A Panchayat at the intermediate level or district level shall be elected by, and from amongst, the elected members thereof". It is clear from above that in the election of the chairperson at intermediate

and district levels, only directly elected members are eligible for choosing the Chairpersons. However, for other matters which were brought out before the intermediate and apex tiers of the Panchayats, all Members whether or not chosen by direct election are entitled for voting. Hence, for explicitly mentioning the intention of the Article 243(4), it may be reworded as follow: "The Chairperson of a Panchayat and other members of a Panchayat whether or not chosen by direct election from territorial constituencies in the Panchayat area shall have the right to vote in the meetings of the Panchayats except in the election of the Chairperson of that tier." Constitution may be amended, accordingly.

Another issue is the election for the office of a Chairperson reserved for the SCs, STs and OBCs categories. This important issue cropped up in West Bengal and Karnataka. The point is whether or not can an SC or ST or OBC candidate elected from general or unreserved seat for any tier of Panchayati Raj System (PRS) be a candidate for the post of Chairperson of such tier reserved for the SC or ST or OBC including women belonging to these groups. In this context, it is suggested that in order to remove doubts, if any, an enabling clause in Article 243D may be added, which may be worded as follow: "A person elected from the general seat at any level of the Panchayati Raj Institutions is eligible to contest for the post of Chairperson of any tier reserved for a category provided the person concerned belongs to that category.

5. Conducting regular and timely elections to the Panchayats is necessary to ensure continuity and sustainability of the PRS in the country. There is a mandatory provision in the Constitution (Article 243E) for conducting election every five years. And there is a mandatory provision (Article 243K) for constituting State Election Commission (SEC) for "superintendence, direction and control of the preparation of electoral rolls for, and conduct of, all elections to the Panchayats". These powers are adequately comprehensive to enable the SEC to conduct elections in a free and impartial manner. However, Clause 4 of the Article 243K limits the SEC's powers in this connection because it states: "...the Legislature of a State may, by law, make provision with respect to, all matters relating to or in connection with, elections to the Panchayats". These provisions

have clipped the wings of the SEC because it can exercise its powers subject to the provisions of such laws as may be legislated by the State Assemblies. It is well-known that the States have powers pertaining to umpteen important poll matters such as issuance of election notification, delimitation of constituencies, reservation of seats, rotation of reserved seats, etc.

Hence, for strengthening the institution of the SEC, powers relating to the delimitation of territorial constituencies of Panchayats and Municipalities, allotment of reserved seats to various electoral wards and rotation of reserved seats among various constituencies, be vested with the SEC through suitably amending the Constitution.

6. It is evident from the experiences gained so far that the State Governments have implemented the mandatory provisions of the Central Act in letter and spirit and the rest of the provisions, which do, in fact pertain to the power and authority of the Panchayats, in a topsy-turvy manner, as per their discretion and/or expediency. Why have Panchayats which have not been given requisite power because the powers and functions given in Article 243G of the Constitution do not make it mandatory on the part of the State Legislatures to do. Hence, clear-cut demarcation of the functions and powers of the three tiers of the Panchayats should explicitly and definitively be given in the Constitution. There is indeed, the need to evolve a separate list of powers and functions for the Panchayats which may be named as Local List. The local list may be further sub-listed for the lowest, intermediate and apex tiers of the PRS.
7. The Panchayats should have their own service cadre on the lines of the Central and State services. Hence, in order to fully implement the Central Act, another amendment to the Constitution is needed to provide for a Panchayat Service in Part XIV of the Constitution. This seems quite logical to do in view of the fact that, under Article 280 of the Constitution, provisions of measures for augmenting the Consolidated Fund of a State to support the resources of the Panchayats has been added. Similar provision for the Panchayat personnel may be made in the Constitution
8. The Panchayati Raj Acts and their rules should be brought in consonance with the letter and spirit of the Constitution. This is

necessary not only for enabling people to participate in the PRS but also for awarding autonomy to these institutions. There are inherent anomalies in the various State Panchayati Raj Acts. For example, in Haryana, as per Section 121 of the Panchayat Act, there is the provision "On the constitution of a Zilla Parishad…, there shall be called the first meeting for the election of the President and the Vice-President by and from amongst its elected members in the manner prescribed, by the prescribed authority." Here, instead of providing to phase in the manner prescribed by the prescribed authority", it should be the phrase "through secret ballot", to avoid any sort of arbitrariness and political influence during the election of the President and Vice-President of the Zilla Parishad.

The Panchayats have, in fact, been made excessively subservient to control by the State Government, as would easily be evident from a perusal of the provisions of the Panchayat Acts of different States. The example of Andhra Pradesh is worth quoting to prove the point. Thus, in the State Act ",...the word 'government' has been used 316 times and the word 'Commissioner' has been used 111 times in the Act. The clause 'subject to' appears 135 times in the Act.... The word 'prescribed', which means that the Government has retained power through the rules it would frame, appears 179 times in the Act....the Government may, by notification and subject to such restrictions and conditions and to such control and revision as may be specified therein, direct the Panchayats." (ISS)

Hence, there is need to properly define and formulate the roles of different institutions such as the SEC, the SFC, the DPC, as well as the election procedures and the jurisprudence of the Panchayats to enable these to function as *de facto* local government.

9. The Central Act has provided constitutional status to the Panchayats endowing them with the powers of preparation and execution of plans for economic development and social justice, including the subjects listed in the 11th Schedule to the Constitution. But no legal framework has been given at the local level to settle disputes, if and when these arise, in the implementation of the schemes and programmes relating to the 29 subjects listed in the Constitution. Although some states have made provision for constituting Nayaya Panchayats, in most of the cases they are dormant and defunct. The existing courts at

district and sub-district levels have already been over-burdened and pending cases are rapidly piling up further. In view of this, it becomes necessary to create an institution of justice at the local level, having representation of the weaker sections, including women, for dispensing speedy and inexpensive justice to the villagers. For this purpose, a constitutional amendment may be carried out for providing a viable and sustainable decentralised judicial system at the local level itself.

10. People's empowerment is basically based on the basic premises that the people, particularly the marginalised and underprivileged ones, have their say in the system of governance for their betterment on a sustainable basis and thereby reduce their dependence on the centralised governance and bureaucracy. This can *inter alia*, be achieved by building the capacity of the elected representatives in a systematic and planned manner. There should be training courses for orientation, sensitisation and attitudinal motivation of the elected members of the PRIs, particularly the new ones particularly those from the weaker and women segments.

Elections of the Panchayats in all the States have recruited more than 3 millions elected representatives at all levels of PRS. Of them, an overwhelming majority comprises the new entrants, particularly from the marginalised sections. The Article 243G of the Constitution expects from the Panchayats to prepare and implement plans for economic development and social justice including the 29 subjects listed in the 11th Schedule to the Constitution. The versatile experiences of the training courses have thrown up two more critical reasons: one, that training should be a continuous process focussing mainly on the conduct of business, powers and functions devolved on the PRIs, micro-planning, financial management and social development; and, two, initially conducting separate courses for women which would later be followed by joint courses so as to facilitate the process of their overcoming the gender taboos restraining them from sitting and participating alongside their male counterparts. The courses of short duration have not much sensitized Panchayat personnel about their economic and social responsibilities. Hence, the need for longer duration course, preferably at the village or a village cluster level, to facilitate and promote greater participation by all members, especially the weaker and women segments.

The findings of the study on the Accounting System and Training in Panchayats in Rural Sector conducted by the Institute of Public Auditors of India in the States of Kerala, Madhaya Pradesh, Tripura, Tamil Nadu and Uttar Pradesh are worth recalling in this context: "It seems that most of the staff members in Village Panchayats are not qualified to keep accounts. They carry out their duties through experience and knowledge acquired in practice... State Planning Boards and State Institutes have been set up for imparting training in Kerala, Tamil Nadu, Tripura and Uttar Pradesh. Their training programmes were inadequate. In Madhya Pradesh, only 18 per cent of functionaries have received some training in budget, accounts, finance ...In Tripura, none of the members of Indira Nagar Village Panchayat was trained except the Panchayat secretary....In all the States there has been complete lack of awareness among the elected members of Panchayats about their democratic rights, duties, entitlement of funds and that actually received, etc. The Panchayat members were not aware of their rights to impose taxes within the Panchayat areas to improve their financial position. The Village Panchayat members were treating the money received from the block Panchayats and other agencies as the money received from the Government and not their own money....The records in various States disclosed that either the meetings of Gram Sabha are not held at all or are held infrequently leading to lack of people's participation in plan formulation. This resulted in selection of unproductive works, creation of non-durable assets, construction of *Kutcha* Link Roads and brick lanes (*Kharanja*) and earthwork by a handful of functionaries, thereby leading to the entire expenditure not being of optimum benefit to the society" (IPAI, 2001:35). In view of above, Panchayats' functionaries have properly been capacitated and enabled to formulate Panchayats' plans and budgets, maintain proper finance and accounts/audits.

After having worked in the Central Ministries including the Planning Commission, Training & Research Institutions and Panchayats, the author of the present paper holds the considered view that training/capacity building of the various stakeholders has to be taken up in the action mode. In this connection, an external agency, may be an NGO, initiates the process of training which covers the holding of the meetings of the Gram Sabha, constitution of the various subject/standing committees, proper conduct and holding the meetings of the Gram Panchayats, and establishing and activating the organic link between and among the different tiers. In other words, the Panchayat Act has to be implemented practically in the Panchayats by activating and involving the various stakeholders. Merely imparting training in the classrooms would not be sufficient.

Conclusion

To conclude, two Hindi sayings aptly describe the present status of the PRIs in the country. One:, '*Panchon kee baat seer matha patnala yahin girega*' (We appreciate your point of view, but the *status quo* will remain). Two: '*Ghar-bar sab tumhara, kothi kuthala ke hath maat lagana*' (Everything of the house belongs to you, but do not touch the money and other precious things). There have been a number of meetings and conferences between Central and the State Governments. All-India Panchayat Adhyakshas Sammelans were also held in the past. The result has been as expected. The Central and the State Governments, instead of initiating such policies which could strengthen the PRS, have used these institutions as conduits for distributing funds to the target groups. The Schemes such as the MP Local Areas Development Scheme and strengthening the District Rural Development Agency have been initiated which are in essence and in practice,anathema to the motto of decentralisation through the PRIs. It was also expected that the constitution of the PRIs would create a new leadership at local levels which would not only abjure obscurantism and backward-looking attitude but also would be instrumental in changing the anti-development outlook of the traditional caste leaders in the villages so that the people could achieve their collective aspiration of developing themselves holistically. But the hopes have been belied because there is no urge for decentralisation at the grassroots. Remedy of the malady lies in creating effective demand on the part of the people at the grassroots in terms of asking more powers from the Governments. For this, it is necessary for Panchayats leaders to get organized and to form their associations at the local, district, state and national levels in order to enhance Panchayats' bargaining powers and enable themselves by learning from each other and by extensively deliberating matters of common interest.

REFERENCES

Department of Rural Development (1986). *Report of the Committee for the Concept Paper on Panchayati Raj Institutions*, 1986, pp. 22.

Gram Sabhas Put Social Develcpment Agenda Before Gram Panchayats: An Analysis of the Outcome of the Gram Sabha Meetings Held in Nilokheri Block of Karnal District in Haryana, *Empowerment and Sustainable Development Initiatives*, 2004, p. 32.

Mahi Pal (2000). 'Panchayats in the Fifth Scheduled Areas', *Economic and Political Weekly*, May 6-12.

Mahi Pal (2001). 'Eighty-Seventh Amendment Bill: Where are We Heading', *Economic and Political Weekly*, May 26, p. 1783.

Mahi Pal (2004). Haryana: Caste and Patriarchy in Panchayats, *Economic and Political Weekly*, August 7, pp. 3581-3583.

Mahi Pal (2004). Panchayati Raj and Rural Governance: Experience of a Decade, *Economic and Political Weekly*, pp. 137-143.

Mahi Pal (2005). *Panchayat Elections in Haryana—A Hurried and Sarpanch-Centric Affair, Panchayati Raj Update*, Institute of Social Sciences, New Delhi, p. 6.

Ministry of Rural Development (2001). '*Report of the Task Force on Devolution of Powers and Functions Upon Panchayati Raj Institutions*', August.

Mor, K.K. (2004). Management of Common Property Resources in *Decentralised Governance: Myth and Reality*, (ed.) *Surat Singh*, Deep and Deep, New Delhi.

Parameswaran, M.P. (2000). Role of Kerala Sastra Sahitya Parishad in the Movement for Democratic Decentralisation, paper presented in the International Conference on Democratic Decentalisation, during May 23-27, Thiruvananthapuram.

'Panchayati Raj Institutions—A Balance Sheet', February, *PRIA* (2000).

Planning Commission (2000). *Mid-Term Appraisal of the Ninth Five Year Plan* (1997-2002), October, pp. 175-76.

Planning Commission (2001). '*Report of the Task Force on Panchayati Raj Institutions (PRIs)*' December, p. 9.

Planning Commission (2001). '*The Report of the Working Group on Decentralised Planning and Panchayati Raj Institutions for the Tenth Five Year Plan* (2002-07)', November, p. 9.

Planning Commission (2001): '*Report of the Working Group on States' Resources for the Tenth Five-Year Plan* (2002-2007)', Volume II, January, p. 93.

Planning Commission (2002). *All-India Conference on the Role of the Voluntary Sector in National Development* (proceedings)', April 20, p. 7.

Planning Commission (2002). *National Human Development Report*, March.

Review of State Panchayati Raj Acts: Note on Preliminary Findings for Discussion, Presented in the Workshop on 27th August, 2005 organised by the Institute of Social Sciences, New Delhi.

Sundaram, K.V. (1997). *Decentralised Multi-level Planning: Principles and Practice—Asian and African Experiences*, Concept Publishing Company, New Delhi, p. 514.

The Constitution (Seventy Third) Amendment Act, 1992.

The Provisions for Panchayats (Extension to Scheduled Areas) Act, 1996.

Varghese, A.V. (1997). *Volunteers to Help People's Planning*, Deccan Herald, May 20.

12

Transforming Village Clusters into Territorial Production Complexes

K.V. SUNDARAM

Introduction

In order to conceive a well-balanced model of rural development, economically, socially and environmentally, the Government of India hopes to evolve an Action Plan for making the Tenth Plan, a People's Plan and Development, a People's Movement. The programme announced for attaining this objective through the establishment of Knowledge Powered Village Complexes, known by the acronym "PURA" (Provision of Urban Amenities in Rural Areas). This is a key element of "Vision 2020" and is intended to give a new thrust to the all-round development of viable village complexes through the twin strategies of 'Effective Planning and Implementation of Rural Development Programmes with People's Participation' and 'Effective Development Communication for People's Participation'. The programme PURA is essentially a Habitat Design to achieve prosperous, peaceful and safe human habitat in rural areas, that would improve the quality of life in rural places and lead to "Inclusive Rural Development".

The objective of this paper is to set the PURA Concept in the wider context of Holistic Village Cluster Development Planning to achieve "Inclusive Development". Given the initial push for the vision by the President of India, what is needed today is nothing less than spearheading a nation-wide Village Renewal Movement, based on a wider understanding of the PURA Concept.

The Concept of a Micro-Territorial Production Complex

Our past experience has shown that conventional development strategies, which emphasize the growth of gross national product *per se*, without at the same time looking into area and community specific potentials and needs cannot alleviate mass poverty and unemployment. It is in this context that we advocate layered models of territorial production complexes in our rural areas to trigger off the full utilisation of natural resources and human potentials.

A Micro-level Rural Territorial Production Complex is a model of spatial organisation, envisaged as contributory to developing agriculture and allied activities as well as small and medium enterprises, effectively in the rural areas and to promote the welfare of rural residents. Ideally, such a Micro-level Territorial Production Complex should lead to rapid agricultural development in the villages and also produce and supply various equipments, materials and consumption goods necessary, with appropriate technology for a basic needs strategy, guiding the local development. It will provide the necessary backward and forward linkages and seek to augment the value addition to the agricultural products emerging from the rural areas and thus enrich the income of the local population. It is the linkages and flows, as well as the managerial improvements brought about to aid and facilitate the production activity in the primary villages, through encouragement and management of the production, processing and marketing activities in the functional area of a village cluster, that constitute the hallmarks of the concept of a Micro Rural Territorial Production Complex.

In such an approach, all incentive and subsidy programmes of the Government, as well as management, extension and training activities will be pooled and administered in a concerted manner to assist private initiatives in local areas. It will lead to a territorial division of labour, leading in due course to effective production relationships, determining the spatial structure of the economy at micro and macro-levels.

The strategy implied here must concentrate, more than in the past, on meeting the requirements of the small farmer, the small entrepreneur and the informal sector producer of services. Such a strategy calls for, and is in turn supported by, a special kind of appropriate technology, which will contain a greater element of technological innovation, governed by selective choice and adaptation applied to existing technologies.

In the context of our rural areas, the Micro Territorial Production Complex, as envisaged here, will be a relatively Homogenous Community of Interest Area, i.e. a Village Cluster.

Why a Village Cluster?

In our present development context, the Indian village, with its small population, has ceased to be an economic and social entity for planning. The knowledge century demands a size suitable for the effective transfer of technology. Besides, people would want higher forms of services and of quality in provisions like health, education, agricultural inputs and services etc. There is, therefore, a need to think in terms of 'viable unit areas of development in rural space'. It is in this context that the concept of a 'cluster of villages' comes as a useful grassroots level unit for planning, for organizing rural development programmes. It is not a new idea and has been discussed, since the middle of the 1960s, though no concerted efforts were made to operationalize the approach. Now, with the PURA concept gaining ground, there has been a revival of interest on this approach to micro-level planning.

Evolution of the Concept

A review of the evolution of the thinking on 'village cluster approach' and of its relevance and usefulness as a tool for rural development would be a useful beginning to this discussion.

An early recognition of the concept was, when the First Master Plan for Delhi was prepared under the guidance of Albert Mayer of Etawah community development fame. The Delhi Master Plan adopted a hierarchical pattern of central villages, service centres and rural towns (based on Christaller's concept), for the limited purpose of bridging the rural-urban gap in the provision of social services[1]. A more concrete beginning towards this approach commenced when the Ford Foundation initiated in 1969, a Pilot Research Project on Growth Centres in collaboration with Government of India, envisaging a 'village cluster', around an identified growth centre, as an intermediate areal unit between the village and the block. It followed the typical spatial-*cum*-regional planner's approach of identifying a hierarchy of service centres (based on techniques like scalogram analysis), for extending a broad range of amenities, services and infrastructure to all settlements in the block. Thereafter, the sectoral scenarios of development are developed for agriculture, industry, education, health and transport to create an integrated spatial plan of development. This approach was also recommended by the Working Group on Block Level Planning, set up by the Planning Commission in 1978.[2]

Around the same time, an International Organization—The United Nations Research Institute for Social Development (UNRISD), Geneva, initiated research studies to examine the relevance of 'Growth Poles and Growth Centres' for regional development in the developing countries.

We (Prof. R.P. Misra and myself) undertook a prestigious project funded by this organization to test the relevance of this concept in a most underdeveloped area of India. This led to our field study in 1963 in the Bastar district of Madhya Pradesh, a most backward tribal district. After this experience, we returned with rich insights and boldly stated our conviction that what are required to stimulate developments in such areas are 'Social Development Poles', and 'People Centered Development', rather than 'Growth Poles'. We viewed with great skeptism, the nature of 'modern' developments taking place in the area by external agencies, e.g. the Bailadilla iron ore mining developments, which were sweeping the economy of the tribals, 'off their feet'.[3]

From the above experience stems our criticism about the mechanistic methodology, promoted through the spatial-*cum*-sectoral strategy, advocated by the growth pole researchers which, while using sophisticated criteria for determining the status of settlement centres (e.g. 'Population Threshold', 'Distance Norms', and 'Movement of People and Commodities'), blatantly ignored the inter and intra-community relations in lower and grassroot level settlements and insights derived from the perception of rural contrasts and uneven economic and social development.

Since the 1970s, the concept of rural development has undergone many changes. It is now "Inclusive Socio-economic Development', which inheres in it, the tasks of 'poverty reduction', 'employment generation', 'empowerment', 'equity', 'sustainable development of the natural resources', besides the provision of basic minimum services. It has to be generated from "within", by unleashing the "Capillary Mechanisms" in the village situation, with people deciding their course of development and Government supporting it. In this conception, the core tasks of rural development will include the fuller use of land and water resources without destruction, modernization of farming by extending processing and marketing activities, building non-farm and allied agricultural activities, setting up institutions like cooperatives and other forms of social capital, empowering Panchayats and women and improving health and education services. We, therefore argue, that in order to bring about 'Inclusive Rural Development, with people's participation nexus', the growth impulses need to be structured and built up at the level of "clusters of Villages", in a highly integrated and effective manner. This line of reasoning implies a new perspective and a much deeper '*economic and social content*', of development addressed to both policy delivery and receiving sides. It is much more than ensuring merely a viable size of settlement for supporting a particular amenity or service. In this conception, the effectiveness of a 'cluster' at grassroots level depends not so much on the size alone of the population brought together, but on the degree of cohesiveness and integration among the constituent villages and social groups, quality of local leadership and capacity of rural institutions to bring the rural weak and poor within their fold. It is here that understanding the village as a part of rural neighbourhood assumes importance in planning for development. This new concept of rural development can be translated into action only through 'social animators' (committed NGOs) and 'change agents' (researchers) joining hands with local bureaucracy/technocracy and local people and their institutions. It has to be seen as a joint endeavour of various "Partners in Progress", with clear goals (Fig. 12.1).

The Delineation of Village Clusters

The first initiating step for creating '*Inclusive Socio-economic Development in the Rural Areas*' would be the delineation of 'Village Clusters (VCs)' as the grassroots level units for rural planning and development. The ideal rural planning units would be the *Cohesive Rural Community*, which can be delineated by taking into account the social, economic and spatial inter-relationships, as well as the power structure of the community. The Village Clusters (VCs) are thus envisaged as 'Community of Interest Areas (CIAs)', in which people's association and social contacts will be of a voluntary nature, with people being drawn together by common interests in economic affairs, trade, social interests for recreation or religion etc. The focal point for each 'Community of Interest Area (CIAs)' will be a 'central place' of viable population size, already performing some area-wide functions such as, school services, medical services, retail services, social and recreational services, business services and transport services. In order to delineate the hinterland of the selected central place, the analysis could proceed both from quantitative data available from secondary sources (e.g. for education and medical services), and combining this with enquiries made in the surrounding villages, so as to obtain a correct knowledge of those rural settlements, which customarily obtained a particular service from the given central place. Such investigations, coupled with the local knowledge of the relative "pulls" of the central places and the transport orientation and consumer travel patterns, will help to determine the hinterland of the selected central places. Mapping of "Desire Lines" for selected services in the rural areas may also be useful. With such methods, decisions about the central places and their hinterlands could be taken with a reasonable degree of accuracy.

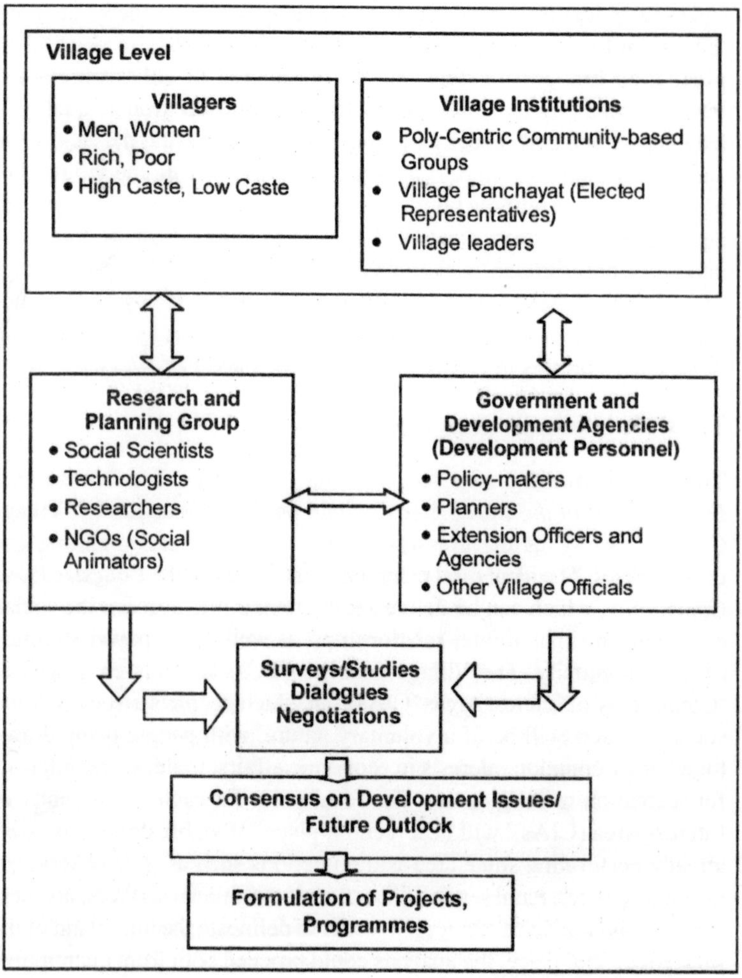

Fig. 12.1: Village Cluster Planning: Partner and Processes

Some of the studies made in micro-level planning in our country have concluded that a viable Village Cluster, which would be the lowest area level unit for planning and development, may cover 5-6 contiguous villages, having a distance span of 4-6 kms and with a total population of about 8,000-10,000. One of our important Committees on Panchayati Raj, which studied the optimum size of the rural unit for planning and development has also concluded that "a large size of rural unit should be appropriate for the villager's own interest for input supplies and provision of services like hospitals and schools, warehousing facilities, availability of higher order technical services and a host of other inter-related functions, which forms part of the design of rural life" (Ashok Mehta Committee, 1978).[4]

PURA in its Wider Context

The case for Village Clusters (VCs) as grassroots level units for planning and development has been made in the previous paragraphs. The VC is only a structural component of rural development. We may now consider as to how to make it a really 'functional unit', surging with dynamism and activities. In the 'structural plan' that has been unfolded with the 'central place' as the 'hub' or 'node', and with the constituent villages as the different 'spokes', dynamism has to be infused through various human activities. Here comes the PURA concept to our aid, as a tool to infuse the 'nodes', with various forms of 'connectivities', 'flows', and 'iterative interactions'.

'Nodes' are 'Nodal/Focal' points, which are endowed with a certain minimum threshold of population, infrastructure and services, and which have the potential to propel activities relevant to development and radiate development impulses to surrounding villages.

The four critical 'connectivities' to induce development are: (i) *Physical Connectivity* or road and transport and power connectivity to bridge distances; (ii) *Electronic Connectivity* to increase communication through media and digital opportunities; (iii) *Knowledge Connectivity* through education and imparting of skills and extension services to provide information, technology, and increased understanding; and (iv) *Market Connectivity* to link rural products to markets. The inadequate presence of these connectivities have been seen as the "major irritants", or obstacles to development in our rural areas and their removal is considered vital for rural transformation.

Simply providing connectivities will not do. They should be rendered *dynamic* through 'flows', which have to be consciously induced in four directions, viz. vertical, downward, horizontal, and circular. The 'Vertical Flows' will provide bottom-up connectivity to higher order centres of activity for processing of commodities produced in primary villages and information on local needs and preferences, area priorities, targets, local problems and constraints. The 'Downward or Top-down Flows' will provide macro-micro linkages and information flows relating to policies, priorities, targets, guidelines, budgetary and technical support. 'Horizontal

Flows' will establish inter and intra-community information flows and interactions. 'Circular Flows' will enhance the recycling-ratio within the villages, adding income to the village and bringing about a circular movement of commodities from the primary villages to their processing centres and the return movement of finished products to consumer centres.

'Interactions' constitute the vital component of this dynamic system and is closely related to both electronic connectivity and knowledge connectivity. It involves (a) the interaction of the people's institutions with Government's institutions; (b) the interaction of people, as individuals, with their own institutions, like the Village Panchayat and Community-Based Organizations (CBOs); and (c) people to people informal interactions through their social organizations.

Thus, the planning intervention envisaged as "Cluster Development" and sought to be operationalized here through the nodes, connectivities, flows, and interactions, in space will ensure participatory development and enable bridging the rural-urban gap, providing urban amenities in rural areas (PURA) and enhancing the quality of life and livelihood opportunities for the rural people.

The Operating Vision, Tactics and Planning

"Where there is no vision, the people perish" is a well-known axiom. It is only with a clear operating vision of objectives, thrusts, tactics, and programmatic interventions that any village cluster (VC) can hope to meet its '*tryst with destiny*'. Initially such a vision will never be totally explicit. It is latent, as a community's attitude towards itself and its aspirations are not easily visible. With iterative interaction with different sections of the community, however, this can be discovered with some clarity. With the modern tools of analyses, the vision that is latent could be objectified.

Tactical Planning

The discussion on 'Nodes', 'Connectivities', 'Flows' and 'Interactions' presented in the fore-going paragraphs relate to the rebuilding of fundamental community structures for the *Holistic Village Cluster Development Planning* in a multi-level framework (Fig. 12.2). They provide a broad framework for tactical planning, i.e. for developing the tactics required to shape the destiny of the village. Thus tactical planning will describe and rationally organize the practical actions required for socio-economic change in the rural areas. The major tactical areas representing the 'overarching tasks' to be accomplished include:

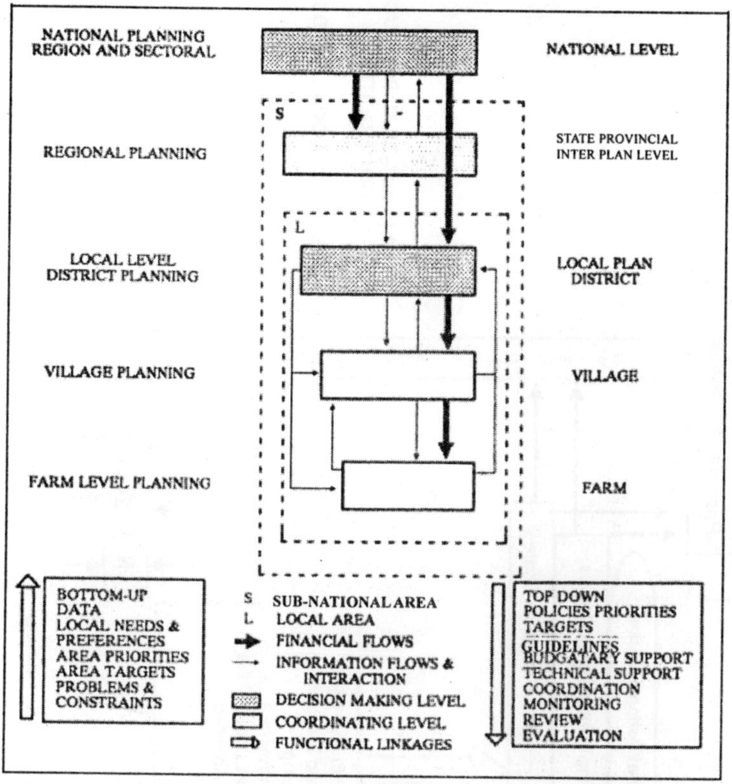

Fig. 12.2: A Hypothetical Multi-level Planning System Showing Levels, Flows and Interactions

(i) Creating an economically viable agricultural and industrial base for the village on the lines of an Auto-Kinetic Model of Development (Fig. 12.3). By auto-kinetic development, we mean a process by which the auto-centric and auto-dynamic forces of development can be unleashed, which will release the inner urges, potentials and energies inherent in the rural community. It is a process of engineering the overall development of productive forces, with the objective of better satisfying the basic livelihood and the material needs of the rural population, based as far as possible on a more intensive utilization of local resources, by means of locally adapted and locally controllable, labour-intensive and small-scale technologies. It is the same as the Gandhian Concept of Village Swaraj or "Development from within", with the capillary mechanisms in place, but points to a definite method for augmenting rural incomes.

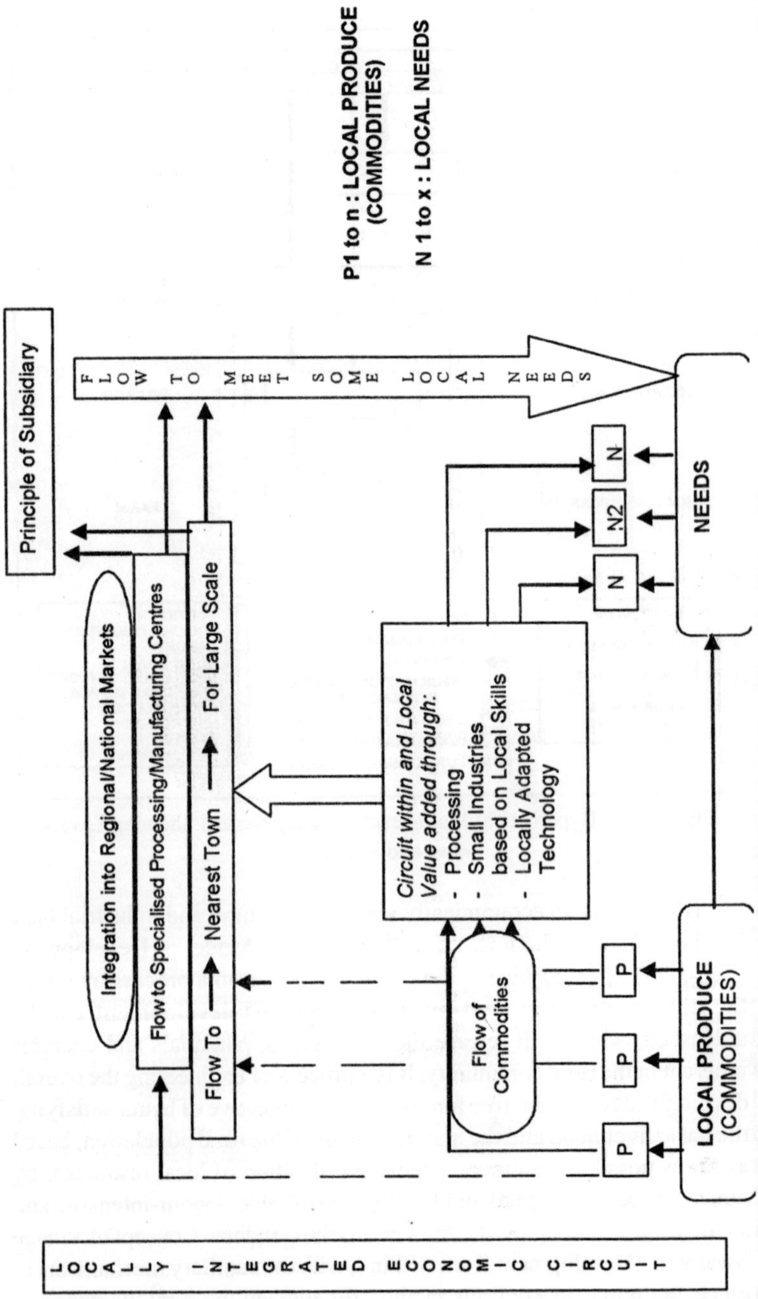
P1 to n : LOCAL PRODUCE (COMMODITIES)
N 1 to x : LOCAL NEEDS
Principle of Subsidiary
FLOW TO MEET SOME LOCAL NEEDS
Integration into Regional/National Markets
Flow to Specialised Processing/Manufacturing Centres
Flow To → Nearest Town → For Large Scale
Circuit within and Local Value added through:
- Processing
- Small Industries based on Local Skills
- Locally Adapted Technology
Flow of Commodities
P
P
P
N
N2
N
NEEDS
LOCAL PRODUCE (COMMODITIES)
LOCALLY INTEGRATED ECONOMIC CIRCUIT

The auto-kinetic development process rests on the hierarchic principle of subsidiarity. That is, we seek to establish processing/manufacturing units at higher level, only when the existing technologies and skills do not allow these to be performed in a satisfactory fashion and at a justifiable cost in the lower spatial level. In the auto-kinetic development approach, the basic objective of achieving the full development of the natural resources and human skills available in the villages can be secured only by establishing "locally integrated economic circuits".

The integrated economic circuit concept implies several other ideas such as engineering circular effects, chain effects and recycling the primary produce of the village into secondary lines of production. The Recycling Ratio is a measure of the extent of rural industrialization. Application of these concepts may lead to specialization and in some areas even to the establishment of a "territorial production complex".

(ii) Guarding the Basic physical well-being of the village society by ensuring food and nutrition security, public health and sanitation and providing health care.

(iii) Developing functional life skills, including upgrading of basic practical literacy skills at all age levels and providing practical training in marketable business and crafts and providing access to agricultural services including inputs and markets.

(iv) Another tactical arena would be breaking the isolation of the village and establishing external support relations. This could be established through fostering the necessary institutional relations with the Public and Private sectors and adapting the development programmes initiated by the Government, to suit the local specificities and needs, forging extended business contacts, e.g. for securing bank loans to fund local projects and promoting the village public image.

(v) Leveraging Information and Communication Technology (ICT) is one of the major tactics for transforming the village into a knowledge-based society. This would call for establishing a "Knowledge Communication Centre (KCC)" and training the rural youth in computer literacy. In this manner, agricultural information for precision farming from extension agencies and price signals from neighbouring market centers could be accessed by the rural community. The 'Hub & Spoke Model' for a knowledge communication centre is illustratively indicated in Fig. 12.4.

The arching dimensions of the operating vision may be visualized in five broad groups, viz. community sustenance; habitat improvement;

sustainable development; institutional support; and social awakening. Each group has been further elaborated to indicate the major thrusts and further detailed into 'enabling programmatic interventions', for actuating the 'Vision'. These are outlined in Table 12.1, which illustrates in a nutshell the planning framework for the study.

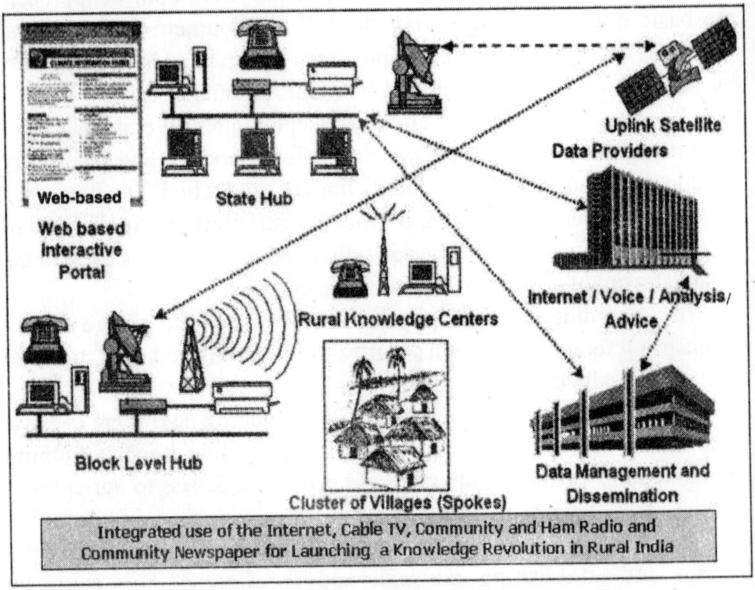

Fig. 12.4: Virtual Academy for Food Security and Rural Prosperity (Hub and Spokes Model)

Programme Content

Successful implementation of the Holistic Village Cluster Development strategy requires complex decentralized matrix structures, with permanent mechanisms for vertical and lateral integration, a combination of specialist and generalist skills, institutional leadership, social intervention capability and systems management. The methodology and programme content would include the building up of the databases for decision-making design is advocated which will aim at (i) Ensuring Sustainability of Resource Base, (ii) Poverty Alleviation, (iii) Empowerment of Women, (iv) Full Employment, (v) Production Systems Planning, (vi) Infrastructure Planning and (vii) Habitat Planning. These are further elaborated in the accompanying Flow Chart (Fig. 12.5). The Bhoovigyan Vikas Foundation hopes to seek to translate the operating vision elucidated in this paper into a few pilot projects in different typological areas.

Table 12.1: Holistic Village Cluster Development Project Operating Vision

Objectives (Broad Groups)	*Major Thrusts*	*Programmatic Interventions*
Community Sustenance (upgrading the economic foundations of village life)	1. Agri-Production Complex	Production and productivity increase, Transportation, Marketing and Credit Linkages, product diversification
	2. Food and Nutrition Security	Grain Golas; and Effective PDS; Livestock/ Poultry/Horticulture Development
	3. Full Employment	SME's through Integrated Economic Circuit
Habitat Improvement	4. Village Reconstruction and Provision of Basic Amenities	Housing rehabilitation, Public Utilities and Facilities, Power, Communication and Transport Systems
Sustainable Development	5. Eco-development (Bio-village)—Promoting conservation and job-led growth opportunities.	Land and Water Management; Bio-diversity Conservation
Institutional Support	6. Poly-centric Planning and Management Institutions	Village Planning Forum, Village Skill Development Centre, Village Knowledge and Communication Centre Co-operatives, User Groups, and CBOs.
Social Awakening	7. Social Rebuilding (through social animators)	Training and education for social effectivity, Literacy Training, Training for women/youth, Land Army, Barefoot Doctors/ Engineers/ Technologists etc.

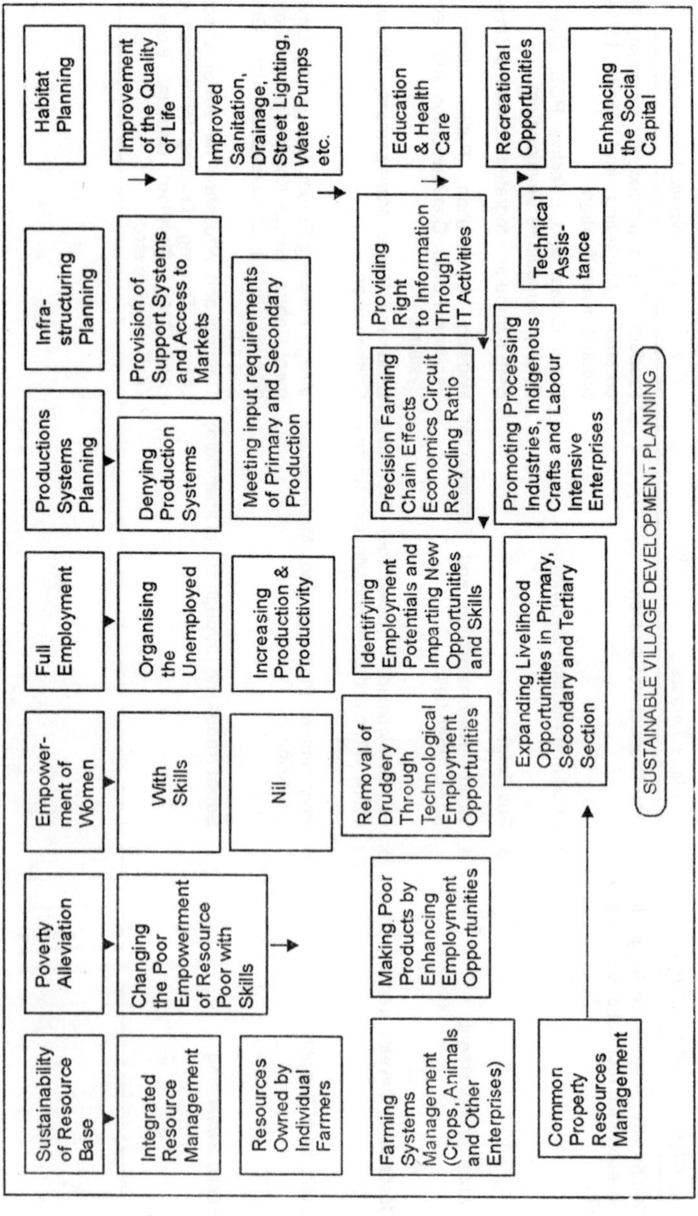

Fig. 12.5: Holistic Planning for a Village Cluster Concepts and Key Dimensions

The operating vision for rural development outlined in this paper and many of its underlying concepts bear a close imprint of Mahatma Gandhi's vision for the reclaiming of the rural villages across our country.

REFERENCES

1. Sundaram, K.V. 'Service Centres and Community of Interest Areas: A Case Study of Rural Delhi' in Urban and Regional Planning In India, Vikas Publishing House, New Delhi, 1977. Reprinted in 'The Trodden Path—Essays in Regional and Micro-level Planning, Anamaya Publishers, New Delhi, 2004.
2. Report of the Working Group on Block Level Planning, Government of India, Planning Commission, 1978.
3. Sundaram, K.V. "Spatial Planning for a Tribal Region" in Urban and Regional Planning in India, Vikas Publishing House, New Delhi, 1977.
4. Report of the Committee on Panchayati Raj Institutions, Government of India, Ministry of Agriculture and Irrigation, New Delhi, August 1978.

13

Socio-Cultural Dimensions of Sericulture
A Village Study from Andhra Pradesh

ESWARAPPA KASI

Introduction

In any kind of developmental activity, socio-cultural factors play a significant role. In this regard, the adopters of innovations attempt to interpret them in tune with their socio-cultural settings. They had to fit into their cultural matrix only then, would the innovations obtain a survival chance. It is in this context sericulture, an innovation for economic upliftment of rural masses has been examined in this paper. The roles of socio-cultural factors such as education, occupational differences, caste and community differences, religion, beliefs, etc., which influence the practice of sericulture or aiding or effecting sericulture are looked into.

These socio-cultural factors are studied in terms of plantation methods, and rearing practices and in relation to marketing, social composition, shift from agriculture to sericulture. Further, the study also attempts to understand the change that has come about in the value system, inter-caste relations, notions of purity and pollution, etc. through the practice of sericulture.

The present paper is an outcome of a study, which is carried out in the Kotha Indlu Village of Chittoor District of Andhra Pradesh, India. This study is a qualitative micro-level study of a sericulture village. The primary data was collected through a structured household schedule administered to all respondents in the village, informal interviews using detailed checklists, key informant interviews, and case studies. Quantitative information with regard to the technological development, land holding, demographic aspects, cropping patterns, irrigation systems were collected by using household schedules and District Statistical Hand Book. The paper has been organised into four sections: The first section of the paper provides the introduction; the second section is about the status of sericulture and its sustenance; the third section describes the location, people and their livelihoods and practice of sericulture and rituals in the village and final section examines the changes that have come in the village while practicing the sericulture activity.

The Background: Status of Sericulture and its Sustenance in India and Andhra Pradesh

In India sericulture has become the most promising allied agricultural activity in rural areas for the reasons of minimum gestation period, less investment, maximum employment potential and quick turnover for the investment. Sericulture generates direct and indirect employment in various ways. Firstly, mulberry cultivation creates employment on farm, and, secondly, cocoon production, which uses mulberry leaves as an input, creates large-scale employment for the family labour of the mulberry growers. There are, instances of non-mulberry growers taking up cocoon production alone as a full-time occupation. They buy leaves from mulberry growers and use them as raw material for cocoon production. Further, the reeling activity is also mainly undertaken in rural areas or semi-urban areas and the employment generated there would help reduction of the rural unemployment in a significant way. In short, sericulture as a whole, by its very nature of activity, creates large-scale employment and income generation opportunities in the rural and semi-urban areas accelerating the economic growth of these areas.

In its long history sericulture in India has experienced many ups and downs. However, during the last 30 years, India has made tremendous progress in the production of mulberry silk for which there is an increasing international demand. There is tremendous scope for the expansion of its production in the country. In recent years sericulture has achieved enormous progress in evolving suitable mulberry varieties and techniques to bring about new silkworm races suitable for tropical climatic conditions. With the evolution and introduction of more productive silkworm races, the productivity has increased and sericulture has become a highly remunerative activity. Attracted by these advantages many more farmers have taken up sericulture and the industry has spread to almost all the States in India (Narasaiah and Jayaraju, 1999).

Karnataka has been the leading producer of mulberry silk accounting for more than 50 per cent of its production in the country. This State is now regarded as the "Silk Bowl of India". Andhra Pradesh comes next to Karnataka in producing mulberry raw silk. In Andhra Pradesh, sericulture has proved to be a money-spinner for many middle class families. Except for Hyderabad district, almost all the districts in the State have taken to sericulture activity. In 1956, mulberry cultivation in the State was undertaken in 1,212 hectares which went up to 90,800 hectares by the end of 1993-94 (Department of Sericulture, Government of Andhra Pradesh, 1993-94).

Once confined to a few pockets like Anantapur and Chittoor districts bordering Karnataka, sericulture has caught up in such a big way that in just 10 years time the State of Andhra Pradesh has come to rank next only to Karnataka, the premier State of silk. Though, almost all districts in the State have taken up sericulture, the main concentration of mulberry silk production is in the four Rayalaseema districts, namely Anantapur, Chittoor, Cuddapah and Kurnool, of which Anantapur, followed by Chittoor, stand first and second in the entire State in producing mulberry raw silk. At present Anantapur and Chittoor show enormous growth rates in sericulture in Andhra Pradesh. Chittoor has been ambitious of achieving first rank in the production of mulberry raw silk in Andhra Pradesh. If the present tempo of growth is maintained, it would not be a surprise if it achieves its desired goal before long. Given these developments, it is worthwhile to study the changes that have taken place due to sericulture among the farmers, especially among the poor, in Chittoor district.

In Andhra Pradesh, mulberry cultivation and cocoon production are mainly concentrated in Chittoor district, which account for nearly 23 per cent of the area under mulberry cultivation. The mulberry plant, being hardy plant adjusts well to low rainfall and drought conditions and responds well to irrigation. The district appears to be gradually finding mulberry cultivation to be a profitable proposition. Prior to the inception of Drought Prone Area Programme (DPAP) in the district, mulberry cultivation as well as cocoon rearing activities were found to be concentrated in the areas adjoining to Karnataka State. The cocoons used to be sold in Bangalore. However, after the inception of DPAP in this district, mulberry cultivation as well as cocoon rearing slowly turned out to be major activity, especially in Palamaner, Madanapalle, and Thamballapalle and Kuppam blocks. In course of time it spread to the adjoining localities of these blocks. Now, at present, there are around six to seven blocks where the rearing activity has become prominent. Overall, cropping pattern in the district has been found to be undergoing a change since the last decade. However, earlier not a single crop, except groundnut, was cultivated consistently for a reasonable period by the farmers of the district. This may be due to the uncertainty involved in the cultivation of such crops. However, what is interesting to find is that silk rearing activity has not only expanded and covered almost two fourths of the district but also it appears to have been stabilising in this region.

What is equally interesting to note is that notwithstanding the increasing cocoon rearing activity, reeling was never carried out in this region until the sixth plan period. The major part of cocoons produced used to find its sales outlet in Karnataka, where reeling activity is traditionally carried out extensively. The same raw silk used to be purchased by the weavers of the Chittoor district at a higher price. Price used to be higher due to the operation of the exploitative network of middlemen. However, with the Government intervention, silk reeling activity came up in the district to a reasonably good level. Subsequently, raw silk market as well as cocoon market got organised with the establishment of silk exchanges in those places where the concentration of silk weaving is relatively higher. As a result, weavers are relatively in a better position as they can purchase the raw silk at the "Government" price.

Now that sericulture from 'leaf to fabric' activity is more organised, a large number of farmers are adopting sericulture. But, there are quite a few serious limitations, which might act as stumbling blocks for the development of sericulture in this district in the years to come. These are found not only in respect of the farmers but also at the Government level. Government assistance for sericulture development in this district can be seen in various forms, such as installation of seed farms, rearing units, grainages, silk reeling units, twisting units, cocoon markets, silk exchange, technical services and financial assistance in the form of subsidy and loan.

Although, infrastructural facilities do not appear to be inadequate, response of the people to reeling activity is not substantial. This may be the reason why production of raw silk has been low. What appears to be true is that a large percentage of cocoons are sold outside the district especially in Karnataka, where the cocoon rearers of this district find it more profitable to sell cocoon.

Another interesting observation one makes is that around one-third of the total cocoon produced is marketed through the regulated markets. There may be two reasons for this, namely price is less remunerative due to less number of buyers who bid and secondly, which is related to the first, there is a thin concentration of reelers in the district. It indicates that two-thirds of the total cocoon produced in the district is sold to the reelers of the adjoining State, namely Karnataka.

Therefore, in order to make "Leaf to Cloth" programme successful in the district of Chittoor what appears to be important at the moment is for the Government to take steps in relation to improving the managerial capacity of the cocoon rearers through better extension mechanism and motivate people, irrespective of caste, to take up silk reeling as well as weaving in the process of development of sericulture in the district.

As observed earlier, sericulture activity is highly concentrated in the four Rayalaseema districts, i.e. Anantapur, Chittoor, Cuddapah and Kurnool. Among them, Anantapur ranks first with 33,190 hectares (36 % of the total area), followed by Chittoor with 18,108 hectares (24% of the total area). Kurnool and Cuddapah occupy the next places in ranking. The Rayalaseema region recorded good progress than other regions in sericulture activity. This is due to the favourable climatic conditions in this region and also rising interest of farmers in this very lucrative activity.

Location, People and their Livelihoods

Kotha Indlu is situated in the Kotha Indlu Panchayat of Kuppam Mandal in Chittoor district, about 15 kilometres from Kuppam, the Mandal head quarters. It is a multi-caste village inhabited by Balija, Vanniar, Scheduled Castes and Chakali (washermen) castes. The Balija is the dominant caste in this village in terms of politics, economy, education and also numerical strengths. While the Balija occupies the highest position in the local caste hierarchy, Vanniar, Chakali and the Scheduled Castes occupy the subsequent positions in that order. Vanniar, who are a backward caste, are competing with the Balija economically and educationally, though they are weak numerically and politically. Next to Balija, the Scheduled Castes are numerically more than that of Vanniar and Chakali castes. The total population of 274 is distributed in 50 families.

On the basis of land holdings, the residents of Kotha Indlu are classified into marginal, small, medium and large farmers, and the landless. The number of small farmers forms the majority, followed by the medium farmers. The numbers of large and marginal farmers are quite few, while the landless are almost insignificant. The crops cultivated are Paddy and Ragi, which are the staple crops and they constitute the staple diet of all communities. Commercial crops like flowers and groundnuts are also grown in the village. Floriculture is specially found among the Balija farmers, while all the castes raise groundnut crop in their lands. Though sericulture occupies the second position after agriculture, it is the major source of income for the farmers in the village. Cattle, goat, sheep, buffaloes and donkeys are the major livestock in the area. Several varieties of grass along with stocked paddy straw; maize stalk, groundnut and *bhoosa* are used as fodder. Poultry is widely prevalent in the village and only the Scheduled Castes raise pig/swine.

The village is very well connected by road and it has good transport facilities. The sericulture farmers use the tillers, tractors and bullock carts to transport their produce to the towns. The village has telephone as well as postal facility. This village has a primary school and a high school. It also has a ration shop under the public distribution system to supply the necessary commodities to the people, who have white ration cards, at concessional rates.

The President of the Gram Panchayat is from the Scheduled Caste community. In fact, this (Panchayat) seat was reserved for the Scheduled Castes in the last elections held for the local bodies on rotatory basis. There is a farmers' club in the village, which takes care of their problems both in the society and outside.

Practice of Sericulture in Kotha Indlu

Farmers in Kotha Indlu follow row system and S1, S5, and S36 varieties of mulberry predominate over the local variety, M5. Before planting, farmers perform certain rituals. Farmers perform *puja* (prayer) to their caste deity, which will be placed in the northern corner of their land, and light incense sticks. They apply *kumkum* (Vermillion) and turmeric powder to the deity and also sprinkle around the place where they put their deity. They start planting *resham kaddi* from that end. The rituals observed here are akin to the ones that are being followed in agriculture in the village.

Farmers in Kotha Indlu maintain thickly populated gardens mostly of local mulberry (M5). They believe that local variety is more drought resistant and requires less water in contrast to the improved varieties, as they are time tested and are passed on to them by their predecessors. Though some of the farmers are willing to raise improved mulberry, they are unable to procure improved variety cuttings. Department of Sericulture, Kuppam, supplies S36 cuttings free of cost to farmers. However, it is not able to supply to all farmers. Farmers also get cuttings from their relatives and friends. According to the Convenor of the farmers' club, there is a scheme wherein the Department of Sericulture gives Rs. 500 to the farmers who plant "Swarnandhra" and some farmers have adopted this improved mulberry in Kotha Indlu. Thus there is receptiveness among the sericulturists to changes in the cropping pattern, though they still tend to continue to follow the ritual processes with great fervour.

In a majority of the sericulture households, women do bed cleaning. This activity is compared with attending to the children and they believe that bed cleaning is part and parcel of women's routine activity. Sericulturists compare worms with children. So, they regard bed cleaning and litter cleaning are women's activities. In some sericulture households, men also do bed cleaning. In Kotha Indlu, sericulturists do bed cleaning once in a day. One can notice some variation during different moulting stages in the village.

Sericulturists in Kotha Indlu say that there are so many reasons for the diseases to silkworms. They are critical of the extension staff of Department of Sericulture. They complain that the staff lack experience with regard to disease control and also identification of diseases at an early stage of rearing. The Sericulture Inspector pointed out that there is inadequate ventilation in the rearing houses and also the farmers' rearing practices are bad. He also says that sericulturists are not taking proper care from the beginning of crop. In this regard, he points out that the farmers after collecting disease free layings (DFL) sheet go to movies and upon reaching home they don't even cover any paper over the DFL sheet to protect the seed. He complains that they are not properly maintaining hygienic conditions in and around the rearing house. On the contrary, some sericulturists from the lower castes complain of discrimination by the extension staff. They aver that the extension staff favours always rich farmers, who rear Bivoltine.

Social composition could be seen in terms of caste differences, occupational differences, family structure, etc., in Kotha Indlu. Here, caste differences are reflected in different aspects of sericulture. For instance, every aspect of sericulture like mulberry cultivation, silkworm rearing, disinfection and upkeep, maintenance of rearing sheds, hygienic conditions in the surroundings of rearing sheds, *chandrikas*, stands and trays, etc., are shaped by caste differences.

Majority of the sericulturists from the upper caste, Balija, have improved mulberry variety, where as the others in the village have local variety. This can be seen in terms of availability of land, irrigation water facility, etc. Moreover, Balija are more receptive to new innovations than the others.

Table 13.1: Distribution of Land among Different Land Size Classes (in Acres)

Land Size Class	*Wet*	*Dry**	*Leased in Wet*	*No. of HHs*
Landless	—	—	0.75	1
Marginal	1.25	1.50	—	4
Small	51.50	15.09	—	31
Medium	26.00	11.33	—	10
Large	18.00	4.00	—	3
Total	96.75	31.92	0.75	50

* Figures under this should be multiplied by 3 to get actual dry land.

There is a correlation between the family structure and sericulture. Sericulture requires an intensive involvement of family labour. The size of mulberry garden has a direct relationship with the family size. Irrespective of caste and land size class, almost all the sericulturists possess mulberry garden between ½ an acre to a maximum of 1½ acres (see Table 13.2). Thus to a certain extent, it is acting as a leveller and the families who have productive labour are better suited to bring more land under mulberry. Kotha Indlu village has more number of nuclear families and hence, their mulberry gardens are also smaller in size. There are totally 50 households in Kotha Indlu villages out of this 39 are involved in sericulture, three in agriculture, one in horticulture, three are in agricultural labour and others form a miscellaneous group of petty businessmen, employees, etc. Balija caste members dominate the sericulture activities.

Table 13.2: Land under Sericulture (in Acres)

Type of Land Holder	*Own*	*Leased In*	*Total HHs*
Landless			1
Marginal Farmers	1.25	0.75	2
Small Farmers	23.50	—	27
Medium Farmers	11.00	—	7
Large Farmers	04.00	—	2
Total	**39.75**	**0.75**	**39**

When it comes to the institution of marriage, Kotha Indlu people prefer marriages within the family. They marry cross cousins. Usually, elders arrange marriages. There are two cases of inter-caste marriage, one between an upper caste woman (Balija) and a lower caste man (Vanniar) and the second was between a Vanniar woman and a SC man. In both cases the women were considered as outcastes. The villagers consider the idea of inter-caste and love marriages as an influence of modern television and films.

Manjula, aged 25, studied up to 7th class and, being the eldest daughter, started helping her mother in sericulture activity. Her mother, Neelamma, did not have any formal education and her father, Ratnappa, is an attender in the school in Santhipuram. He studied up to 10th Class. Manjula eloped with a lower caste man and, following this, Ratnappa and other caste elders disowned her. Balija in the village felt that this is their caste issue and no outsider should involve in this matter. Balija in the village are afraid that through inter-caste marriage their caste status/ prestige get declined. Manjula has left her home nine-months back and started staying with Srinivas in the village. Even though, it is an inter-caste marriage, Vanniar have welcomed this marriage. The Vanniar in fact "look up to" such type of marriages in near future. They think that, this would enhance their caste position in the society in general and Kotha Indlu village in particular.

Kotha Indlu village has a high school. Girls are usually drop-outs before or on reaching the high school level. Very few boys go out for pursuing higher education. In the Anganwadi Centre, pre-school children are given nutritious food. The teacher in Anganwadi Centre is a Scheduled Caste Woman. The adult literacy programme has not made its inroads in Kotha Indlu.

The Practice of Rituals and Worship in the Village

The chief deity of Kotha Indlu village is Timmaraya Swamy. After Timmaraya Swamy they worship Anjaneya Swamy. The local deities are Agnimaremma, Kunti Gangamma and Gangamma. The local deities are worshipped on special occasions. For instance, Gangamma is worshipped when children have problems; Agnimaremma is worshipped for health of women; Kunti Gangamma is regarded as a protector of cattle life in the village and is worshipped during *Sankranti* (in January). Untouchability is still in vogue. For instance, Scheduled Caste members are not allowed into the premises of Timmaraya Swamy Temple.

In Kotha Indlu village people give more importance to rituals. Rituals are associated with all their activities. Before starting every activity they perform some kind of ritual, e.g. while doing *chawki*, they perform *puja* and offer *samrani*, coconut and *kumkum* (Vermillion) to their family deity. Another example is, when a girl reaches puberty, certain rituals are performed. During every menstrual period, women are kept outside daily work for three days. Even rearing houses are kept out of their reach.

In Kotha Indlu majority of the people are Hindus. Some from the Scheduled Castes follow Christianity. Every caste has their own deity as well as common deity in the village. For instance, the chief deity of Balija (Family deity) is Venkateswara Swamy and Vanniar family deity is Muneswera Swamy or Muneppaswamy temple, which is situated in Mallappagutta, which is 2 kms. away from Kotha Indlu. Vanniar from Kotha Indlu go to Muneppaswamy temple for the first hair shaving ceremony (tonsure ceremony) of their first child only. Dharmarajulu is their caste deity. They claim him as their ancestor and claim themselves as belonging to Kshatriyas caste. They perform Dharmarajulu panduga (festive occasion) once in three years. They visit Dharmarajulu temple situated in Pedda Bangaru Nattam, 3 kms. from Kotha Indlu village. During this ceremony, they perform animal sacrifice and have a communal feast. Scheduled Caste people have their own religious ceremonies in Kotha Indlu. Hair shaving ceremony is practiced at the time of Munidevara panduga. They also perform animal sacrifice (offering pig, goat or buffalo) attended by their relatives and neighbours. Other caste people are not allowed to attend this ceremony.

Agnimaremma is a common deity for all the villagers. But each caste has its own rituals and in this way every caste preserves its integrity and purity. For instance, Balija have the practice of inviting Guruvulu from Kanchi (Tamil Nadu) once in every five years. When Guruvulu arrives, he makes sacrifices for the betterment of Balija. He also does purification ceremony in case of any inter-caste marriage. Teertham or holy water is used for this purpose. Kotha Indlu villagers believe in magical spells and practices. For instance they believe magical spells can even save a person after snakebite.

Rituals are part of people's personality and life in Kotha Indlu Village. Even before a child arrives into the world, when the child is in mothers womb rituals begin. So, when a woman is pregnant, a ceremony called "Sreemantham" is performed. After the child is born naming ceremony, "Namakaranam", is performed. When the first child reaches 3 years, hair removing/Tonsure ceremony is performed in the village. For this ceremony Balija go to Tirupati, Vanniar go to Dharmarajulu temple (Pedda Bangaru Nattam) and Scheduled caste people go to Mallappagutta, 2 kms. from Kotha Indlu (Yellamma). When a girl reaches puberty it is celebrated within the caste members. Marriage ceremony takes places for three days. Death ceremony is performed in the village after 2 or 11th day of death. Those under the pollution period by custom are not allowed to enter into rearing shed. In case they want to enter rearing shed, they have to have a ritual bath and pray to their family deity. Women are engaged in domestic callings and also work in their rearing sheds. But when it comes to performing rituals, men dominate.

The physical structure of the village itself set boundaries between castes in the village. This means every caste maintained its own identity by performing its own rituals. In terms of labour distribution, Scheduled Castes do agricultural labour. In case of those who have lands, they also hire Vanniar for labour. The payment is in the form of cash or kind. The upper caste Balija and Vanniar if they have rearing sheds, allow Scheduled Caste labour into the rearing house. The lower caste people (SCs) are not allowed into temple premises. But now-a-days they do allow SCs to perform *Puja* outside the temple.

Goddess Agnimaremma is the chief deity of the village. On festival occasion, she is taken in a palanquin and taken around the village. For the convenience of lower caste *darshan*, the deity is kept in a common place and lower caste people are allowed to perform *Puja*. However, unlike the earlier times, now-a-days the deity is even taken up nearer to their dwellings.

The concepts of purity and pollution have special meanings for Kotha Indlu villagers. Purity is a matter of both faith and hygiene. For instance, purity is equal to hygiene in the maintenance of rearing sheds. They are very particular in maintaining the hygiene of rearing sheds with cowdung and bleaching powder. The concept of pollution is equally important for Kotha Indlu villagers. In most cases it is a question of faith or even taboo. For instance, the women's menstruation is considered a polluting phenomenon. When she is in periods she cannot enter rearing sheds and dwellings for a period of three days. In case there are no women for cooking and rearing, they are allowed to participate only after taking ritual bath. After taking ritual bath, urine of cow is taken as Teertham and sprinkled in and around the house for purification. Many sericulturists in the Kotha Indlu village still believe that allowing a polluted person into rearing house may cause crop failure (Mulberry). Kotha Indlu villagers remember their elders through following these beliefs. Death is also considered a pollutant because of its strange nature. When in case of a death in a family, the persons are not supposed to participate in social gathering or rearing. They have to undergo the same process of purification for de-polluting themselves.

Krishnaveni, leader of a DWCRA group, says that generally women do not participate in religious activities in Kotha Indlu. Usually men do all the activities. With regards to pollution and purity, she says that they practice these very strictly. Polluted women do not enter into rearing house in general and dwelling house in particular. If any body dies in the village, they consider it also as pollution. After taking ritual bath then they allow them into their house. With regard to purity, following their custom, they clean the rearing house and dwelling house with cowdung on every Friday.

Krishnaveni believes that her lifestyle has changed a lot after becoming a member of the DWCRA group and sericulturist. She also says that whenever they get leisure time they go on tour to Tirupati or go for watching movies. Regarding education, she wanted to send her children to convent school and she has definite idea about her children's future. She feels that, whatever they have now they got it because of sericulture. She has radio, wall clock and got gas cylinder during Janmabhoomi period.

The economic condition of Kotha Indlu village is largely dependent on the level of land holdings. There are big landholders, small landholders, marginal landholders and landless labourers. The big landholders have their impact on small landholders. For instance, the landlords of larger land holdings hire tractor and other machinery for getting better yield. This has led to denial of labour to the lower caste/class members, thus leading to unemployment. The distribution of land holdings also reflects the social status of each caste in Kotha Indlu village. The big farms are held by the upper caste and the landless labour is usually from the lower castes. If we observe, the pattern of land holdings, in terms of number, we can see the impact of development programmes in Kotha Indlu.

Fifteen households hold big land holdings, 27 households hold small landholdings and 6 households hold marginal land holdings and two are landless (see Table 13.3). The development schemes of the Government strengthen and maintain such a divide. Supplying DFLs, promotion of drip irrigation, construction of rearing sheds, as part of development, benefit the big landholders rather than landless labourers.

Table 13.3: Distribution of Landholding by Caste (in Acres)

Size Class	*Balija*		*Vanniar*		*Chakali*		*SCs*		*Total HHs*
	Wet	*Dry*	*Wet*	*Dry*	*Wet*	*Dry*	*Wet*	*Dry*	
Landless	0.75*	—	—	—	—	—	—	—	2
Marginal	0.75	1.50	0.50	1.00					4
Small	25.00	17.50	6.50	3.50	7.00	2.00	12.00	22.25	31
Medium	22.00	28.00	2.0	4.0			3.00	2.00	10
Large	18.00	12.00							3
Total HHs	**30**		**6**		**3**		**11**		**50**

In terms of labour, Kotha Indlu villagers have their own wage structure and rates of payment. Landless labourers are paid both in terms of cash and kind. The cash payment differs between men and women. Men are paid Rs.35 and women paid Rs.25 per day. The miscellaneous services like running of tiffin stalls, teashop and general stores also fetch reasonable amount. For instance, Sridevi gets Rs.100 per day from running a tiffin stall. For women and landless labourers, there is hardly any capital available. For women DWCRA serves as a chief source of capital. Non-sericulturists either go for agriculture or economic services like running of tea stall or rice milling for livelihood.

The economic organization of Kotha Indlu village is sericulture based (See Table 13.3). There are no formal banking services for capital generation and economic organization, except for DWCRA groups. The ratio between land, labour and organization is based on caste and other social factors than economic factors. Marketing of sericulture products is done through Cocoon Marketing Centres of Government as well as private agencies. However, private merchant agencies also exist for marketing.

The level of production depends on the quantity and quality of inputs given for production. For instance, if the production level in sericulture has to go up, it requires good quality seed, high breed mulberry plants, pesticides, labour intensive technology and clean atmosphere. If we consider supply of seed, upper caste people who already have proper infrastructure appropriate high quality "Swarnandhra" seeds. So, their income level also goes up. This in turn gives them enough purchasing power of labour. The lower caste people are left to practice conventional practices with non-hybrid varieties like cross breed DFLs. Such economic disadvantages perpetuate the connection between community and production differences.

Changes

Kotha Indlu has a home for families who have consistently been progressive. The chief factor for such a change is primarily attitudinal. If there are any political and economic factors behind the growth, it is only due to the positive attitude of the people in absorbing such factors. To go back to the stories of sericulture at first, the introduction and growth of sericulture itself is an attitudinal change. The first person who brought sericulture is a woman, who introduced it in the year 1977. She had faced lot of resistance in making sericulture rooted in Kotha Indlu.

The villagers turned the adversity into an opportunity when their village got burnt accidentally. Some of the older generation persons say that it is like rejuvenating oneself by burning all the bad thoughts and things in the "Bogi" fire. Now sericulture is the main occupation of Kotha Indlu village. This village also stands as an example of better participation of women in economic development along with men. Women not only do family labour but also contribute for increase in production levels. Literacy rate of Kotha Indlu village is more than the literacy rates of Andhra Pradesh in general and Chittoor in particular (See Table 13.4). Women also are enterprising. For instance, they run tiffin stalls and rice milling independently. They have also organized themselves for economic benefits as the case in DWCRA.

Table 13.4: Distribution of Education by Caste

Level of Education	*Balija*	*Vanniar*	*Chakali*	*SCs*	*Total**
Illiterates	28 (18.54)	06 (24.00)	06 (28.57)	22 (32.35)	62 (23.39)
Primary	67 (44.37)	08 (32.00)	09 (42.86)	26 (38.23)	110 (41.51)
Secondary	45 (29.80)	10 (40.00)	04 (19.05)	17 (25.00)	76 (28.68)
College	11 (07.29)	01 (04.00)	02 (09.52)	03 (04.12)	17 (6.42)
Total	**151 (100) (56.98)**	**25 (100) (9.43)**	**21 (100) (7.93)**	**68 (100) (25.66)**	**265 (100)**

* There are 9 infants who are excluded from the total.

Note: Parentheses indicate percentages.

Sericulture as an economic and social activity has brought a major change in Kotha Indlu life. They have given up looking at silkworm as worms and started looking them as babies and a lucky crop. People who are in sericulture get more benefits in terms of economic and social status. Agriculture and miscellaneous activities do not even match anywhere to sericulture. Dairy farming also complements sericulture activity. The leaves are not wasted and the hybrid cows (Jersey) are made to browse the mulberry leaves that are left behind by silkworms. This is something akin to mixed cropping in agriculture.

Sericulture has brought a change in the marriage pattern of Kotha Indlu villagers. Even though, they practice cross cousin marriage, earlier they were not taking dowry. But now, dowry has become essential in most of the marriages. For instance, according to one key informant, the convenor of the "Farmers club", who got married two years ago, took about Rs.50,000 as dowry. Sericulturists in the village think that, giving or taking dowry enhances their status in the society. Sericulture enhanced the financial status of the farmers and now they spend more money on marriages, as it demonstrates the improved economic as well as social status.

Traditional occupational pattern in Kotha Indlu village is largely dismantled due to arrival of sericulture. Caste based occupations have been disturbed. Caste hierarchy has increasingly been questioned by the lower castes. The Vanniar now claim an equal, if not superior, status with the Balija. Sericulture as a commercial activity has brought all caste groups to compete with each other in the open market. For instance, the people of Chakali caste have given up their age-old occupation of washing clothes. They have adopted sericulture as their main occupation. But, when it comes to utilisation of benefits like procuring of seeds, irrigation facilities, etc., upper caste people still maintain their traditional advantage over the others.

Economic independence of lower castes was made possible in Kotha Indlu due to sericulture. For instance, Scheduled Caste members traditionally used to do jobs such as agricultural labour, drumming (Dappu), etc., which were not done by the others. Since they did not have any powerful economic base, they felt highly insecure. But now they have sericulture to support. As one key informant says "we use to beat Dappu and we did not have money, and we were frightened, what to do everyday. But now we have sericulture".

The developmental activities like Catalytic Area Development Programme, implemented by the Government, have a significant role in complementing the growth of sericulture in Kotha Indlu. For example, they give loans for drip irrigation, offer new hybrid variety seeds and so on.

In spite of improved economic status of farmers from different castes, the lower caste members are still denied entry into public places like temples. Caste of a person continues to have a bearing on the patterns of land holdings, distribution of labour, appropriation of loans, etc. Most of the sericulturists have TVs and scooters/motor vehicles. The entry of cable television and other communication network have not brought about any change in the attitudes of the villagers with regard to their traditional caste attitudes. Thus, there is a greater need to understand the role of NGOs and Government developmental programmes in bringing about the desired change among the villagers.

REFERENCES

Eswarappa Kasi, 2000, *Development and Change Due to Sericulture—A Study in Chittoor District* (An unpublished M.Phil Dissertation submitted to the University of Hyderabad for the Award of M.Phil Degree).

Government of Andhra Pradesh, 1993-94, *A Note on implementation of National Sericulture Project in A.P.* (Summary and Progress), Department of Sericulture, Hyderabad.

Narasaiah and Jayaraju, 1999, Sericulture, Its Place in National Economy in Indian Silk, May.

14

Empowering Rural India with SERVIS

AMARNATH KSHATRIYA

Introduction

President of India, Dr. Abdul Kalam, in his 2004 Republic Day speech suggested that,

> "Our scientists should become civic scientists and contribute towards societal transformation. Civic means concerning or affecting the community or the people. In the new capacity, scientists step beyond their campuses, laboratories, ministries and institutes and move into the centre of their communities to engage in active dialogue and action with their fellow citizens.
>
> They should ask themselves a question, how their knowledge can make an impact on the common man's life. Our civil servants and others in the service sector should become fearlessly people-friendly, have a positive attitude, and provide responsive, proactive, transparent and unbiased administration and service to the billion people...."

We applaud President Kalam's appeal to the scientists/civil servants to move into the communities and share their knowledge and establish rapport with the soul of India that toils in the villages. SERVIS is poised to inaugurate this movement.

Our Proposal

On November 26, 2002, H.E. Mr. V.K. Nambiar, Permanent Representative of India to the United Nations, following his contribution to the "Outcome of the International Year of Volunteers and its Follow-up" observed that:

'Volunteerism' has formed an integral part of the Indian ethos—known as "*shramdaan*"in Indian society ..." This social wealth was sought to be harnessed towards national development through National Service Scheme.

The Theme of SERVIS (Scientists/Engineers/Researchers/Visitors in Schools) is in essence a replacement of "*Shramdaan*" (donation of labour) by "*Gyandaan*" (Sharing Knowledge and Wisdom). The dormant spirit of willingness to volunteer to build a strong India can be catalysed into action with the genuine support of all the "stakeholders" in the nation. After all, from the Ambanis of the Coroprate India to the lowly Zaduwala in the village have at least one thing in common and that is the longing to belong to a nation proud and strong (*Mera Bharat Mahan*).

SERVIS seeks to promote the spirit of volunteerism among the academics and other professionals in India and to induce them to participate in a Science/Technology Out Reach Initiative (STORI) Camps in rural villages. SERVIS volunteers will act as a catalyst or facilitator helping school age children to understand natural phenomena and technology with emphasis on the ecology of the environment and the human body. School age children in the village will be encouraged to attend the STORI Camps and engage in fun, interactive group activities—through experimentation, exploring, designing and creating—to learn about the Five Elements [Earth, Water, Fire (Energy), Sky (Environment), Wind (Sustainability)] and how these five elements affect their daily life.

These were the five elements [Earth, Water, Fire, Sky, Wind (*Kshit, Jal, Pavak, Gagan, Sameera Punch rachit yeh adham Sharira*)] mentioned in *Ramayan* as the components that comprised the mortal human form.

The Beginning of SERVIS

I conceived the idea of forming SERVIS in May 1997 following my retirement from British Columbia Institute of Tecnology in Vancouver, Canada. Over a period of 10 years, I have visited many elementary schools through out British Columbia and have spoken with over 3000 students. I engaged the students in a range of fun and exciting hands-on clasroom activities, using laboratory demo equipment from British Columbia Institute of Technology, physics department to illustrate/explore concepts of force, momentum, magnetism, inertia, energy, sound, laser light and to emphasize that it is the human endeavour that invents.

Prior to each visit I requested the teachers to pose three questions to their students. One of the question was: what is one question that they will like to ask a Scientist? Below is a sample of some of their responses.These responses were most revealing to me as they reflected, in my view, the level of curiosity the respondents had about their 'universe'.

1. What is a fiber optic cable? What is the most information that it can carry?
2. Why is the sky blue? What comes first? The chicken or the egg?
3. How much energy does a beam of light have?
4. What is the purpose of life?
5. How do clams make pearls?
6. How fast does my hair grow?
7. When did trees start growing?
8. When does earth end (going into space)?
9. How fast does light travel?
10. How do you make nuclear energy?
11. How far does light travel until it loses light?
12. How fast is the speed of light?
13. What is the speed of sound? How is sound made? Why does sound don't make sound?
14. How can sound vibrate?
15. How do snowflakes make their shape and size?
16. How was the earth made?
17. How does salt lower the freezing point?
18. How do cell phones work?
19. What is the most inventive technology on earth?
20. How was science or technology made?
21. How were the galaxies made?
22. Why do planets orbit?
23. How many earths can fit in all of the planets?

I was also delighted with the positive responses I received following the demonstrations/presentations. Here is a sample of a set of responses from the students and their teacher.

E-mail from Mrs. Hehn, Sir James Douglas Elementary School, Victoria BC dated March 30, 2001

To Amarnath Kshatriya,Thank you so much for coming to our school and giving science demonstrations to all of our classes. You gave so much teaching time to us as well as spending all that time setting up the classroom with the equipment that the children certainly enjoyed. All the teachers raved about what an excellent experience it was for their students. My grade 6/7 class certainly enjoyed the experience and we talked a lot about it.....

P.S. I didn't censor the student's writing; I let them use their overly, friendly, casual style.

Dr. K.

Thanks so much for giving Mrs. Hehn's 2nd class and many other classes the time to show us the science experiments. I think the most interesting thing was the red laser light. I also really liked the paper you put to hold the water in the tube.

—Courtney

Dear Dr. K

I thought the science lesson was awesome!!! I liked the lazer pointer with the comb the best; it was so cool how it made a noise when you put the comb through! Where did you learn that? Where did you grow up? What school did you go to? That sound wave thing was pretty cool as well... you should have put it to the max level and it would have hurt our ears. That would have been really funny. Are you ever going to come back to SJD? I definitely think you should and teach us something every week. I am really interested in science and I thought your little show was AWESOME! You should come again. Peace-Out Doctor K.

From your biggest fan Eliot.

Dear Dr. K.

Thanks very much for your presentation that you did for us. It was really cool especially with that light/laser and that sound wave detector(?) that you used. I found it really interesting and I hope you come out and show us again. Thanks again,

Shohei

Dear Dr. Kshatriya,

Thank you for the presentation you did for our class. I really enjoyed it, and I think all of the other students liked it too. I think the sound vibration prongs were really neat, especially the one that sounded like a mosquito. I also liked how the paper held the water in the tube. My favourite was the red laser light experiment. It was really cool!!! Thanks again!

Eric

The impetus to form SERVIS, was a direct result of the stimulating experience of interacting with the young minds and the realisation that curiosity is nature's gift to the yougsters to help them expand and build their conscious mind. Unfortunately that very invaluable gift of curiosty is rapidly stifled in the Indian home environment and eventually school education (rote learning driven by tuitions, no culture of children asking question of elders, etc.) almost extinguishes the level of curiosity. I know that curiosity can be re-kindled. An enabling learning environment can be a 'nursery' for re-kindling curiosity among the school age children.

In concert with the eight sub-themes suggested for discussion at this conference, it is clear to us that there is an urgent need to create an enduring culture of interest in science and technology which focusses on education and training for sustainable development among school-age students at the village level and beyond.

SERVIS can meet this need. SERVIS volunteers will set out to create an excitement about science and technology by making it relevant to the childrens' daily lives.

This implies the use of hands-on, fun, yet informative activities, not requiring expensive equipment, to discover/explore elements of:

I. *Life Sciences*: Characteristics of living things: different habitat of plants and animals; exploring the human body Nose to Toes; topics in human health, personal hygiene, diseases, drugs, population.

II. *Physical Science*: Properties of objects and materials: explore concepts of measurement, matter, volume, temperature, time, force, magnetisim, power (energy); mechanical concepts; making things, e.g. create a simple machine to harness the energy of the wind.

III. *Earth Science*: Elements of weather, the atmosphere (pollution), food production, sustainability, irrigation, water table, rivers and canals, evaporation,..,..

IV. *Technology*: Computer, how it may be used, ...

STORI Camp facilitators will not adopt the role of traditional teachers but will act as model/mentors/facilitators of learning to help the youngsters develop questions, communicate it and seek guidance from the facilitators.

STORI Camps will not replace the traditional schools in the village, such as they are, but will act as a parallel modality for nurturing curiosity, imparting skills of comprehension, interpretation of assimilated knowledge and analysing and synthesizing the information presented to them, albeit, at an elementary level.

As the SERVIS programme grows and begins to sustain itself, new programmes may be added and SERVIS can be expanded to include SERVIS Science Mentors, Ask-A-Scientist, etc.

Volunteering to make the MDG goals a reality

Volunteering is increasingly seen as an effective tool of social transformation in India. Without the involvement of the ordinary citizens and the dedication of volunteers, the intended social transformations that we are seeking to institute at this conference will be more dormant than enduring.

I understand that over the past 15 years there has been a prolific growth of Volunteer NGOs in this country and quite a few are doing good work. SERVIS will aggressively seek to connect with those that are truly dedicated and willing to join hands to accelerate the empowering of the Rural India.

The Indian coporate sector is also contributing. As an example, Wipro and Intel have jointly launched, *i-shiksha*, a low-cost solution and technology to meet the needs of education segment in India.

The *i-shiksha* is a network of computers that is pre-loaded with learning and assessment software that enriches the learning experience in an interactive learning classroom environment.

Also "*e-choupal*", the unique web based initiative of ITC's International Business Division, offers the farmers of India all the information, products and services they need to enhance farm productivity. Farmers can access latest local and global weather information, scientific farming practices as well as market prices at the village itself through the web portal.

I am optimistic that we will be able to entice well-established and reputable scientists, engineers, researchers, physicians, social scientistis, educators, entrepreneurs, etc., to lend their support to the SERVIS initiative. And that the corporate sector and high tech companies will be prepared to invest in the required start-up funding to support the implementation of the SERVIS proposal (administrative support, purchasing, storing, maintenance and repair of demonstration equipment).

I believe we will also be able to count on Indian Colleges and Universities' support. This will be needed to provide access to expert faculty, required demonstration equipment and advice on the best way to engage the students in learning about and using ICT. All those departments in the various Government levels in India with the responsibility of Human Resource Development, Literacy, Rural Development, etc. can count on SERVIS STORI CAMPS to ensure that the benefits of various schemes will reach the intended "grassroots level and do not remain only on paper",[1]

---

1. a comment made by the former Union Minister for Human Resource, Dr. Murali Manohar Joshi, during a "Conference on Convergence of Social Sector Programmes for Population Stabilization", held in Delhi in 2001.

Summary

Early this September 2005, the gathering of the World leaders at the UN treaffirmed its commitment to the MDGs "to substantially reduce poverty, hunger and ill-health for the world's poorest citizens by 2015."

On September 26, 2005, Kemal Dervis, Head of the United Nations Development Programme (UNDP), addressing a World Bank Conference in Washington DC, commented that "deployment of resources itself will not solve the problem of poverty". Adding that a new emphasis in development focuses on national efforts to achieve political and social freedoms. MDGs have been a "tremendous outreach effort" and have "energized" the world on giving attention to development issues.

SERVIS is ready to launch the next step of "outreach" to activate the volunteer spirit of dedicated professionals to help build Rural India. SERVIS has the formula for success in accomplishing its mission given:

(a) It is an organisation of people in the nation who are dedicated to the cause of empowerment of rural India and are willing to contribute their "*Gyandaan*" to do that;

(b) It is a free assembly of people each of whom is prepared to share his/her best to build up the potential at the grassroots level and help build a strong, caring and compassionate Nation with robust democratic values. Each participant can feel good about the success of the collective effort.

Table 14.1 summarises various phases envisioned to implement the SERVIS project.

Table 14.1: SERVIS—Project Phases

Phase I — Pilot Project	— Create SERVIS volunteer database — Conduct a STORI Camp — Monitor/evaluate pilot project outcome
Phase II —Programme Development	— Solicit programme sponsors/stakeholders — Solicit active support for programme (Government sanction, authority to promote/market programme to schools, volunteer marketing support, financial support, programme implementation support. Gather/analyze info on the differing perceptions of Rural and Urban students on science and technology. To serve as a resource for preparing presentations and demonstrations.

(Contd.)

(*Contd.*)

	— Create database of volunteers for SERVIS Speaker's Bureau — Create profiles of SERVICE volunteeıs to be used as marketing collateral — Secure financial support for programme implementation
Phase III —Programme Implementation	— Development of sample units and lesson demonstrations including the full implementation of the Visiting SERVIS Speaker's Bureau and a gradual introduction access to third-party learning resources. — Develop website dedicated to providing marketing information, registration services, Visiting SERVIS Speaker's Bureau requests and links to relevant resources.
Phase IV —New Activities	Full implementation of the Visiting SERVIS Speaker's Bureau and a gradual introduction of new programme activities such as SERVIS Science Mentors, Ask-A-Scientist Website, Virtual Field Trips and Student Exchanges and Teacher Workshops.
Phase V —SERVIS Railcar Program	— The goal of this programme is to reach communities and schools spread out along India's vast railway system to deliver the highest quality Science and Technology education available in a Visiting SERVIS Speaker's Bureau format. We anticipate the donation and retrofitting of 7 railway cars: (2) Railcars will be retrofitted to contain a modern and technology-enabled classroom-on-wheels which will illustrate how ICT access and e-governance can be used (4) railcars, each being sponsored by a multi-national corporations/local high tech performers, will be used to exhibit the latest in science technology relating to the Four Elements: Earth, Water, Energy and Environment; the seventh railcar will be retrofitted to include sleeping accommodation for 6 SERVIS staff/volunteers, kitchen and eating facilities, bathing facilities, office and storage.

STORI

Science/Technology/Out reach/Initiative

1. **Who will be a Sponsor?**
   Sponsor may be a government agency, a corporation, a philanthropic association, group of individuals/Government Departments—Education/Rural development/MHRD/.
2. **How will the Mission of STORI be accomplished ?**
   By way of two week STORI CAMPS IN RURAL AREAS, where the volunteers will act as facilitators:
   (a) Encourage and sustain curiosity among the learners and teachers;
   (b) Encourage class participation, promote and reward discussions/learning made relevant to the 'real world';
   (c) More 'Why' and 'How' in the Classroom;
   (d) Teachers as Facilitators in Learning;
   (e) Develop a scientific approach to learning (by way of selecting/doing projects in the real world outside the class-room, communicating their results to the class/learning to look at all sides of an argument,...)
3. **Who will be the SERVIS Volunteers?**
   Volunteers in the community who are dedicated teachers and researchers, Volunteers from outside the community who are dedicated teachers and are willing to act as "pollinators".
   We all get together for a week or so (Workshop with the teachers) and share our best practices among one another, following that and in response to a school/village head/a local resident in the village etc. request Volunteers go to a school and try out.
4. **What are the Required Resources?**
   (a) Equipment needed for demonstrating the concept and technical application;
   (b) Equipment for projects/audio-visual aides/tapes etc./ computer access;
   (c) Off site resources may act as web contacts/science technology development communicators;
   (d) Logistics of getting the volunteers to the STORI CAMP, accommodation, board, best times for visit,..,..,
5. **What will be my role? (or for that matter that of any of the Board Members/Directors?**
   (a) To establish an association of local dedicated researchers/ educators who subscribe to the mission of STORI.

(b) Following that a one week workshop is conducted where the local volunteers along with school teachers get together and identify concepts that are difficult to teach/get the students interested in. In the workshop we all share our best practices to engage students' interest/an inventory of equipment we will need along with appliances etc. which clearly demonstrate the technology in action/principles of science underlying/How the computer can entice the students and help the teacher to act as a facilitator?

(c) Contact person dealing with questions/resource material via internet. I expect to visit the Pilot Project site once a year for 4 weeks.

6. Monetary Requirements

(a) An honorarium to the local community volunteers who go to a school on per diem.

(b) Cost of room/board for the week long conference for all participants. Cost of developing relevant equiment/etc. from Local Corporations/Processing Feedback.

7. Evaluation of the Project

SERVIS Board Member, Students, Education Deptartment/ Sponsor of the Project.

PART III

BRIDGING THE DIGITAL DIVIDE

15

Cost Effective Solution for Effective e-Governance-e-Panchayat

C.S.R. PRABHU

1. Introduction

The Governments all over the world are interested in enhancing the quality and quantity of services to the citizens in a cost effective manner. The total cost of ownership for all the stake-holders should be minimum during the entire period of services/ownership of the system. Another major problem the Governments are facing today is to create more employment opportunities for its citizens at least cost. The system proposed shall address these issues also very effectively with a well defined strategy.

2. Institutions of Governance in the Field

The State Governments have created several Panchayati Raj Institutions to ensure grassroots level development. At village level the Gram Panchayats are there, these are grassroots level institutions. At the intermediate level, i.e. Tehsil/Block/Mandal level there are nodal executive authorities, they coordinate the implementation of welfare schemes and services. In the case of Andhra Pradesh, there are Mandal Parishads. Similarly, at the District level the common nodal administrative authority exists, in Andhra Pradesh at District level there are Zilla Parishads. Some places there are Divisions. At the State level there is State level Administration to prepare policy guidelines and ensure the smooth functioning of the Panchayati Raj Institutions.

3. ICT Application at Grassroots Level for e-Governance

How to streamline the administrative processes and avoid delays and conserve resources in the country. Is there a feasible solution for this problem. How to implement the solution in a cost effective manner. Answer to several such questions was e-Panchayat. The main idea is if the adminstrative processes are streamlined and resources are conserved and citizens are empowered with efficient and responsive local administration at every village, then it results in development of districts, states and subsequently the entire nation. So grassroots level problems were taken up for ICT application under the e-Panchayat project.

4. EKPANCH Architecture and Features

EKPANCH's architecture is web-based n-tiered. It functions like Applications Services Provider for all Panchayat level digital services to all the stake-holders. The stake-holders are Citizens, Elected Representatives, Gram Panchayat Officials, the Government and the Knowledge Workers. At Panchayat level an ordinary client with internet access is sufficient to implement e-Panchayat. Most of the computation is done locally, the reporting and association communication takes place through available network for a very brief period of time. Thus field level units can conserve a lot of communication cost and at the same time ensure the uniformity and quality of a e-Governance services.

The EKPANCH Application Server Software has four main modules:

1. Administration Module,
2. Management Information System Module,
3. Citizen Services Module,
4. Elected Representative Module.

The Administration Module is going to help the Gram Panchayat officials in executing their duties and responsibilities delegated to the Gram Panchayat Officials. The MIS Module with a few hundred reports is envisaged to facilitate the State level Gram Panchayat, Planning, Finance and Revenue Department Officials. The Citizen Module with a few hundrend applications is going to be the one-stop and non-stop provider of all the services to the citizens. The Elected representatives module is going to render services to the elected representatives. The multi-media based information dissemination software, the bio-metric based attendance monitoring software etc., are also part of this EKPANCH to facilitate empowerment through simplicity, morality, accountability, responsiveness and transparency (SMART) at the Panchayats.

5. EKPANCH Implementation Model

The main issue is how to ensure successful implementation of EKPANCH in all the Panchayats, in shortest possible time with least amount of resources and overheads by the Government.

The following strategy is suggested. The Commissioner Panchayati Raj and Rural Employement need to buy only the servers. At Gram Panchayats, on Build Own and Operate (BOO) basis the systems can be facilitated through a self-employment generation scheme so that in each village one or two families shall be able to get their livelihood by rendering services to the Citizens and the Gram Panchayats Officials by operating and implementing EKPANCH. For rendering the services, the service charges can be fixed. It will be the enterpreneurs responsibility to collect and feed data, and generate reports. The amount of data fed and the number or reports generated are monitored through software and accounted for making payment. The system maintenance, procurement, internet connectivity and associated logistics shall be taken care by the enterpreneur. Following is the comparative chart of BOO model and normal model.

Sl. No.	*Normal Model*	*BOO Model*
1.	Government has to procure the systems.	Government need not buy systems. It is enterpreneurs responsibility. Government recommends a soft loan through Bank or through Government's economic upliftment welfare schemes.
2.	Government has to prepare site.	Government need not prepare site.
3.	Government has to arrange for internet connectivity.	Government need not arrange for internet connectivity.
4.	Maintenance is Government's responsibility.	The village level enterprenur's takes care of the system, as it is the means of livelihood for him/her.
5.	Citizens may or may not get prompt response.	Citizens get prompt response, as every transaction is chargeable, at the rates fixed by Gram Panchayat.
6.	May get requests from non-serious users.	Non-serious users are eliminated.
7.	systems may fail frequently, as sense of responsibility is low. thus whole system may come to a standstill.	Systems failure rate is less, as a person is there to take care of it 24 hours.
8.	The Gram Panchayat Secretary or other officials may have to feed the reports to the central server. The existing staff may not be willing to operate and	The data will be collected and fed by entrepreneur, as for each report entered and sent, he gets a fixed amount of revenue. So from villages reports come to Government

(Contd.)

(*Contd.*)

Sl. No.	*Normal Model*	*BOO Model*
	feed the data, as they may have to attend to other duties. Even if a person is posted, he may be on leave, or may learn and resign or may not respond properly, perhaps due to lack of motivation or may render services to selected persons only, who have power.	Entrepreneur has to do business with everyone and has to keep everyone happy. So he will serve all sections of society. Based on the size of Panchayat, the GP may decide about the number of self-employed entrepreneurs to be given licences to operate e-Panchayat in the village.
9.	Gram Panchayat has to bear the overheads of system maintenance and communication and stationery and consumables charges. It will be difficult to monitor the expenditure on consumables.	Gram Panchayat need not bother about the recurring expenditure and need not keep track of the associated expenditure.
10.	Operations may be slow, timely reporting may not take place.	Operations shall be faster, as Gram Panchayat can remove slow and non-responsive enterpreneurs. Or if load is more, may issue licence to additional enterprenuers and thus create competition.
11.	Difficult to encourage and create competition.	Prompt services results in quick revenues and thus speed and promptness are encouraged. And Scope exists for creating a healty competition for operating and utilising the EKPANCH system.
12.	Gram Panchayat has to incur expenditure.	Gram Panchayat may also may get a share of revenue.
13.	Revenues may not go up as all payments may not be accounted.	Revenues go up as all accounts are recorded, even otherwise, people shall demand a receipt from a powerless businessmen for the amount paid rather than a powerful Government servant.
14.	Takes time to implement in all the villages, due to overheads.	Implementation can take place simultaneusly in all villages, as overheads are almost nil for the Government.
15.	Difficult to sustain the model.	Easy to sustain the model, so runs for longer time.
16.	The Government shall be under pressure to ensure successful implementation, as overheads are more and expenditure is involved.	Government shall not have any pressure, as there is no expenditure for the project implementation in the field from the Department. The pressure is distributed to all the village level enterprenuers. So problem is getting divided and solved.

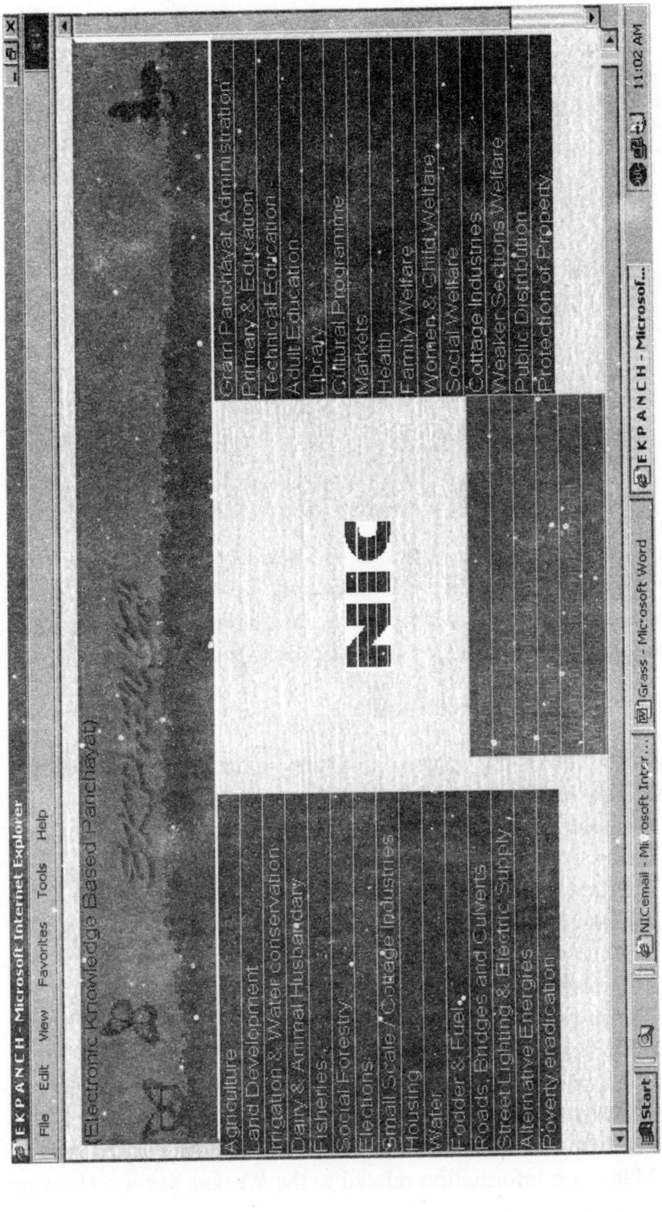
E K P A N C H - Microsoft Internet Explorer
File Edit View Favorites Tools Help
(Electronic Knowledge Based Panchayat)
Agriculture
Land Development
Irrigation & Water conservation
Dairy & Animal Husbandary
Fisheries
Social Forestry
Elections
Small Scale / Cottage Industries
Housing
Water
Fodder & Fuel
Roads, Bridges and Culverts
Street Lighting & Electric Supply
Alternative Energies
Poverty eradication
NIC
Gram Panchayat Administration
Primary & Education
Technical Education
Adult Education
Library
Cultural Programme
Markets
Health
Family Welfare
Women & Child Welfare
Social Welfare
Cottage Industries
Weaker Sections Welfare
Public Distribution
Protection of Property
Start
NICemail - Microsoft Inter...
Grass - Microsoft Word
E K P A N C H - Microsof...
11:02 AM

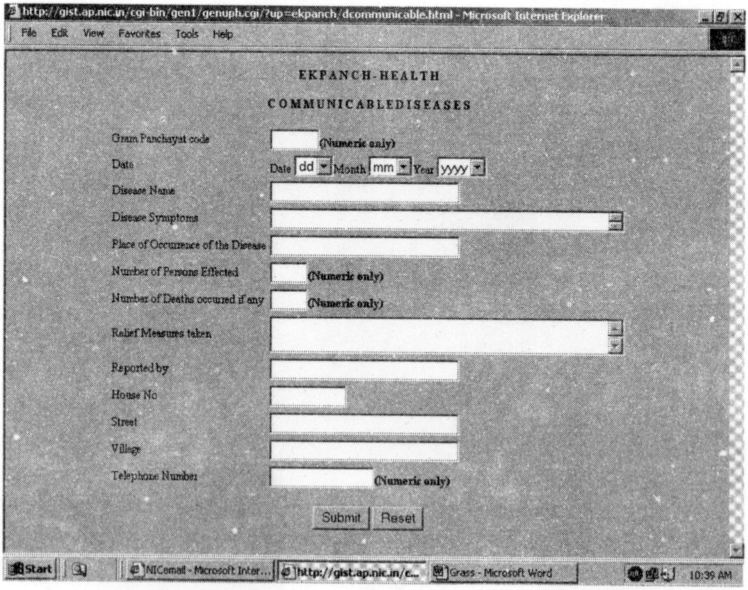

A model in the similar lines is successfully functional in West Godavari District. The software is developed by NIC and the Central server infrastructure is created by the State Government and the field level resources are facilitated on lines similar to the BOO Model.

6. Features of EKPANCH Software

The main features of the various software modules of EKPANCH—(Electronic Knowledge Based Panchayat) are:

1. Bio-metric Attendance Monitoring,
2. Below Poverty Line Enumeration,
3. Village Level Planning,
4. Meetings Management,
5. Self-Help Groups related Information Management,
6. Monitoring Attendance in Gram Panchayat Meetings,
7. Managing Information related to the Meetings at Mandal/ District Officials,
8. Managing Information related to the Assignments of the Superior Officers related to the Village,
9. Managing the Information related to the Gram Panchayat Property.
10. Managing Information related to the Weaker Section Housing Programme,

11. Encroachments related Information Maintenance,
12. Managing Information related to the Atrocities against Women and Children (if any),
13. Gram Panchayat Registers Computerisation,
14. Communicable Diseases related Information Management,
15. Tax demands, receipts generation, revenue accounting,
16. Anti-Malarial Spraying related Information Management,
17. Crops related Information Management,
18. Anganwadies, Balwadies related Information Management,
19. Generation of Ceritificates for Citizens,
20. Tree Plantation and Protection related Information Management,
21. Maintain Sanitation related Information,
22. Maintaining information related to the Atrocities against SC, ST (if any),
23. Planning and Implementation related Information Management
24. VTDA related Information Maintenance,
25. Relief and Rehabilitation related Information Management.
26. Untouchability Eradication Programmes related information Maintenance,
27. Assistance to AP TRANSCO related Information Maintenance,
28. Enrolment related Information Management,
29. Minimum Wages Act voilation related Information Maintenance,
30. Maintaining Information for organizing Literacy Classes,
31. Birth-Death Registration related Information Management,
32. Crop Coverage Information Management,
33. Marriage Registration Information Management,
34. Extension Information Management for Dissemination,
35. Grama Sabha—Beneficiaries Information Management,
36. Managing Land Information for Azmoish of Crops,
37. Property Details Information Maitenance,
38. Maintenance of Agricultural Statistics,
39. Managing Information for Serving Legal Notices (as and when required),
40. Maintaining information to ensure up-keep of Display Boards and Assets,
41. Managing Information related to Village Level Events to Inform Public,
42. Maintaining Information related to Black Marketing of Inputs for Reporting,
43. Managing Information to Assist in Loan Recovery,

44. Maintaining Weavers Information for Reporting,
45. Maintaining Information related to Panchanamas conducted,
46. Managing Information for Generating and Issuing Dependancy Certificate to Weavers,
47. Managing Information to ensure safety of Government Attached Property,
48. Managing Information to ensure better Co-ordination among various Agencies operating in the Village,
49. Managing Information for Performing Election Duties,
50. Managing Information related to Government Programmes for Public Participation,
51. Maintaining Information to inform Police about Unlawful Incidents,
52. Managing the information related to Village Teachers for better liaisoning,
53. Managing Information to Organize Meetings for Community Work,
54. System to Report Accidents immediately to Higherups,
55. System to ensure Co-ordination among all Departments right from hamlet level to the State level,
56. Manage Information related to the Old Age Pension Scheme,
57. Maintenance of Information to ensure Clean and Green,
58. System to Report Epidemics Outbreak in Time,
59. System to provide Marketing Services,
60. Managing Information related to the Community Welfare and Development,
61. Other Information, Counselling, Transaction Processing (including payment), Certification, Authentication, Grievance Redressal etc., Services are included in EKPANCH.

7. Current Implementation Status and Conclusion

For sustainable socio-economic development at the grassroot level, a wide involvement of the communities and businesses would need to be ensured. This would automatically warrant forging of sound business relationships based on transparency, mutual goodwill and benefit. NIC's e-Panchayat offers a unique opportunity for establishing inter as well as intra stakeholders partnerships at the grassroots level. Realising this fact the Government of Andhra Pradesh has decided to implement the e-Panchayat system in all the major Gram Panchayats in the Andhra Pradesh. It went one step ahead and to ensure e-readiness of the Panchayats, training,

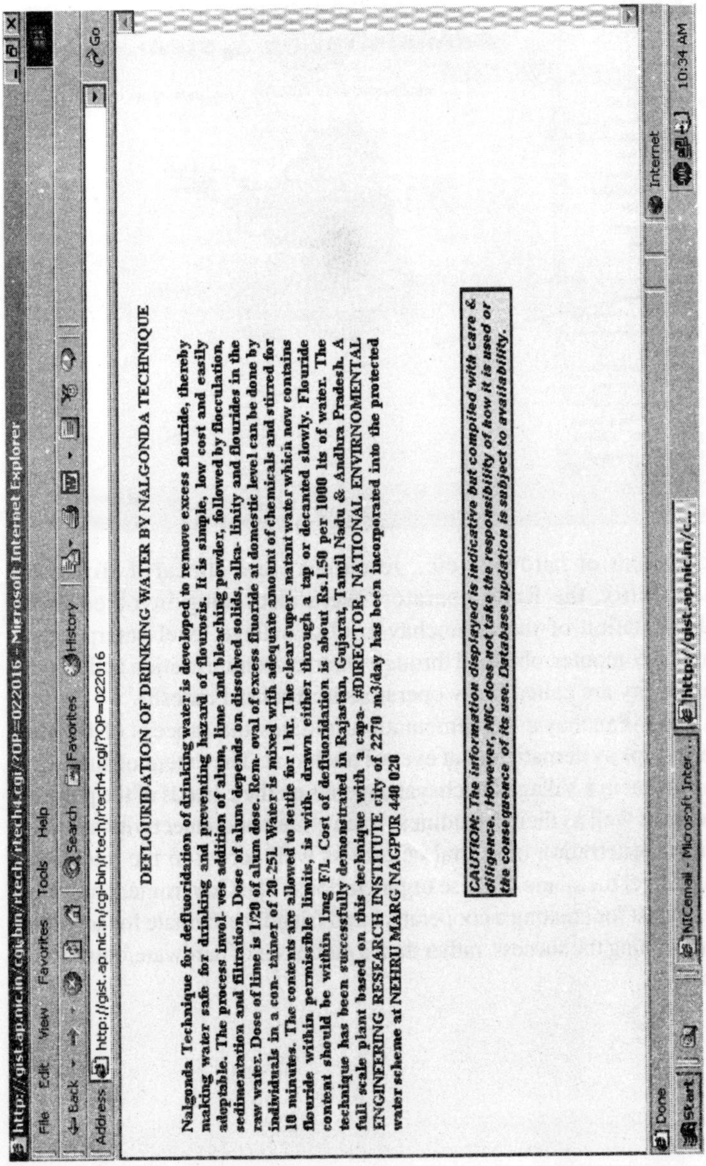
http://gist.ap.nic.in/cgi-bin/rtech/rtech4.cgi/?OP=022016 - Microsoft Internet Explorer
File Edit View Favorites Tools Help
Back Search Favorites History
Address http://gist.ap.nic.in/cgi-bin/rtech/rtech4.cgi/?OP=022016
Go
DEFLOURIDATION OF DRINKING WATER BY NALGONDA TECHNIQUE
Nalgonda Technique for defluoridation of drinking water is developed to remove excess flouride, thereby making water safe for drinking and preventing hazard of flourosis. It is simple, low cost and easily adoptable. The process invol- ves addition of alum, lime and bleaching powder, followed by flocculation, sedimentation and filtration. Dose of alum depends on dissolved solids, alka- linity and flourides in the raw water. Dose of lime is 1/20 of alum dose. Rem- oval of excess fluoride at domestic level can be done by individuals in a con- tainer of 20-25L Water is mixed with adequate amount of chemicals and stirred for 10 minutes. The contents are allowed to settle for 1 hr. The clear super- natant water which now contains flouride within permissible limits, is with- drawn either through a tap or decanted slowly. Flouride content should be within 1mg F/L Cost of defluoridation is about Rs.1.50 per 1000 lts of water. The technique has been successfully demonstrated in Rajastan, Gujarat, Tamil Nadu & Andhra Pradesh. A full scale plant based on this technique with a capa- #DIRECTOR, NATIONAL ENVIRNOMENTAL ENGINEERING RESEARCH INSTITUTE city of 2270 m3/day has been incorporated into the protected water scheme at NEHRU MARG NAGPUR 440 020
CAUTION: The information displayed is indicative but compiled with care & diligence. However, NIC does not take the responsibility of how it is used or the consequence of its use. Database updation is subject to availability.
Done
Internet
Start
NICemail - Microsoft Inter...
http://gist.ap.nic.in/c...
10:34 AM

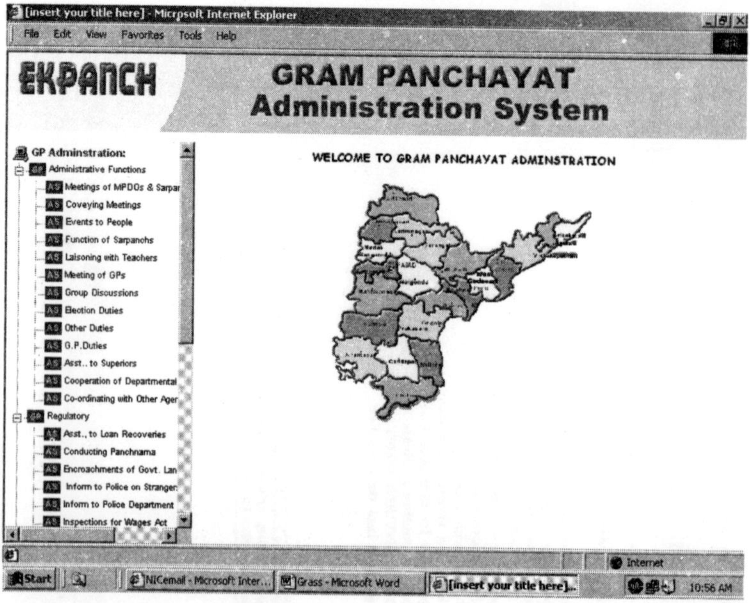

procurement of hardware etc., resources were allocated. To ensure sustainability, the Rajiv operators were also being involved in the implementation of the e-Panchayat. (The village level enterpreneurs owning a computer, obtained through District Administration and internet connectivity are called Rajiv operators in Andhra Pradesh).

The e-Panchayat implementation initiates the process of building partnerships systematically at every Panchayat. The extent of success of e-Panchayat in a Village Panchayat depends on the e-needs of the potential partners as well as their e-readiness to an extent of connectivity available to them, penetration of digital culture as well access to the internet by village level members of these organizations. The Government has to act as a catalyst for creating a cooperative and friendly e-climate for achieving and sustaining the success, rather than a purchaser of hardware/outsourcing agency.

16

Mainstreaming ICT for Fostering Agricultural Growth, Poverty Reduction and Sustainable Resource Use to Usher in Rural Prosperity

MADASWAMY MONI

1. Introduction

Agriculture, including crop and animal husbandry, forestry and agro-forestry, fisheries, and agro-industries, provides livelihoods to over 70 per cent of the Indian population. Agriculture is decentralised but small scale manufacturing (small and marginal farmers) units. In addition, India is a land of diversity with different types of terrain, various agro-climatic conditions, different levels of socio-economic conditions, and varied levels of regional development. There have been concerns about persistent rural poverty, unemployment and inequality, and resulting social tensions at grassroots level in India. The strategies and policies developed by the Planners and Policy-makers, adopted two approaches: one focusing on the overall economic development (through percolation, trickle down and spread effect), and the other on poverty alleviation (direct intervention). Though these two approaches reinforce each other, there has been no attempt to integrate them.

At the beginning of the new millennium, 260 million people in the country did not have incomes to access a consumption basket, which defines the poverty line. Of these, 75 per cent were in the rural areas. Agricultural wage earners, small and marginal farmers, casual workers engaged in non-agricultural activities, rural women (especially women-headed households), among the others, constitute bulk of the rural poor.

After decades of limited success in eliminating rural poverty, new ideas about rural development (i.e. viable society in a rural space, livelihood approach, sustainable livelihood approach, sustainable agricultural and rural livelihood approach, sustainable community concept, Multiple Livelihood Opportunities, etc.) are emerging so as to reduce the vulnerability of the rural poor. Poverty Alleviation, Livelihood Opportunities and Gainful Employment are closely linked.

There have been both national and international efforts (DOT Force of the UN, the UN/ESCAP Committee on Poverty Reduction, the Millennium Development Goals, Poverty Net of the World Bank, etc.)[1] to improve information flows and communication services to eliminate poverty (ICT for Poverty Reduction), which are a necessary but not sufficient condition. In poor rural areas, where agricultural productivity is low and unreliable and there is food insecurity, better information and knowledge-exchange can be important in lessening poverty.

The growing populations need food, clothing, shelter, fuel and fodder for their livestock. As market oriented economic development proceeds, Indian farmers in rural areas continue to experience great disparity in income compared with other sectors, and they are forced to revert to natural resources as the most accessible sources of livelihood. As the natural resources which are critical inputs for providing sustainable livelihood to the rural poor are in a degraded condition, it is difficult for the SMFs to generate gainful employment in rural areas and to come out of poverty. Lack of infrastructure, inadequate financial resources and poor managerial capabilities have further shaken their confidence to fight poverty[2]. Degradation of natural resource is a key threat to socio-economic development, and to global environment (e.g., climate change and loss of bio-diversity).

Since the Rio Earth Summit in 1992, agriculture remains high on the international agenda because it brings together critical issues like water, poverty, hunger, and health. "Reaching the Rural Poor" makes broad-based economic growth its primary objective, and treats agriculture as the leading productive sector within the rural economy and closely linked to non-farm activities. Science and Technology (S&T) offers tremendous opportunities in simultaneous achievements of the goals of sustainable agriculture and improving the rural livelihoods[3]. "Doubly Green Revolution" adheres about "growth in agricultural production (GAP)" and "improved livelihoods" through:

- Access to land, capital and knowledge,

- Dissemination and sharing of knowledge and technologies (IPRs),
- Improvement of user training and qualification,
- Affordable and accessible technologies,
- Gender equity,
- Improve/enable risk/benefit assessment at national, institutional and private level.

Demands for agricultural technology are changing and diversifying[4]:

- Technologies to break the stagnation in the yield growth in intensive irrigated agriculture,
- Technologies for dryland farming which accounts for two-third of the country's cultivated land,
- Technologies for diversification of farming systems for employment and income generation and in high value commercial crops, and development of sustainable agricultural practices that preserve natural resources while enhancing the productivity.

Informatics Networks, besides Computer Networks, are increasingly considered as development tools for achieving (a) Reaching the Unreached—Public Services; (b) From Digital Divide to Digital Opportunities for sustainable development and economic growth; (c) Fostering agricultural growth, poverty reduction and sustainable resources use; (d) Sustainable Development and Earth Care Policies—Water, Energy, Education, Health, Agriculture and Rural Development, Bio-diversity; and (e) a Cluster of Villages—Sustainable Societies in Viable Rural Space. Emergence of Information and Communication Technologies (ICTs) on the national agenda and the announcement of ICT policies by various State Governments, since 1995, have recognised the "Convergence of Core Technologies and e-Governance" as the tool for sustainable development and globalisation of economy.

Good Governance and institutions are indispensable for sound agricultural and rural development in developing countries. Models of e-Government (i.e. digital Government) are continuously evolving and improvising to harness the potential offered by the Information and Communication Technologies (ICTs) and deal with new realities in the area of governance, throughout the World. Generic Models of e-Government (www. digitalgovernance.org), viz. Broadcasting/Wider-Dissemination Model, Critical Flow Model, Comparative Analysis Model,

E-Advocacy/Lobbying and Pressure Group Model, are relevant while discussing "design of an e-Government for Poor". Studying and influencing the "Geometry of Information Flows" facilitates direct benefits rather than trickle-down benefits for the disadvantaged community (i.e. the Poor).

Digital opportunities have emerged as "power tool" for fostering agricultural growth, poverty reduction and sustainable resources use in developing countries. The National Agricultural Innovation Project (NAIP) of the ICAR envisages, among the others, to transfer appropriate technologies and to develop innovative models for promoting agricultural production, particularly farmers belonging to weaker sections of the society and located in disadvantaged regions, through the components of:

(a) Collaborative Research Models for Sustainable Improvement of Agricultural Value Chains;
(b) Collaborative Research Models for Livelihood Improvement in Disadvantaged Regions.

Rural India desires to take advantage of "knowledge-intensive" techniques in sustainable crop farming and animal husbandry, to get better value-added for their products via information links to markets, and to obtain up-to-date, accurate information on entitlements, viz. health and welfare benefits, employment opportunities. Innovation is the need of the hour in India's rural sector.

2. Viable Society in a Rural Space: An Epitome for Rural Livelihoods

Our economic and industrial growths are dependent on production and productivity in agriculture. Agriculture, with its large dependent population, has to thrive and flourish in order to secure rural prosperity. Agricultural development, along with village and cottage industries, tiny and micro-enterprises, are the cornerstone for promoting sustainable rural livelihoods[5]. Agricultural productivity is a function of seed quality, nutrient adequacy, irrigation and adoption of appropriate package of practices. Agricultural growth is an important factor in poverty reduction, especially in agrarian economies like ours, in view of its "multifunctionality"[6].

The constraints faced by the Indian Agricultural sector, for achieving its sustainable growth in many parts of the country, are well documented as follows:

- Extensive pressure on land due to urbanisation,
- Fragmentation of land holdings,
- Frequent failure of monsoon and uneven distribution of rains (climatic aberrations),
- Depletion of ground water due to over-exploitation and lack of recharge,
- Increasing areas under fallows,
- Declining nutrient status of soil and soil health due to intensive cultivation—Degradation of land (problem soils),
- Infrastructural inadequacy,
- Inadequate market support,
- Weak linkage between farmers and R&D institutions,
- Inadequate post-harvest infrastructure, processing facilities, research and extension support,
- Paucity of resources for large investments with long gestation periods,
- Ineffective pre-harvest and post-harvest Supply Chain Models and CRM Models.

In areas where investments in agricultural infrastructure have lagged (viz., State of Bihar, State of Orissa, State of Assam and State of Madhya Pradesh), rates of growth in agricultural productivity and poverty reduction also lagged[7]. Most parts of the Eastern Region (ER)[8] lags rest of India on some social indicators like: (a) Poverty Ratio (highest in Bihar); (b) Infant Fertility (highest in Orissa); (c) Overall Literacy (lowest in Bihar) and Female Literacy (lowest in Orissa and Bihar), and (d) Poverty incidence varies within the Eastern Region. Given the strong linkage between output growth (particularly agricultural growth) and reduction in poverty, accelerated agricultural growth is critical, to reduce poverty in the Eastern Region (ER) States.

The Green Revolution has been "supply-driven" whereas the Livestock Revolution is driven by "demand-driven". Livestock are a renewable natural resource but Livestock Production requires natural resources such as land, feed and water. Mishra (2004)[9] has claimed that the mixed farming system, in which livestock and agriculture are integrated or interdependent, is environmentally most benign, and also open for sustainable development in future. Studies have shown that dairy enterprises as against crops in rural areas gave large profit margins in marginal, small and medium holdings. Livestock farming is major player in dryland and hill regions. Sustainable rapid growth and development in this sector can be only ensured if the livestock owners, service providers, veterinarians and planners: become knowledge-based and acquire the ability to absorb, assimilate and adopt the spectacular development in the veterinary sciences and related technologies. Bhat (2004)[10], therefore, envisages an urgent requirement of "Animal Production and Health Information System (APHNET)" in the country.

The experience shows (Acharya, 2004)[11] that rural livelihoods can be improved in a sustainable way if the following conditions hold true:

- Availability of appropriate technology;
- Availability of inputs, services and credit to farmer and rural families;
- Availability of market support to farmers and rural households;
- Adequate local institutions and social capital;
- Security of tenure or rights to resources;
- Absence of poor non-friendly macro-policies; and
- Good governance.

Our vision is to realise an economically and socially vibrant agricultural industry, i.e. to accelerate commercialization of innovative, alternative and value-added agricultural and farm related enterprises, in rural areas. This development strategy, *inter alia*, facilitates skill improvement, providing employment in rural areas, transfer of technology, rural industrialization and promoting self-reliance among the people and to build up a strong rural community base. That is to build a Rural India, which shines. Farms, households, and other components of rural systems need a minimum bundle of rural infrastructure services to function efficiently, which include:

- An adequate supply of safe drinking water;
- Health and education infrastructure;
- Transportation infrastructure and services;
- Information systems; and
- An adequate and dependable energy supplies.

Agriculture in India needs a productivity and quality revolution, i.e. a Total Quality Management (TQM) and a Total Factor Productivity (TFP), which can be brought out through the much-needed agricultural reforms. Agricultural reforms mean land reform plus a package of support services such as credit, marketing, pricing, and rural infrastructure (water, roads, bridges, hospitals and schools). Agricultural research, education, extension and training are the essential four pillars of sustainable Agriculture and induction of "Knowledge Triangle" (Fig. 16.1) is essential. Information and Communications Technologies (ICTs) have tremendous role to play in these four components.

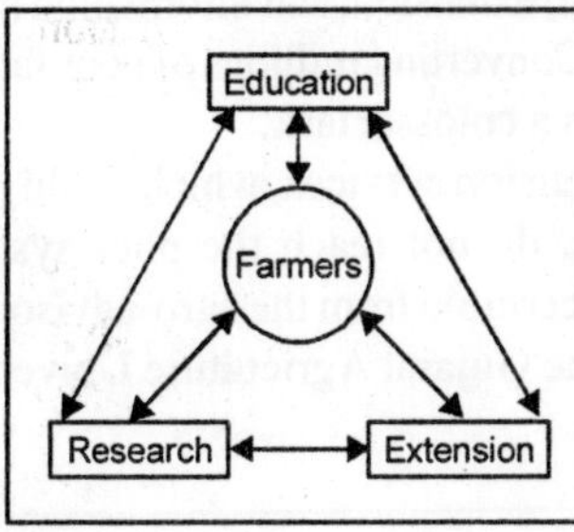

Fig. 16.1: Knowledge Triangle (www.fao.org)

3. Agricultural Extension: An Agricultural Knowledge and Information System for Rural Empowerment and Improved Livelihoods—A Situation Analysis

Agriculture is different all over the world and hence the 'one-size-fits-all' model (i.e. "Green Revolution Model") of agriculture is inappropriate and cannot reflect the environmental, social and cultural diversity that exists around the world. Rural households at various agro-eco-regions practice several farming systems. While each system might bring benefit to an individual household, the conflicts in resource utilization often nullify the effort of the whole community. Societies have also different attitudes to intensification and new technologies, and with regard to traditional forms of farming, e.g. maintaining small-scale mixed farming, specific types of produce, knowledge and skills etc. Extension strategy for rainfed farmers and farming is different than irrigated rich resource and progressive farmers[12]. Agricultural practices have, however, a major impact on the use of natural resources, eco-systems and bio-diversity. Therefore, a better approach to sustainable natural resource management must be designed to help small and marginal farmers (SMFs) in less privileged, fragile areas. Competent techniques of transferring technologies to the farming communities are of prime importance towards all round development of the agricultural sector.

In developing countries, the productivity and incomes of the Small and Marginal Farmers (SMFs) have stagnated or even decreased. SMFs, traditionally the backbone of much of developing countries agriculture, are under threat[13]. Farmers, in particular SMFs, must be able to choose agricultural practices and technologies from the full range of approaches available for tackling their problems: agro-ecological methods, conventional research methods, and molecular biology research methods. The slack in adoption of technology by the SMFs is largely due to inaccessibility or unavailability of credit, extension, and information rather than any inherent incapability in them. Converting millions of poor farmers to the use of new technologies has been a colossal task.

Agricultural extension services, which could help reduce the risk of on-farm investments, do not reach the poor systematically. What the farmers require is discernible from the agro-advisories in farm operations being offered from the Gujarat Agriculture University (GAU)—Anand, as follows:

Sl. No.	Farm Operations	Percentage
1.	Pest and Disease Information	32.4
2.	Sowing time Information	22.7
3.	Crop Information	16.3
4.	Weather Information	13.1
5.	Irrigation Scheduling	4.4
6.	Tobacco Crop Information	4.3
7.	Rice Crop Information	2.3
8.	Potato Crop Information	2.2
9.	Harvesting Information	1.2
10.	Fertiliser Application	1.1

There are four types of actors involved in the agricultural extension: public agencies, private agencies (input suppliers etc.), Production Organisations (POs), and NGOs. Agricultural extension and its services involve basic functions such as informal, dissemination, advisory, and educational. The Agricultural Extension System Infrastructure in India includes:

- Agricultural Extension Officials (Full time) in States — 90000 (approximately)
- Extension Education Institutions (EEIs) including MANAGE — 5
- Krishi Vigyan Kendras (KVKs) in 252 Districts (Proposed to have one KVK in every district) — 261
- Advanced Training Institutes (ATIs) — 15
- Farmer Training Centres/Kisan Vidyapeeth — 137

- Rural Development Institutes (mainly in the State of Uttar Pradesh) — 52
- Trainers Training Centres — 8
- ICAR Institutes/Directorates — 85
- State Agricultural Universities — 40
- ICAR Deemed Universities — 5
- Agricultural Colleges — 300

The ATMA (Agricultural Technology Management Agency) Model, which was the extension component of NATP implemented in 7 States, during the Ninth Plan period (1997-2002), was considered as a "way forward" to increase farmer's input with respect to the followings:

- Farming system innovations (intensification or diversification),
- Farmer organisations,
- Technology gaps both in crop production systems and livestock production systems,
- Natural Resources Management (Soil Conservation and Water Conservation, Reduction of Pesticides use through IPM Programmes), and
- Marketing and Agro-processing linkages.

The Agricultural Extension Service is a medium-term and long-term investment in the same way as education and research. The farm is a business and the farm manager (farmer) needs to optimize the returns from their agricultural investment. The interactions between agricultural systems and practices, economics, society and the environment are highly complex. Human Resources Development among farming communities is very essential, as they have to face the challenges at ground level: creation of technology (IPR Divide), diffusion of recent innovations (Digital Divide), diffusion of old innovations (Extension Divide) and diffusion of human skills (Educational Divide).

The ongoing scheme of "Agri-Clinics"[14] intends to advise farmers on cropping practices, marketing their products, technology dissemination, crop protection from pests and diseases, market trends and prices of various crops in the markets. Agri-clinics are expected to augment the existing agricultural extension network, which continues to be the only farm support facility in most areas in India. Establishment of a "Network of Agri-Clinics and Agri-Business Centres" is a new dimension in agricultural extension services to farmers.

The Government initiative of AGMARKNET[15] Programme during the Ninth Plan with the roadmap to network 7,000 Agricultural Produce Wholesale Markets and 32,000 Rural Markets is a step towards "market-led extension" in India and is expected to have impact on the supply-chain of pre-harvest and post-harvest segments of agricultural operations. This has definite impact on farm investment. This Government initiative (AGMARKNET) and Private initiatives [e-choupal by ITC Ltd., Agriventure, and Mahindra Krishi Vihar (MKV)[16]] ushers in "market-led extension" in the country, facilitating farmers to (i) enhance farm productivity, (ii) improve farm gate price, and (iii) cut transaction cost. Demonstrations of Mahindra Krishi Vihar (MKV) at Madurai district in Tamil Nadu have shown a significant increase in productivity (as against an average paddy yield of 1.8-2 tonnes per acre, farmers have achieved yield ranging from 3-3.6 tonnes per acre). The consultancy and technical support package was at a cost of about 10 per cent on the increase in productivity achieved[17]. Community Agro-biodiversity Centre (CAbC) of the M.S. Swaminathan Research Foundation (Chennai) aims at protection of livelihood and bio-diversity in villages.

According to Preeti Ramanathan (2002)[18], Agri-Clinics, in its present form and without the supporting commercial and legal mechanisms, would remain yet another well-intentioned government intervention in agriculture and rural development, without sustainable growth and without any relations to macro-level planning, for development. The rural economy is larger than the yearly 12,000 graduates or so in agriculture and related subjects, and the ills are far beyond and over encompassing. The government expects that "Dial 1551 Call Centre" which rides on the fact that six lakh villages have phone connections, could by-pass a clearly inadequate system of extension workers and NGOs.

Agricultural Extension Programmes need to be oriented towards enhancing the decision-making abilities of rural people, especially farmers, by expanding their exposure to effective organization and management skills, not just new agricultural production technologies alone, i.e. "knowledge-intensive". The task of extension would become more challenging in the wake of post WTO era, which demands a "system of market-led extension" with specific focus on diversification, post-harvest management, and export orientation. There are about 600 million farm workers, most of whom are not agricultural science graduates, yet possess inherent knowledge to carry out commercial activities. What they lack is capital, access to trade, agricultural resources, market information system, connectivity among others of their ilk, decision-making support systems and market mechanisms to convert their labour into profitability.

Mohan and Moni (2001)[19] proposed an exclusive Agricultural Extension Information System Network (VISTARNET) to reach Technology to the Small Holders (Resource-Poor-Farmers) in India, by networking the existing Agricultural Extension Infrastructure, given above, and also undertake "capacity-building" through Human Resource Development Programme for using ICTs. However, The Government Extension Machinery had been promoting "WAICENT" for retrofitting in India during the Ninth Plan. With the implementation of DACNET Infrastructure (An e-Governance Infrastructure for Development information Exchange—E-GOV4D and also see http://www.dacnet.nic.in) in all Directorates of the Central Department of Agriculture and Cooperation (DAC), and the AGMARKNET, the VISTARNET has been realised in India. "Dial 1551" to DACNET VSAT Nodes can generate a synergetic response to farmers.

In the Indian context, emergence of an e-Farmer[20] is the need of the hour. Synergetic collaborations among Cooperatives, Agricultural Produce Markets, Agri-Clinics, Agri-Business Centres, and AGMARKNET could become the "pathways" to rural prosperity. The Agricultural Extension System is expected to be "an Agricultural Knowledge and Information System, using Geomatics Technology" in India. Development of Agricultural Extension XML (AeXML) is an essential requirement.

4. Developments in Digital Governance (e-Governance)

E-Governance is more about putting together the IT infrastructure to make the average citizen's interface with the Government easier, while e-Governance has more to do with the laws and regulation of the Internet. Models of Digital Governance (e-Governance) are continuously evolving and improvising to harness the potential offered by ICTs and deal with new realities in the area of governance. There are no rigid and finite models of digital governance. In fact, several developing countries are putting into practice innovative e-Governance models that may technologically be simple but are changing the way information is distributed in the society (http://www. digitalgovernance.org). All these models share in common, the inherent characteristics of the new technologies, which are: (a) enabling equal access to information to anyone who is a linked to the digital network and (b) de-concentration of information across the entire digital network.

Studying and influencing "geometry of information flows" is important because the aim is to use ICTs for development purposes—those that bring real, significant changes in the lives of disadvantaged communities (i.e. farmers), rather than simple embedding of ICTs in the society. Somewhat detailed steps, as available from published materials, for e-Governance projects are given below:

- Who are our Target Groups that we want to reach out to, through ICT for Development Projects?
- Small farmers with less than 1 acre of land.
- Farmers who have land away from roads and markets.
- Farmers farming in ecologically fragile areas.
- Newly turned farmers, young and women farmers.
- Farmers lacking credit, tools to enhance land productivity.
- What are the key information needs of the disadvantaged community?
- Information on identifying and dealing crop pests and livestock diseases.
- Technical inputs on how to carry contour bunding, land-levelling, water harvesting activities, composting to increase productivity.
- Information on Government subsidies and schemes on seeds, fertilizers, horticulture and minimum support price.
- Information on new crop varieties, irrigation frequency, setting up farm-based enterprises.
- Information on market prices of the crops, availability of credit, agriculture fairs, soil-testing laboratories and training programmes.
- What are the existing channels by which information reaches to the disadvantaged community?
- Through other farmers, progressive farmers, money-lenders, teachers, public phone operator, postman and health workers.
- Through Government Officials, agricultural extensionists, agriculture fairs, agricultural universities and NGOs.
- Through radios, televisions, folk songs and newspapers.
- What is the weakest link in the chain of information flows: from source to the disadvantaged community?
- Information may be available at local agricultural centres or in markets but these are not easily accessible by farmers.
- High levels of illiteracy prevent farmers to benefit from available information.
- Agricultural extensionists are knowledgeable but do not visit farmlands away from roads or in remote areas.

- Agricultural extensionists and local agricultural centres do not have updated knowledge of new crop varieties, pest control and Government schemes and subsidies.

Here the disadvantaged community is the resource poor farmers (land farming and landless farming).

Information, Knowledge and Communication (IKC) are the lifeblood of economic and social interaction. Agriculture (genetic modification), Medicine (Genome Research and Bio-informatics) and Information and Communication Technologies (ICTs) are the three fields where diffusion of technology holds particular promise for the poor (*The Economist*, November 2001). The poor lack:

- Access to information that is vital to their lives and livelihoods;
- About market prices for the goods they produce;
- About health;
- About the structure and services of public institutions;
- About their rights;
- Political visibility and voice in the institutions and power relations that shape their lives;
- Access to knowledge, education and skills development that could improve their livelihoods;
- Access to markets and institutions, both Governmental and societal, which could provide them with needed resources and services;
- Access to, and information about, income-earning opportunities.

Harnessing the information revolution for economic development, social cohesion and poverty alleviation in the 21st century is the theme for various National and International Conferences. The present day Digital opportunities (0s and 1s) will assist us in realising the concept of "Sustainable Communities", which is one where all stakeholders, as partners in progress on the road to economic development, will be enabled to achieve "Sustainable Production" and "Sustainable Consumption". There is a "need to bridge theory and reality (i.e. grassroots)" (Fig 16.2). An ICT based Gateway to farmers is essential.

ICT Diffusion and Infusion in agricultural sector provide the necessary "digital opportunities" or "advantages" for productivity increase, income generation and decrease in regional disparity. ICT has opened considerable opportunities for the rural poor, both in terms of direct employment and in improving their linkages with the market. Markets are now the driver for agricultural growth. Marketing today depends on information system, i.e. on adequate information about what people want, at what price, and who can supply it (*http://www.agmarknet.nic.in*). It is essential to look into the sustainability aspects of "technology transfer" to grassroots level development. ICT diffusion and Infusion should induce developments in the areas of:

- Physical connectivity to bridge distances;
- Electronic connectivity to increase communication;
- Knowledge connectivity to provide information, technology, and increased understanding;
- Market connectivity to link rural products to markets.

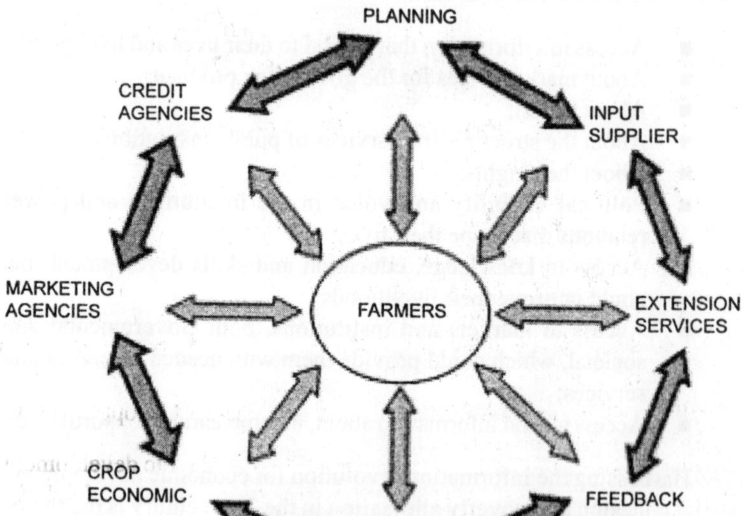

Fig. 16.2: Gateway to Farmers—Interaction of Farmer with Various Components of Agricultural Input-Output Systems

"Networking of People" and "Networking of Information" are essential. One of the major problems of using ICT for poor is language barrier. Content development in local languages is challenging task ahead. Grassroots level Information access (Contents) and Grassroots level access to Information (Networking) are the two essential components for grassroots level development strategies through ICT.

5. Mainstreaming of ICT for Collaborative Research Models for Ustainable Improvement of Agricultural Value Chains

Agricultural Resources Information System—A Needed Domestic Strategy for Sustainable Rural Livelihoods: Proof-of-Concept Projects

The scientific management of agricultural resources becomes a very important part of modern agricultural production in the 21st Century and is a new direction in the domain of agricultural engineering. Agriculture is highly dynamic in nature, because of the changing phenomenon of agricultural crops, which is further complicated by the interaction of crops with environment. A wide range of agricultural practices could be affected including potential environmental benefits that could come from minimizing adverse impacts, by reducing external inputs and greater use efficiency. Farmers, Land and Natural Resources (Supported by the Land) have intrinsic and dynamic relationship.

Resources Application and Agronomic Practices are to match with soil attributes and crop requirements, as they vary across a site (McBratney and Pringle, 1997). The site-specific management (or precision farming) is farm-size and production-system neutral, and will make agriculture "information intensive". This will impact rural economies. Despite potential of economic and ecological benefits, adoption of precision technologies is very slow throughout the World. The reasons for limited implementation of site-specification management (or precision farming) in Asian Countries are due to: small land holdings, cost-benefit aspect, heterogeneity of cropping system, lack of local technical expertise, and knowledge and technological gaps. In India, about 57.8 per cent of operational holdings have size less than 1 hectare.

Proper analysis of the agricultural sector requires that it be seen as a system of functionality inter-related and inter-dependent elements, each of which contributes to the existing and potential level of performance of the sector. A stocktaking and diagnostic survey is needed early in the planning process to provide information about the wide range of factors, among the others, influencing agricultural performance:

- Agro-climatic data,
- Agro-economic data,
- Agro-forestry Resources,
- Animal Resources,
- Capital Resources,
- Crops and Cropping Systems,

- Environment Data,
- Fisheries Resources,
- Forestry Resources,
- Institutional Resources,
- Land owners data,
- Plant Resources,
- Socio-economic and infrastructure data,
- Soil Resources,
- Spices Resources,
- Water Resources.

Mollet (1984) provides details on the contents of Agricultural Resources Information System and recommends data and information on basic resources such as (i) soil resources, (ii) water resources, (iii) climate resources, and other data sets (collated from Remote Sensing as well as conventional means) such as (iv) basic data on crops, (v) basic data on animal husbandry and fisheries, (vi) basic data genetic (plant, animal and fisheries) materials, (vii) basic data on land ownership, (viii) socio-economic data, (viii) data on infrastructure for agricultural development. Zhu Zesheng and Sun Ling (1996) has proposed a seven-layered architectural model for agricultural resources management: application layer, management layer, decision models layer, decision data layer, production data layer, weather data layer, and environmental data layer. This layered approach is essential to facilitate development of metadata and application of Open GIS model for optimal utilisation of agricultural resources.

Development of Metadata (data about data) is required, as overall rate of collection of data increases rapidly with advances in technologies such as high resolution satellite-borne Imaging Systems and Global Positioning System (GPS), and with the growing number of people and organizations who are collecting and using data (Ground Truths). Synchronization of Geo-processing Technology (GPS—Global Positioning System and RS—Remote Sensing), with the emerging Information Technology Standards based on "Open Systems", "Distributed Processing", and "Componentware Frameworks" provides Open GIS Architecture (*see http://www. opengis.org*) to facilitate Information Systems and Solutions, over INTERNET/INTRANET for developing DSSs (or GDSSs) on sustainable development issues. The Componentware Framework includes GIS Tools, Imaging Tools, Expert Tools, and RDBMS Tools; and each Tool has algorithms, data and interfaces to services in the distributed computing environment.

One of the central innovations of the World Wide Web (WWW) was the advent of a platform-independent graphical user interface markup language. Modern markup languages, such as *SGML* (Standard Generalized Markup Language) and *XML* (eXtensible Markup Language), which were initially conceived for modelling texts, are now receiving an increasing attention as formalisms for data and knowledge modelling. Geography Markup Language (GML) is an XML encoding for transport and storage of geographic information (both spatial and non-spatial properties of geographic features) is a significant milestone in the development of truly interoperable architectures for the use of spatial information between commercial applications. The progress of GML is an example of the growing momentum for the acceptance and use of Open GIS Consortium (OGC) specifications around the world, and represents one of the most visible steps taken by the geo-spatial community towards the creation of widespread spatial interoperability. Geomatics technology facilitates building up of the following DSSs for the benefit of farming community:

- Crop Suitability based on Factor Endowment;
- Land Suitability Assessment;
- Land Productivity Assessment;
- Population Supporting Capacity;
- Land Evaluation and Land Use Planning;
- Land Degradation Risk Assessment;
- Quantification of Land Resources Constraints;
- Land Management;
- Agro-ecological Characterization for Research and Planning;
- Agricultural Technology Transfer;
- Agricultural Inputs Recommendations;
- Farming Systems Analysis and Development;
- Environmental Impact Assessment;
- Monitoring of Land Resources Development;
- Livestock (cattle, buffalo, goat, and sheep) Farming Systems;
- Water Allocation in an Irrigation System;
- Fodder Resources Development;
- Water Bodies (Basin) Planning Systems using Watershed and Agro-Eco Region Planning Concepts.

The Central Ministry of Agriculture, on the basis of Report of the Core Group-V (AgRIS Report, 2000) of the Standing Committee on Agriculture and Soils, has planned to take up "Agricultural Resources

Information System" in all districts (regions) of the Country, "irrespective of past or future growth regions" (Fig. 16.3). Development of Agricultural Resources Information Systems, using Geomatics Technology in pilot districts with public funding, has been sanctioned as a scheme during the Tenth Plan (2004-07), in the following district typologies:

- A tribal district,
- A hill district,
- A dryland farming district,
- A socially backward district,
- A Green Revolution district,
- A district dominated by cash crops,
- A coastal district,
- A district in a mining/industrial belt,
- A district dominated by forest economy,
- A dairy farming district,
- A district dominated by one or two urban centres,
- A district in arid-zone,
- A district which is flood prone but having vast wasteland that could be used to generate forest cover.

This pilot project demonstration in districts of above mentioned typologies, in each state, will facilitate development of Decision Support Systems (DSSs) on "production practices and systems" which need to be adapted to respond to new market demands and export opportunities, poverty alleviation or growing labour shortages, depending on the setting (NATP, 1998). Agricultural Resources Information System (AgRIS) is a "way-forward" to improve agricultural productivity in rural areas, and a much "needed domestic strategy" for sustainable rural livelihoods. During the Tenth Plan, the Department has planned to:

- Initiate pilot projects on "Agricultural Resources Information System (AgRIS) in districts in order to work out the cost and efficiency of this project and then expand to the entire country;
- Develop a comprehensive database on various parameters related to land use, inputs (seeds, fertiliser, agricultural technology, agricultural credit) use, and water use;
- Development of Decision Support Systems (DSSs) packages for strengthening advisory services to farmers; and
- Capacity-building through Human Resources Development.

The AgRIS Project is expected to emerge as the richest "agricultural resources information system" (Fig. 16.3 and Fig. 16.4) covering all aspects of agricultural, natural resource, and food systems, to:

- Enable farmers to locate needed information to improve yields,
- Plan for weather contingencies,
- Access research,
- Calculate treatments and runoff,
- Simulate the growing season,
- Visualize precision data,
- Manage finances,
- Buy inputs and sell outputs, and
- Monitor prices in local as well as world markets.

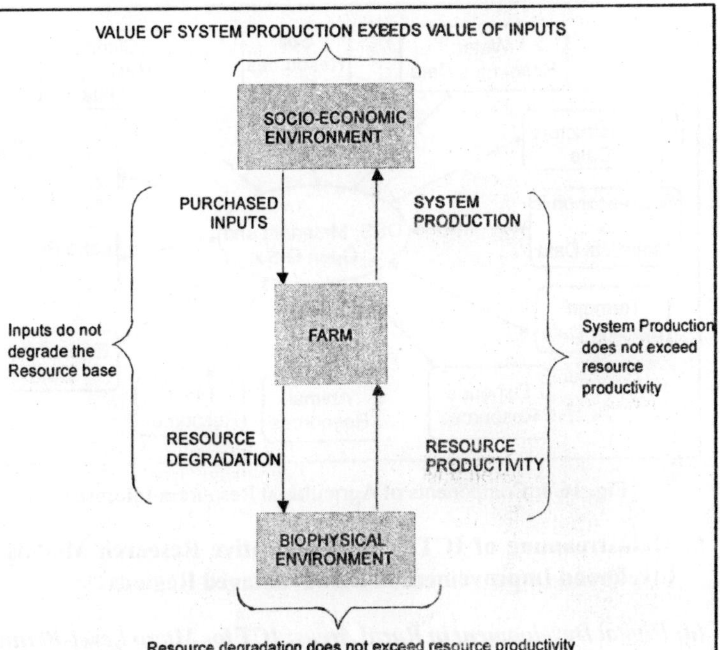

Source Proceedings of an International workshop on "Sustainable Land Use Systems Research" held at New Delhi on Feb. 1216, 1990

Fig. 16.3: Sustainable Land Use System

Guiding Principles of designing AgRIS will be as follows:

- Focus on the Disadvantaged Communities, who otherwise will be excluded;
- Provide that information or service which otherwise will not be provided;
- Focus on utilizing and where possible building upon what is existing rather than thrusting a new intervention;
- Create an outcome which in absence of ICT, will not be produced efficiently or timely;
- Understand the difference between direct benefits and trickle-down benefits for the disadvantaged community.

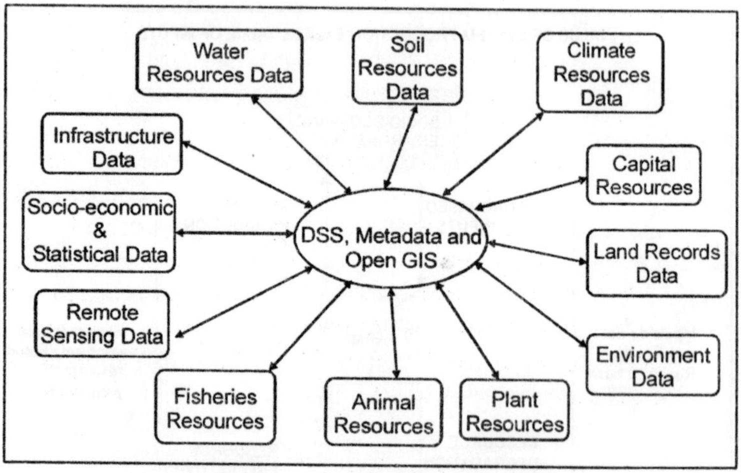

Fig. 16.4: Components of Agricultural Resources Information

6. Mainstreaming of ICT for Collaborative Research Models for Livelihood Improvement in Disadvantaged Regions

(a) Digital Development in Rural Areas: ICT for Micro Level-Planning

Indian village is a cognizable unit located in a specific agro-ecological and sociological environment. Its potentials and constraints for development are well known. As of today, the development planning is a highly compartmentalized activity managed by atomized Government departments handling agriculture, rural enterprises, forests, fisheries, water, health, education, culture, technology and livelihoods—almost in isolation of each other through different projects that rarely converge. Village problems are inter-related and the resources (natural and human) are integrated. People are both the "end and means" of development and also bound by a common space, history, culture and know-how. Indian villages are still complex, intertwined and multi-faceted.

In India, "district" is the basic administrative unit at the sub-state level and also consistent with the decentralized planning process prevailing at the grassroot. With the establishment of NICNET in districts numbering about in 1985-87, National Informatics Centre (NIC) has launched its "district information system (DISNIC)"[21] in about 27 sectors viz., agriculture, animal husbandry, education, health, industries, rural development, micro-level planning, etc., as an informatics tool for development planning and responsive administration.

The database project to facilitate micro-level planning (DISNIC-PLAN) was one among the 27 sectors in 1990s. This project has received national level appreciation in 1990s and has been included as one of the recommendations of the National Task Force on Information Technology and Software Development (1998): "Citizen IT Interface—Empowering people through the use of IT and information availability: Decentralised Planning and Implementation: DISNIC-PLAN programme should be made widespread and the database updated online should be made available to the public, the Panchayat among others."

In the era of e-Governance at grassroots, the relevance of the DISNIC Programme was felt necessary and the Planning Commission has desired to re-establish the DISNIC-PLAN Project, with institutional linkages of grassroots level organizations, to provide continuous support to development agencies in districts. This project will facilitate to arrive at "an Informatics Blueprint that covers Villages" for economic and social development; derivation of various development indicators which will act as pointers to understand the development potentials as well as constraints for sustainable development; use of data and sensitization at the lower level spatial units; relevance of geospatial intelligence, research, education, training and extension for micro-level planning; institutional linkages and plan of action for implementation, etc. The Informatics Blueprint includes, among the others, parameters related to the following sector at village level:

- Village Identification and Location,
- General Characteristics,

- Public Utilities,
- Meteorology
- Geology
- Soil
- Water Resources and Irrigation
- Environmental Problems
- Bio-Resources and Forestry Resources
- Household Details
- Agriculture
- Cooperation
- Livestock and Animal Husbandry
- Dairy Development
- Fisheries Development
- Industry
- Transport
- Irrigation
- Agricultural Machinery and Implements
- Transport and Communication
- Energy
- Markets
- Financial Institutions
- Education
- Health and Family Welfare
- Public Distribution System
- Drinking Water and Sanitation
- Development Schemes

The draft dataset is available at http://www.disnic.nic.in.

This DISNIC-PLAN new initiative, during 2004-07, will support building up databases and decision support systems, and communication systems to facilitate: sustainability of resources, poverty alleviation, empowerment of women, full employment, production systems planning, infrastructure planning and habitat planning. Production potentials of village through "circular-flows" and "chain-effect" should be understood. This project is slated for pilot implementation in 28 districts throughout the country (one district per state). After successful implementation, it will be scaled up to cover all the districts in the country.

(*b*) *Government's Initiatives and Agenda on Agricultural Informatics and Communication in India*

According to the National IT Task Force (1999) recommendation (No.79), "the Government shall take all necessary steps to boost IT for Agriculture and Integrated Rural Development". The Ministry of Agriculture and National Informatics Centre (NIC) emphasized informatics for Agricultural development in the National Conference on "Informatics for Sustainable Agricultural Development (ISDA-95)", Many follow up actions (ICT projects: AGMARKNET, DACNET, etc.) were taken up, to provide relevant agricultural information in rural areas, helping farmers to improve their labor productivity, increase their yields, and realize a better price for their produce. India is expected to become a "Knowledge Society" by 2008 and by which time, any farmer in a remote village can demand and get the following information:

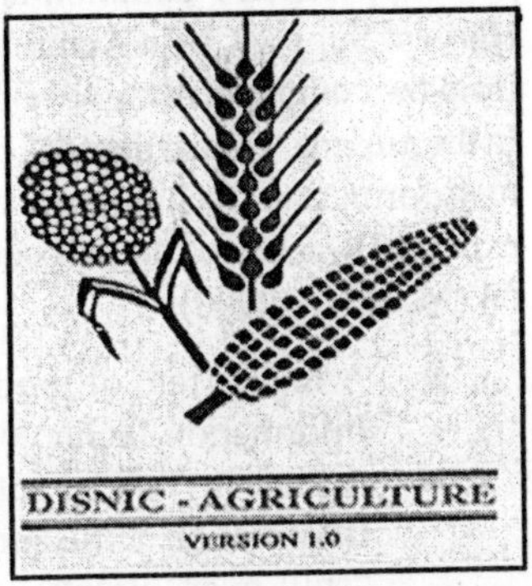

- Landuse planning for cropping strategy for farmers fields based on integrated information on soil, water, weather, fertiliser and pest management models;
- How and where to get proper seeds or good quality nursery plants;
- Prevailing prices of farm equipments, agricultural produce, products and series of such set of information, which can lead to high productivity and optimum cost benefit to the farmers.

To achieve "knowledge society" in agriculture, the following things should happen:

- An agriculture information centre in each village;
- Interactive exchange of information for planning and day-to-day operations by farmers;
- Availability of all the extension and advisory services on demand.

The Vision (2020) Document of the Department of Agriculture and Cooperation envisages that "the tools of IT will provide networking of Agriculture Sector not only in the country but also globally and the Centre and State Government Departments will have reservoir of databases"; and also "will bring farmers, researchers, scientists and administrators together by establishing "Agriculture Online" through exchange of ideas and information". Karnataka State Agricultural Policy (1995) and Gujarat State Agricultural Policy (2000) "Agrovision-2010" have envisaged Information Technology applications to globalize their agricultural sector in a big way.

Various study results strongly support that the "payoff" effect of ICT on economic growth can be achieved only through a robust National Information Infrastructure (NII), i.e. AGRISNET for the agricultural sector that supports ICT adoption and applications in Pre-Harvest and Post-Harvest Supply Chain Activities. ICT diffusion derives economic force from the complementary development of a knowledge-intensive society[22]. In the present "crucial decade" of this millennium, a high rate of investment in Information Technology capital and a supportive environment are expected to achieve "digital economy". Its rapid growth, however, depends on (M. Moni, 2000)[23]:

- A higher rate of productivity growth related to investment in Information Technology;
- A rise in Total Factor Productivity (TFP) growth due to Information Utilisation across the economy and resulting in "spill-over" effects;
- An increase in factor utilisation; and
- A decline in the non-accelerating inflation rate and rate of unemployment.

ICTs Diffusion and Infusion have many potential applications spanning the breadth of the agricultural industry, at all scales of organisation from farmer, to cooperative and professional bodies, from farm machinery vendors, fertiliser and chemical companies, insurance, regulators, and commodities, to agronomists, consultants, and farm advisors. This facilitates farming systems to change in response to economic, technological and social needs. The proposed strategy reinforces commitment for higher agricultural productivity in India, which has been blessed with excellent agro-climatic conditions and water resources. This "informatics-led development" in Agricultural Sector provides enough scope for reduction of new risks of marginalisation and vulnerability of small farmers, in view of the two on-going processes (i) economic reforms and liberalisation (1991) and WTO/GATT Agreements (1994) at macro-level, and (ii) decentralization reforms at micro-level, operating in India.

As a step towards "reaching" technology to the Small Holders (Resource-Poor-Farmers) of the Country, the Central Ministry of Agriculture have taken initiatives to build up an Informatics Network—AGRISNET: A NICNET based Agricultural Informatics and Communication—during the Tenth Plan. During the Ninth Plan, this proposal did not materialize due to bureaucratic interference. This ICT Network envisages to facilitate an integrated approach of 'Internet Technology" and "Sustainable Agricultural, Rural and Backward Area Development" with its farm and non-farm linkages. During the Ninth Plan, the AGRISNET was suggested to have nodes upto "block level", as "block" is the planning unit for agricultural development. However, in the Tenth Plan, the AGRISNET has been envisaged to have Server Access nodes at 35 State Agricultural and Cooperative Departments, and 600 District Agricultural and Cooperative Offices. To usher in "Agricultural Governance" in the country, it is essential to make the AGRISNET as the "rural infrastructure" reaching upto 6.5 lakhs villages.

This Initiative is based on the recommendations of ISDA-95 Conference (Informatics for Sustainable Agricultural Development), which includes, among the others, the following Informatics Networks:

- **AGRISNET**—An Infrastructure network upto block level agricultural offices facilitating agricultural extension services and agribusiness activities to usher in rural prosperity.
- **AGMARKNET** with a road map to network 7,000 Agricultural produce wholesale markets and 32,000 rural markets.
- **ARISNET**—Agricultural Research Information System Network.
- **SeedNET**—Seed Informatics Network.
- **CoopNet**—to network 93000 Agricultural Primary Credit Societies (PACS) and Agricultural Cooperative Marketing Societies to usher in ICT enabled services and rural transformation.
- **HORTNET** –Horticultural Informatics Network.
- **FERTNET**—Fertilisers (Chemical, Bio and Organic Manure) Informatics Network facilitating "Integrating Nutrient Management" at farm level.
- **VISTARNET**—Agricultural Extension Information System Network.
- **PPIN**—Plant Protection Informatics Network.
- **APHNET**—Animal production and Health Informatics Network networking about 42,000 Animal Primary Health Centres.

- **FISHNET**—Fisheries Informatics Network.
- **LISNET**—Land Information System Network linking all institutions involved in land and water management for agricultural productivity and production systems, which has now evolved as "Agricultural Resources Information System" project during the Tenth Plan being implemented through NIC.
- **AFPINET**—Agricultural and Food Processing Industries Informatics Network.
- **ARINET**—Agricultural and Rural Industries Information System Network to strengthen Small and Micro Enterprises (SMEs).
- **NDMNET**—Natural Disaster Management Knowledge Network in India.
- **WeatherNET**—Weather Resource System Information Network of India.

(c) *e-Cooperatives and CoopNet: An Internet Enterprise Development Programme for Fostering Agricultural and Rural Industries*

Rural Connectivity is the lifeline of rural economy. There are about 5.5 lakh cooperative societies with a membership of more than 236 million and working capital of more than Rs. 3,400 billion[24]. Cooperatives have made remarkable progress in agriculture, banking, credit, agro-processing, storage, marketing, dairy, fishing and housing (Table 16.1). This network covers 100 per cent villages and 85 per cent of rural households, and occupies a key position in agricultural development with respect to resources use, inputs use, harvesting of water resources, marketing channels, storage facilities, distribution channels, value addition, market information, and a regular monitoring network system. Cooperatives are also engaged in economic activities like disbursement of credit, distribution of agricultural inputs (seeds, fertilisers, and agro-chemicals), and arranging for Sanitary and Phyto Sanitary (SPS) measures of farm produces.

Cooperatives have inherent advantages in tackling the problems of poverty alleviation, food security and employment generation. Cooperatives are also considered to have immense potential to deliver goods and services in areas where both the State Sector and the Private Sector have failed (Verma, 2004)[25]. By and large, cooperatives (except a few large cooperatives) are "local institutions", addressing "local needs", employing "local talent", and lead by "local leaders". Situation is emerging wherein there is a possibility of people migrating back to rural areas, as there is a vast potential in cooperatives for generating self-employment

Table 16.1: Outreach of Cooperatives

Sl. No.	*Outreach of Cooperatives*	*Coverage/ Contribution*
1.	Rural Network (Villages Covered)	100%
2.	Coverage of Rural Households	71.0%
3.	Agricultural Credit Disbursed	42.8%
4.	Fertiliser Distributed (6.049 M. Tonnes)	36.17%
5.	Fertiliser Production (3.509 M. Tonnes—N&P)	25.0%
6.	Sugar Produced (10.164 M. Tonnes)	50.5%
7.	Wheat Procurement (6.926 M. Tonnes)	33.5%
8.	Animal Feed Production/Supply	50.0%
9.	Retail Fair Price Shops	20.3%
10.	Handlooms	54.0%
11.	Fishermen Cooperatives	21.0%
12.	Storage Facility (Village level PACS)	64.5%
13.	Direct Employment Generated	1.15 Million
14.	Self-Employment Generated	14.79 Million

Source: Indian Cooperative Movement—A Profile, 2004: Developed by National Resource Centre, National Cooperative Union of India).

opportunities at grassroots. The contributions of Cooperatives in the Green Revolution and the White Revolution are significant. The Cooperative movement is facilitated through (a) 19 National Level Cooperative Federations, (b) 367 State Level Cooperative Federations, and 2,890 District Level Cooperative Federations.

The Primary Agriculture Cooperative Societies (PACS), which are about 100,000 in number, are the major player at the grassroots level helping farmers with credit, inputs, marketing, advisories etc. Credit is a very powerful instrument for empowerment of the resource-poor people. Credit can generate "accelerated economic growth", when loans are easily available, properly utilised and repaid in time. A cumulative upward movement of "capital supply—increased productivity—higher real income—higher capital supply" is necessary for sustainable rural development (Tapan Chakraborti, 2004)[26]. These PACS can play a major role in revitalization of cooperative credit structure, provided if they are networked using ICT so as to facilitate "streamlining the Cooperative Credit Structure for hassle free flow of credit". Urgent steps to revitalize Cooperative Credit Institutions are required.

Marketing and Commercial Production have been a longtime part of the life of Indian farmers. A network of cooperatives at the local, state, and national levels assist in agricultural marketing in India, by handling foodgrains, jute, cotton, sugar, milk, fruits, vegetables, and areca nuts. Global competition and new Information and Communication

Technologies (ICTs) are forging new relationships within and between, different layers of agribusiness, transforming the industry from a chain to a complex web. Internet Commerce (or e-Commerce) is growing fastest among businesses and facilitate companies to integrate and maximise changes (i.e. restructuring, business-process standardisation, enterprise resource planning, etc). The Industry developments provide an insight into trends, potential impacts and prospects, as given below:

- e-market place/Neutral e-Hub
- e-distribution sites/Distributor model
- e-procurement sites/Aggregation model

With a view to ensure "appropriate positioning" of cooperatives in the emerging liberalised competitive economy, it is essential to:

- promote cooperation among cooperatives,
- develop coordination among cooperatives,
- strengthen Primary Agriculture Cooperative Societies (PACS) through business development plan for their self-reliance,
- enhance involvement of rural poor in the Cooperative Movement through Self-Help Groups (SHGs),
- develop cooperative insurance system,
- promote international cooperative trade,
- undertake human resources development,
- develop agro-processing industries,
- to widen the business in marketing of food products, seeds and fertiliser,
- provide market support to farmers,
- make available all essential "agro-inputs" to the farmers,
- undertake livelihood generation projects in tribal areas for marketing development of tribal products,
- facilitate cooperative planning,
- transfer of technology to its members and common public,
- assess and disburse of credits to its members, and also promote economic interests of the members,
- provide all types of banking facilities,
- provide godown, cold storage and processing yard facilities,
- develop market intelligence related to supply, demand, price trends, supply/market chain, value addition and processing facilities, product quality specifications, product applications, market trends,

- disseminate information to the members as well as planners, researchers, and associate organisations and business circles etc.,
- develop and establish sound "information network" among cooperatives sector-wise,
- develop ICT enabled services in cooperatives, and
- On-line service to its members, and many more quality services.

To begin with, it is suggested to network about 1.46 lakhs Primary Agriculture and Credit Cooperatives, 19 National Level Cooperative Federations, 367 State Level Cooperative Federations and 2,890 District Level Cooperative Federations. Networking of Cooperatives through state-of-the-art technology is essential, which will facilitate to get connected to a National Data Centre (24/7 Infrastructure) for achieving the following, for which an initial investment of Rs. 2,500 crores is required:

- Build relationships and alliances faster,
- Re-engineer and integrate their processes,
- Develop more and better value-added products and services,
- Share knowledge and experiences,
- Enhance innovation,
- Promote Web-based business trading.

(*d*) *Warana Wired Village Experiment in 1990s: True Reflection of a Public-Private-Participatory (PPP) Model of NIC and Warana Nagar Cooperative Society in the State of Maharashtra*

Warana is a well-developed rural area located 30 kilometres northwest of the city of Kolhapur, is one of the richest states of India, Maharashtra. The main economic activity in Warana has been sugarcane growing and processing. About 50,000 farmers live in 70 villages spread in the 25,000 sq. km area covered by the cooperative. Another Rural digital initiative is the Warana Wired Village in Maharashtra State, which had set up information kiosks in 70 villages to enable villagers to access agricultural, medical and educational information through the Internet, in their local language (Marathi), under a Public-Private-Participatory (PPP) Model of NIC and Warana Nagar Cooperative Society.

Farmers consider the Internet a better source of information than traditional sources such as traders, field officers, television, radio and the print media. By providing neutral information, this ICT service also minimizes cheating by unscrupulous traders quoting the prices of farm

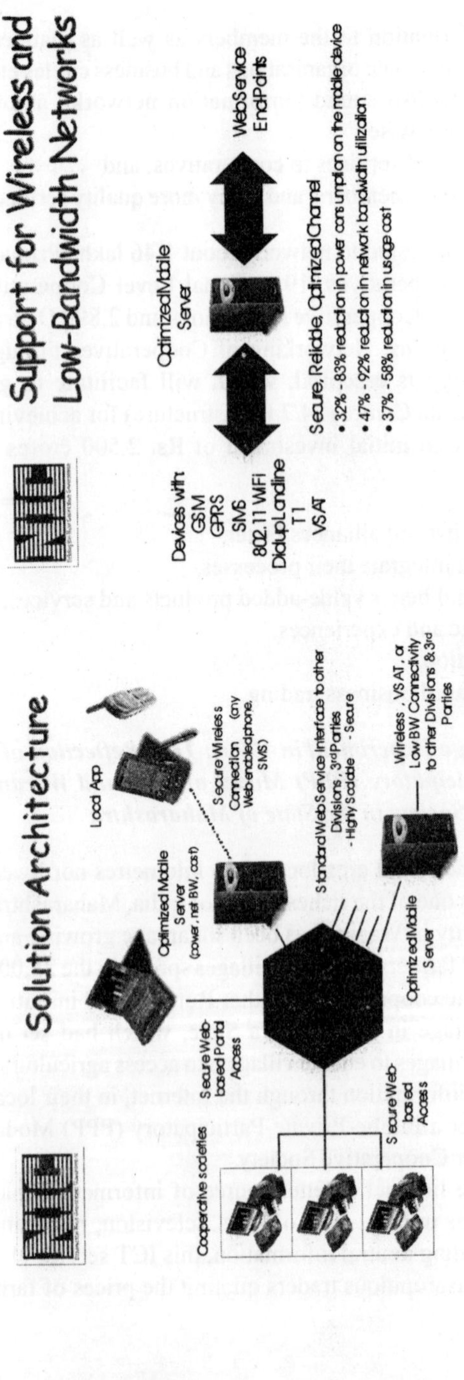

Network Solution Architecture

products. Community-based Telecentres offer a way of providing affordable access to ICT services in rural areas.[27]

(*e*) *AGMARKNET*

A Powerful Government ICT Initiative for Rural Empowerment, a Warehousing of "Data-for-Development", and a "Free-trade-zone" on Internet.

As a step towards to globalisation of agriculture, the Union Ministry of Agricuiture has embarked upon an ICT project: NICNET-based Agricultural Marketing Information System Network (AGMARKNET)"

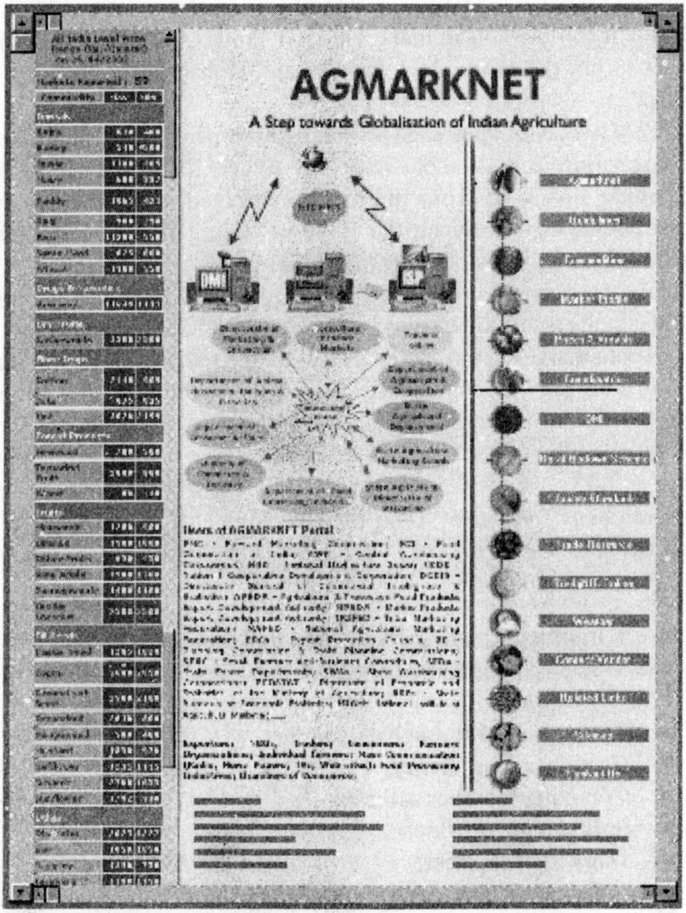

Fig. 16.5: A Synoptic View of AGMARKNET Portal

(Fig. 16.5) in the country. This AGMARKNET project has already networked 735 Agricultural Produces Wholesale Markets (APWMs), during 2000-02 and embarked upon additional 2000 Markets during the Tenth Plan Period (2002-07). As on 31 March 2005, 1347 APWMs have been brought under the AGMARKNET Programme, which has facilitated in developing the database on 300 commodities and about 2000 varieties. The Government initiative of the networking of agricultural produce markets (AGMARKNET[28]) and the AGMARKNET Portal would facilitate the development of B2B and B2C e-Commerce Model in the Country (Moni, 2001)[29]. This project has the potential of expansion to about 7000 Wholesale Markets located throughout the country and further to 35000 Rural Markets in India. This ICT Project is a "farmer-centric" project to put the progressive farmers on "global free trade zone on Internet".

AGMARKNET appears to be filling a huge gap by providing access to information at reasonable cost. The AGMARKNET venture is a heartening initiative from the much criticized and slow-to-react Government, especially on the issue of easing the infrastructural constraints on agriculture (*Times of India*, 2002)[30]. The advantages of this database accrue to the farmers, as they are not forced to sell their produce in the nearest market at uneconomical prices. The challenge, if the full potential of such ventures have to be utilized, is to take IT to rural India in a big way. Constraints/Challenges are (a) connectivity in rural areas, (b) training the stakeholders, and (c) ensuring data updation in real time frame (http://www.eapf.net/ casestudies/in/agmarket.asp). Anand Sagar (2003)[31] considers AGMARKNET, a step towards "Rural Empowerment" and also "Data for Development". Kari Holland of Washington University (USA)[32] categorizes AGMARKNET (India) as follows:

- Nation-wide market information for wholesale produce,
- Project supported by various departments and State Boards of Agriculture,
- Access mainly through the Internet,
- Customers pay (some),
- Computer facilities at the markets,
- Software for download,
- Daily market prices,
- Information collected by nodes in the various markets,
- Weekly trends,

- Information on loans, policies and regulations,
- Income has increased (for some), and
- By-pass middlemen.

The Inter-Ministerial task force on Agricultural Marketing Reforms (2002) has suggested creating an "Atlas of Agricultural Markets" as well as "e-Commerce" on AGMARKNET Portal so as to enable producers (farmers) directly transact business with the buyers. This Digital Advantage Project has already achieved the following:

- "Reaching the Unreached, i.e. resource poor farmers",
- "reduction of distress sale",
- "right to information",
- "base for production planning",
- "base for marketing led agricultural extension",
- "increased competition",
- "reduced marketing margins",
- "vertical linkages in export crop markets that connect multi-national traders to domestic traders".

The country is now witnessing a unified "agricultural marketing price information system" for the entire country. The general Hypothesis is "the more farmer produces the less he gets"—i.e. not getting adequate returns for his toils. This AGMARKNET venture benefits the farming communities from the new global market access opportunities and also strengthened the internal agricultural marketing system in India. There have been requests for AGMARKNET venture in various developing countries (ASEAN), Countries in Africa region, etc., in view of its operational efficiency in India. The Regional Workshop of Asian Productivity Organisation (APO) at Manila in 2004 recommends the AGMARKNET System for its member countries for replication. AGMARKNET is an effort to bring rural people into the mainstream economy and emerges as model of informatics-led-grassroots development in the country.

(f) Digital Opportunities

A Way Forward for Reduction of Spatial Disparities and Enhancing Opportunities for Development of SMEs to Penetrate into Export Markets.

Small and Medium Enterprises (SMEs) have always been the engine of growth in developing as well as in transition economies. Many developed economies (i.e. Japan, USA) owe their industrialization to

SMEs. SMEs are proven innovators and the driving force behind a number of technological breakthroughs. SMEs have been identified as high potential sector for employment generation and source of livelihood to millions of people in Asian, African, and Latin American countries.

In India also, SMEs are the second biggest employment generators after agriculture, providing jobs to over 9.2 million people, and accounts for 39 per cent of industrial production and 34 per cent of exports. It is estimated that there are around 350 Urban SME clusters and approximately 2000 Rural and Artisan based clusters (Rural SMEs) in India. This does not include the Small and Marginal Farmers (SMFs), who constitute about 67 per cent of the farming community in India. The process of globalisation has impacted SMEs much more than larger business enterprises, and SMEs are at crossroads now. While developed countries have already taken a lead in this direction, developing countries have yet to fully exploit this system particularly to invigorate Small and Medium Enterprises (Jürgen Bischoff, 2001)[33]. There is a need for serious global thinking on how to ensure greater stability of SMEs, how to provide greater social security for those working in this sector, and, at the same time, how to retain the dynamic force that drives this sector (Vajpayee, 2002).[34]

Areas of business that are targeted for improvement of SMEs are: User Profiling, Supply Chain, Value Chain, Customer-Relation-Management (CRM), SME Networks, and Suppliers Cooperation. The challenge for entrepreneurs is to think about creating solutions for the twin engines of future growth—Rural India and SMEs. Cluster Approach, Industrial Estates, Industrial Growth Centres, Export Processing Zones, Industrial Parks and Integrated Infrastructure Development Centres etc., have been the spatial policies for infrastructure development of SMEs. The Hypothesis: "the proximity of a web of businesses lowers the unit cost of infrastructure, leads to accretion of skills and is a source of informational economies" holds good in respect of SMEs.

Rapid emergence of knowledge-based economies required a paradigm shift in technology transfer services from traditional methods (i.e. horizontal company-to-company transfer of commercialized technology through a world-wide network of technology brokers) to Internet and vertical transfer of technology from R&D labs to industries. SMEs face information asymmetries in two ways: (i) in their own access to market—and business-related information, and (ii) as providers of services to poor people. Attention is required to SMEs evolution and to the "external" factors, related to "digital economy" that affects SMEs success and development in the era of globalization, addressing:

- Structural changes in SMEs organization and management related to the impact of digital economy;
- Changes in the competitive scenario that may affect the SMEs evolution;
- Interventions and policy instruments that have contributed to foster the SMEs adaptation to digital economy.

The focus is on SMEs' best practices in exploiting the opportunities provided by ICT to improve performance and visualize a "digital SME" in the process (http://www.etw.org), characterized as:

- An SME whose products and services are predominantly digital;
- An SME which uses digital methods as the primary means of carrying out core operations—marketing, sales, service etc. (e.g. a pure dot.com);
- An SME that exploits the benefits of digital methods to a significant extent.

An effective strategy to overcome this inherent weakness of SMEs is to network enterprises with institutions and organisations active in the area of SMEs development, information networks, and data bases as well as with technology generators. Supply Chain Management (SCM) is important for SMEs, as the supplier, the customer, the logistics partner and the manufacturer together involve in the material flow, the information flow and the fund flow. Establishing such an integrated service package on Internet for SMEs is a very challenging task (Vadim Kotelnikov, 2001)[35]. Indian SMEs have realized the importance of productization and specialization to survive in an ever-changing market. There have been attempts, as listed below, to provide Information Services through ICT for SMEs

- Technology-sme.com and Business e-Coach for Asia-Pacific SMEs, established by UN APCTT—an Asia Network of Model Technology Transfer Agencies (MTTAs).
- SMEs in India (www.smallindustryindia.com) with over 50 categories and 30 lakhs records.
- TIFAC (Technology Information, Forecasting and Assessment Council)—http://www.tifac.org.in—for technology and business links, partner with TIFAC;
- Indian Patent database—http://www.indianpatents.org.in

According to (Frank Wilson, Tony Swash and Stephen Anderson)[36], there has been a gap between the strategic level of support for business development (regional focus) and the SME level of development (specific business development). It has been further argued in (Wilson and Pardo *et al.*, 2001)[37] that the threat of marginalization faced by SMEs in the increasingly digital economy can be reduced by systematic support for transformation of traditional business approaches, sectors, and networks. Knowledge Management for regional business development may become a more critical issue, as the digital economy becomes more established.

In India, Internet among the Customers and SMEs is very limited. The Scenario of "internet connectivity is slow, access costs are high and connections are unreliable" has been removed by the National Broadband Policy, 2004: "Always On" data connection with a minimum download speed of 256 kbps. But the key to its success lies in meeting the content development challenge driven by Governments and right regulatory environment (Neel Rattan, 2004)[38]. Availability of broadband services at affordable price levels shall have a significant impact on India's GDP, facilitate new investment, create job opportunities, provide access to new and improved services, and increase productivity through infrastructure creation (Dayanidhi Maran, 2004)[39]. Case Study Reports show that rural GDP in South Korea has gone up due to availability of broadband connectivity.

The evolution of Network Based Computing relates to Distributed and Mobile Computing (DMC) and two new directions of computing; Grid Computing[40] and Pervasive Computing[41] will impact SMEs for emerging as "digital SMEs" and improving their competitiveness. Internet Commerce (e-Commerce) is growing fastest among businesses and four types of economic activities drive its growth (i.e. formation of Free Trade Zone on Internet):

- Building up the Internet,
- e-Commerce among Businesses,
- Digital delivery of goods and services,
- Retail sale of tangible goods.

ICT diffusion derives economic force from the complementary development of a knowledge-intensive society (Eunice Wang, 1999)[42]. The diffusion of ICT throughout all sectors (primary sector, secondary sector and tertiary sector) is, therefore, far more important than the production of ICT industries *per se*, to usher in economy growth and development-based on "digital technologies". India is expected to gain from the "Emerging Digital Economy", as it has:

- affordable access to core information resources, cutting edge technology and to sophisticated telecommunication systems and infrastructure;
- the capacity to build, operate, manage, and service the technologies involved;
- policies that promote equitable public participation in the information society as both producers and consumers of information and knowledge; and
- a work force trained to develop, maintain and provide the value-added products and services required by the information economy.

A synergetic collaborative relationship among Small Industries Development Organisation (SIDO), Khadi and Village Industries Commission (KVIC), National Informatics Centre (NIC), Indian Postal System (IPS), and Bharat Sanchar Nigam Ltd. (BSNL) can invigorate SMEs in India, through Value Chain, transfer of technologies, and utilise their IPRs. This will also facilitate SMEs to access both national and international business networks and emerge as "Digital SMEs".

ICT diffusion for economic growth and sustainable development of SMEs, thus, emerges as a powerful tool for reducing spatial disparities and enhancing opportunities at grassroots level. However, it is essential to have an "Integrated Policy Framework for Community and SMEs Development in Digital Economy" in the country.

7. Mainstreaming ICT for Sustainable Rural Livelihoods: A Way Forward

Rural India should be given a chance through Digital Networks for Farmers (DNF), DISNIC Programme, e-Cooperatives, and Digital SMEs to usher in "digital inclusion" for fostering rural prosperity and grassroots level development, and also provide a broad-base for upliftment of rural poor. Grassroots level information access and grassroots level access to information are the required components of "information and communication" strategy for grassroots level development in India. There is a "need to bridge theory and reality at grassroots". Converting millions of poor farmers to the use of new technologies has been a colossal task.

Science and Technology offers tremendous opportunities in simultaneous achievements towards the goal of sustainable agriculture and improving rural livelihoods. Digital opportunities have emerged as "power tool" for fostering agricultural growth, poverty reduction and sustainable resources use in developing countries. The PovertyNet in India. takes the form of the Government's Digital Initiatives and Agenda (viz., DISNIC, AGRISNET, AgRIS, AGMARKNET, DACNET, VISTARNET, APHNET, FISHNET, HORTNET, SeedNET, PPIN, COOPNET, FERTNET, ARISNET, AFPINET, ARINET, NDMNET, etc.), as a step towards "reaching" agricultural knowledge and technology to the rural Poor. Synergetic collaborations among cooperatives, Agricultural Produce Markets, Agri-clinics, Agri-Business Centres, and AGMARKNET could become the "pathways" to rural prosperity. This "digital development in rural areas" of India facilitates rural prosperity, rural empowerment, and a warehousing of "data for development" is a step towards digital inclusion to foster rural enterprise in India. Strategies for Effective Communication and Public Awareness to be adopted are as follows:

A. Suggested Action Plan

- QMS—Quality Management System to be adopted,
- Intranet for Collaborative Research,
- Portals in 22 officially recognised Indian Languages—Localisation,
- Portal Models to be adopted,
- Broadcasting/Wider-Dissemination Model,
- Critical Flow Model,
- Comparative Analysis Model,
- E-Advocacy/Lobbying and Pressure Group Model,
- Integrated Services Model,
- Geometry of Information Flow to be documented,
- Capacity-Building of all Stakeholders involved in the Collaborative Research Models,
- Participation of all Agricultural Colleges (about 300 in number) which generate more than.12,000 Graduates per annum,
- Establishment of Agricultural Resources Information System (AgRIS) in all Agricultural Colleges, State Agricultural Universities, ICAR Deemed Universities, and ICAR Institutes/Directorates,
- Development of Proof-of-Concept Projects in pilot districts (Typology identified),

- Agricultural Resources Information System,
- AGMARKNET led Market Extension System,
- Digital SMEs,
- E-Cooperative and COOPNET of Agricultural and Rural Credit Societies,
- ICT for Micro level Planning,
- Networking Departments of Geography (~ 215), Departments/ Faculties of Agricultural Marketing, Centres for Regional Development Studies, and Agricultural Colleges (~ 300) for spatial data analysis.

B. Possible Output: Creating Geo-Spatial Intelligence for ustainable Development at Grassroots Level

- Location Specific Databases,
- Location Specific Decision Support Systems,
- Location Specific Information Communication Systems based on Geometry of Information Flow,
- Community Action Plan.

Partial List of Outputs

- An e-Governance Model: G2G, G2B, G2C and C2G Framework,
- Quality and Reliable Information Services,
- Information Modelling of Village,
- Parameterization of Information System: Dataset—Informatics Blueprint for a Village,
- Grassroots Information to Access for Development,
- Sampling Framework for Impact Study,
- Transfer and Utilization of Appropriate Technology,
- Database on,
- Village Identification and Location,
- General Characteristics,
- Public Utilities,
- Meteorology,
- Geological Conditions,
- Soil Characteristics,
- Environmental Problems,
- Bio-Resources and Forestry,
- Household Details,
- Agriculture, Horticulture, Floriculture,

- Agricultural Inputs:
  - Agricultural Machinery and Implements,
  - Seeds,
  - Fertilizers,
  - Credit,
  - Pesticides,
- Land Use Pattern,
- Water Use Pattern,
- Cropping Pattern,
- Seed Consumption and Constraints,
- Fertilisers Consumption and Constraints,
- Pests and Diseases Incidences,
- Agricultural Services,
- Livestock and Animal Husbandry,
- Fisheries,
- Industry,
- Irrigation and Water Resources Related Problems,
- Transport and Communication,
- Power and Energy,
- Marketing Infrastructure and Price Trends,
- Financial Institutions,
- Educational Development,
- Health and Family Welfare,
- Public Distribution System,
- Drinking Water,
- Cooperatives,
- Social Welfare,
- Women and Child Development,
- Development Schemes,
- Sports and Cultural Activities,
- Public Grievances—Constraints and Problems Prohibiting Growth,
- ICT Facilities at District Planning Unit and Identified Blocks,
- Local Language Interface,
- Workshop, Seminar and Conferences for and with Stakeholders,
- Problem Specific Portals in Village,
- Information Access through Distributed and Mobile Computing Services,
- Virtual Call Centre on Village Level Problem,
- Capacity building of Officials of Districts and Sub-districts,

- Target Group Based Information Retrieval and Advisory System: Spatial and Non-spatial,
- Enriched Cadastral Map through GIS Process,
- Facilitates Skill Improvement,
- Raw Material Availability/Requirements,
- Agricultural/Industrial Waste Details.

C. *Possible Outcome: Creating Geo-Spatial Intelligence for Location Specific E-Government Programme*

- Planning for Development and Utilization of Resources,
- Improve Government Programmes and Services through the use of Modern Scientific Methods and Technology.

Partial List of Possible Outcome

- Right to Information,
- Digital inclusion for Fostering Rural Prosperity and Empowerment,
- Improvement of Quality of Life at Grassroots Level,
- Need Based Research and Technology Generation,
- Grassroot Access to Information for Development,
- Facilitate more scientific and transparent way to locate/plan for facilities in the rural areas,
- Addressing Ecological Restoration and Sustainable Livelihoods,
- Promotion of Public Participation and Appraisal in the Planning Process leading to Effective Citizen Interface,
- Better Rural Livelihoods Opportunities,
- Improved Production System and Planning,
- Improved Integrated Macro/Micro Nutrient Management System,
- Stakeholders Satisfaction—Researchers, Planners, NGOs etc.,
- Interoperability of Information System,
- Content Generation and Content Management for Grassroots Level Development,
- Rural Infrastructure Development and Capital Formation (Education, Health, Road and Bridges, Communication etc.),
- Effective Land Resources Management and Environment Protection,
- Improved Habitat Planning, Social Development and Economic Development,
- Better Disease Surveillance and Control,
- Better Understanding of Nutrition Security,

- Promotion of Intellectual Property Rights in the era of Globalisation at Grassroots Level,
- Natural Resource Management and Conservation Leading to Sustainable Development,
- Development/Formulation of Location Specific Schemes, based on Community Needs,
- Optimizing Government Expenditure on Developmental Schemes (i.e. Public Utilities and Services) leading to effective bottom up Monitoring and Evaluation (M&E),
- Strengthening Cooperative Development for Boosting Rural Economy,
- Development of Schemes Facilitating Deceleration of Out-migration,
- Better Disaster Management and Relief Plan Modelling,
- Employment Opportunities in Untapped Areas and Income Generation,
- Enhancing Women Empowerment and Integrated Child Development,
- Strengthening of Rural Extension and Advisory Services (agriculture, horticulture, fisheries, floriculture, dairy, veterinary, health, education etc.),
- Identification and Promotion of Agro and Rural Industries,
- Widening Market Reach for Household and Rural Products and Services,
- Documentation of Traditional Knowledge and its use,
- Identification of Entrepreneurial Development Programme,
- Promotion of Global Practices with respect to the Local Market,
- Planning and promoting forward (post-harvest processing of Primary Products and Marketing) and Backward (input chain) linkages for Agricultural Development in Rural Areas,
- Better Planning, Management and Conservation of Water Bodies and Sustainable Consumption and Water use, in View of Impending Climate Change,
- Exploitation of Non-conventional Energy Sources,
- Application of Technologies such as Database Technology, Web Technology, GIS Technology, Remote Sensing Technology, GPS Technology and Local Language Information Processing Techniques for Strengthening and Enriching Cadastral Mapping content so as to facilitate Land Use Planning and Management,
- Minimizing/Removal of Inter-sectoral Disparities through Spatial Planning.

Trustworthy Networking and Information System reinforces Commitment for Productivity and Sustainable Development.

NOTES

1. Submitted to National Agricultural Innovation Project (NAIP) of ICAR funded by the World Bank, as a Member of the Working Group.
2. Narayan, G. Hedge (2004). "Management of Natural Resources for Sustainable Livelihood: BAIF's Approach", in *Natural Resources Management and Livelihood Security: Survival Strategies and Sustainable Policies* (Eds.): K.V. Sundaram, M. Moni and Mrityunjay M. Jha, Bhoovigyan Vikas Foundation, New Delhi, 2004.
3. World Bank (2003). Multi-Stakeholder Regional Consultation for International Assessment on Role of Agricultural Science and Technology in *Reducing Hunger, Improving Rural Livelihoods and Stimulating Environmentally Sustainable Economic Growth*, organised by the World Bank, 12-13 May, 2003, New Delhi.
4. Rita Sharma (1999) in 18th National Workshop on Planning and Management of Agricultural Extension Training, 5-6 March, 1999, Ministry of Agriculture, New Delhi.
5. Madaswamy, Moni (2005). *Digital Developments in Rural Areas: Government's Initiatives and Agenda,* Presented at Symposium on Information and Communication Strategies for Grassroots Development, March 4-5, 2005, Organised by School of Information and Communication Studies, University of Madras, Chennai (India).
6. Multifunctionality—An Economic Activity (Producing Goods) and a Service to the Community (Food Security, Rural Employment, and Environmental Obligations such as Soil Conservation, Sustainable Natural Resource Management and Bio-diversity Protection)—is thus crucial to sustainable agricultural and rural development in our country.
7. Roy, B.C. and Suresh Pal (2001). Incremental Capital-Output Ratio in Indian Agriculture, *Agricultural Economics Research Review*, 14(1): 34-46.
8. Eastern Region (ER) of India comprises of 130 districts covering States of Orissa, Bihar, Chhattisgarh, Jharkhand, Eastern Uttar Pradesh, West Bengal and Andhra Pradesh.
9. Mishra, S.N. (2004). "Sustainable Livestock Development in India: A Strategic Approach", in *Natural Resources Management and Livelihood Security: Survival Strategies and Sustainable Policies* (Eds.): K.V. Sundaram, M. Moni and Mrityunjay, M. Jha, Bhoovigyan Vikas Foundation, New Delhi, 2004.
10. Bhat, P.N. (2004): "Livestock Revolution as a Sustainable Alternative for

Poverty Alleviation", in *Natural Resources Management and Livelihood Security: Survival Strategies and Sustainable Policies* (Eds.): K.V. Sundaram, M. Moni and Mrityunjay M. Jha, Bhoovigyan Vikas Foundation, New Delhi, 2004.

11. Acharya, S.S. (2004). "Sustainable Agriculture and Rural Livelihoods" in *Natural Resources Management and Livelihood Security: Survival Strategies and Sustainable Policies* (Eds.): K.V. Sundaram, M. Moni and Mrityunjay M. Jha, Bhoovigyan Vikas Foundation (New Delhi), Published by Concept Publishing Company, New Delhi.
12. Prasad, C. and Das, P. (1991). "Extension Strategies for Rainfed Agriculture", Indian Society of Agricultural Extension, New Delhi, India.
13. International Food Policy Research Institute (2002). *Achieving Sustainable Food Security for All by 2020—Priorities and Responsibilities*, May 2002.
14. A one-stop shop for all services for the agro industry and also see (www.agricoop.nic.in).
15. http:www.agmarknet.nic.in
16. Mahindra Shublabh Services Limited (MSSL), which is a private sector initiative, have planned to open Mahindra Krishi Vihar (MKV) in about 180 districts to facilitate bridging the lab-to-land gap in TOT for enhancing productivity, and also by tapping the as-yet-unexploited-potential of available technologies. Each MKV would be equipped with a VSAT for information exchange.
17. Balaji, R. (2001). "Mahindras Foray into Farm Consultancy", *Business Line*, October 22, 2001.
18. Preeti Ramanathan (2002). "Commercial Support Vital for Agri Clinics Role in Rural Development", *Financial Express*, 1 April, 2002.
19. Mohan, A. and Moni, M. (2001). "VISTARNET—A NICNET-Based Agricultural Extension Information System Network to reach Technology to the Small Holders (Resource-Poor-Farmers) in India", presented at the National Workshop on "Cyber Extension: Application of Information and Communication Technology in Agriculture" held at National Institute of Agricultural Extension Management, Hyderabad (India), February 21-22, 2001.
20. e-Farmer: A farmer, who knows the market situation world over, has unhindered access to the markets and technology and also possesses access to meteorological information, and extension advice can be considered as adequately equipped to enter the world of global partnership. He will then be free to deal with the problem of, at least operative consolidation of fragmented land holdings, backlog in technology and feeble infrastructure.
21. DISNIC—District Information System Programme of NIC, National Informatics Centre, Government of India, 1987.
22. Eunice, Wang (1999). "ICT and Economic Development in Taiwan: Analysis of the Evidence", *Telecommunications Policy*, 23(3,4), April/May 1999, pp. 235-43.
23. M. Moni (2000). "New Productivity Paradigms and Strategies in the e-Age—Government Initiatives on IT-led Development in India",

International Conference on Productivity in the e-Age, Asian Productivity Organisation, 22-24 November, 2000, New Delhi.
24. (see http://ncui.nic.in/stat.htm).
25. Verma, S.K. (2004). "Cooperative Centenary in India", *New Sector Magazine*, Issue No. 61, April/May 2004.
26. Tapan Chakraborti (2004). Speech of Mr. Tapan Chakraborti, Minister for Agriculture (Government of Tripura) in the meeting of State Agriculture Ministers held on 21st December, 2004 at New Delhi.
27. (http://www1.worldbank.org/publicsector/egov/warana.htm).
28. "AGMARKNET: NICNET Based Agricultural Marketing Information System Network—A Sponsored Project of the Directorate of Marketing and Inspection (Ministry of Agriculture)", February 2001 and also see to http://www.agmarknet.nic.in
29. Moni, M. (2001). "Leveraging Information Technology for Development of Agri-Business—Agribusiness: B2B or B2C e-Commerce Model?", presented at the National Seminar on "Managing Agri-Business in the New Millennium: Challenges and Opportunities", organised by All India Management Association (AIMA), 20-21 July, 2001, New Delhi.
30. *Times of India*: "Website to help farmers bargain better", dated 31st May, 2002.
31. Anand Sagar, K. (2003). "Agmarknet-Rural Empowerment: Data for Development", *Geospatial Today*, Vol. 2(2), 2003 and also www. GeospatialToday. Com/ journal/Vol2Issue2
32. Www.cis.washington.edu/courses/pbaf537d
33. Jürgen, Bischoff (2001). "An Overview of Successful International Technology Business Incubator Programmes", The First International Workshop on Technology Business Incubators in India (ITBI India 2001), 29-31 January 2001, Bangalore, India and Director, Asian and Pacific Centre for Transfer of Technology (APCTT), New Delhi (India).
34. Shri Atal Bihari Vajpayee, Hon'ble Prime Minister of India, addressed the Global Summit on Small and Medium Enterprises (SMEs) in New Delhi on 20-12-2002.
35. Vadim Kotelnikov (2001). *Development of Web-aided Technology Transfer Services for SMEs in Asia*, Asian and Pacific Centre for Transfer of Technology (APCTT), New Delhi, India.
36. Frank Wilson, Tony Swash and Stephen Anderson. *Promoting SME-Business through Business Support Networks*; www.usherproject.org.uk
37. Wilson, F., Swash, T., Pardo, A., and Crump, M. (2001). "Structured Support for S-M-E-Business", In *Proceedings of Human Factors in Telecommunications-HFT*, 2001 Conference, Bergen, Norway, 2001.
38. Neel Rattan (2004). "Broadband Growth Hinges on Content, Pricing", *Business Line*, 19 January, 2004.
39. Mr. Dayanidhi Maran, Union Minister of Communications and Information Technology, in *The Economic Times*, 3 June, 2004.

40. Grid computing coordinates network resources and mimics electrical power grid by bringing remote computing power uniformly and transparently to the users.
41. Pervasive computing emphasizes everywhere, all-time services and "human-centered" view of computing.
42. Eunice Wang (1999). ICT and Economic Development in Taiwan: Analysis of the Evidence", *Telecọmmunications Policy*, 23(3,4), April/May 1999, pp. 235-43.

17

Bridging the Digital Divide: Information Kiosks in Rural India

Challenges and Opportunities

V.P. SHARMA and K.V. RAO

Introduction

"Information Technology revolution is unfolding, and has very high visibility. However, its benefits have remained confined primarily to the urban areas. Rural communities have not been able to gain to the same extent from ICTs. As a means of agricultural technology transfer to farmers, information technology has had a limited impact. Even the vast potential of the broadcasting network has been tapped only minimally for extension". "The World Bank recognizes that making effective use of knowledge in any country requires developing appropriate policies and institutions to promote entrepreneurship and efficient use of knowledge." Michael Carter, World Bank Country Director for India Dahlman C. Utz A. (2005). Dahlman and Utz recognize that India has made tremendous strides in its economic and social development in the past two decades and is poised to realize even faster growth in the years to come. They note that the time is very opportune for India to make its transition to the knowledge economy. Defining knowledge economy as 'an economy that creates, disseminates, and uses knowledge to enhance its growth and development,' this report claims to provide a "big picture" assessment of India's readiness to embrace the knowledge economy and highlights some of the key constraints and emerging possibilities confronting India on four critical pillars of the knowledge economy: (i) Strengthening the economic and institutional regime, (ii) Developing educated and skilled workers, (iii) Creating an efficient innovation system, and (iv) Building a dynamic information infrastructure. According to Dr. R.A. Mashelkar, Director General, Council of Scientific and Industrial Research of India, "India is already gaining international repute for its innovations in areas ranging from pharmaceuticals to software. IT will achieve even more as it improves the efficiency of public R&D, increase private R&D, and encourages greater university-industry linkages. It is leveraging traditional knowledge with modern science and exploiting public-private partnerships to support grassroots innovations which can improve the quality of life for the poor." Dahlman C. Utz A. (2005).

Indian Villages

Very little has changed in villages of India in past decades. Schools have been built but many still lack of teachers and appropriate teaching methods, there are phone lines in over one lakh villages, but getting a dial tone is still a challenge. Electricity supply is at the best intermittent. Health care is still limited in its availability. Indian villages are dependent on agriculture for much of their sustenance. As a result, the villagers mostly remain a poor lot—per capita income of villages is no more than Rs. 12,000 to Rs.18,000 per annum as compared to national average of Rs. 25,000. Even as one India (Urban India) races ahead with optimism in new millennium, there is another India (Rural India), which seems to be stuck in the past. If India Nation as a nation has to progress, there is little doubt that rural India has to progress. Transforming rural India is the challenge that should focus the best of Indian minds—it is perhaps the single barrier to making India a developed country, and achieving the double digits growth rate, our Hon'ble President Dr. A.P.J. Abdul Kalam talks about. Rural India needs a very strong and innovative intervention to take a giant leap forward. "Connecting India", with focus on rural connectivity is one such intervention, which can help, ignite the rural creativity to achieve its potential by integrating them in global market. "With the kind of awareness and opportunities available in ICT, it will soon become a reality wherein every one of our villages will have computers and connectivity. These would be the windows to the world of knowledge for our villages and also reap the benefits of our e-governance, tele-education, tele-medicine, e-commerce and e-judiciary initiatives. In spite of the all-pervasive nature of computers, they would still be far from being a truly friendly access device for our villagers. We would need in such cases, a human intermediary who would act as the village information officer. He will be the extended eyes and ears of the villager to the world of knowledge. India has approximately 2.3 lakh village Panchayats. I visualize the establishment of village knowledge centres in these Panchayats to empower villagers with knowledge and to act as a nodal centre for knowledge connectivity for villagers. These centres will act as an interface for economic activity among village clusters. The knowledge centre from which the villagers would access information through the village information officer can also be used for collection, digital storage and dissemination of village specific information pertaining to any relevant information to villagers. The development of village knowledge centres has become a reality in the Budget proposals of 2005." (Abdul Kalam, A.P.J., 2005).

"We live in a world where human knowledge is increasing at an unprecedented phase, something that was not imaginable even two decades ago. Thanks to modern developments in Science and Technology, it is possible, as never before in human history, to ensure that chronic poverty, ignorance and disease do not have to be the lot of a majority of humankinds. India lives in villages and therefore unless we take the benefits of modern science and technology to our villages, we cannot get rid of the mass poverty, which has afflicted millions and millions of our people for centuries. This is a most innovative programme—'Village Resource Development Programme' and what I have seen today, I think it convinces me that we can look forward in the years to come a vigorous national effort to improve agricultural productivity, to improve agricultural yields, to launch a new programme for control of pests and diseases, which have remained unamenabless to any realistic solutions, and at the same time to ensure that through distant education we can not only modernize our school system of education but also make a more frontal attack on adult illiteracy" (Singh, Manmohan, 2005). The above views sum up the optimism and also indicate a road-map for improving the livelihoods of rural communities, using the emerging new information and communication technologies (ICTs) to empower the rural India. The methodologies being suggested by Government, non-government and civil society organizations are also converging. The concept of "Knowledge Centre in Every Village: Mission 2007" is a novel thought conceived and being vigorously promoted by M.S. Swaminathan Research Foundation in collaboration with over 80 organizations.

The idea that the Internet and related technologies might have an important role in aiding developmental efforts has captured a central place in international policy debates. Over the course of the last three years, statements affirming the need to close the so-called 'digital divide' between social groups with and without access to the internet have been made through several UN agencies, at the G-8 summit, and at meetings of developmental organizations around the world (Sood, A.D., 2003). In India also the Village Knowledge Centre (VKC) has captured the attention of policy-makers at the highest level. The VKC movement has been joined by many organizations—government, non-government and private sector organizations working for the development of rural areas. "The National Commission on Farmers has recommended the establishment of Rural Knowledge Centres all over the country using modern Information and Communication Technology (ICT). Mission 2007 is a national initiative launched by an alliance comprising nearly 80 organizations including civil society organizations. Their goal is to set up a Knowledge Centre in every village by the 60th anniversary of Independence Day. Government supports the goal, and I am glad to announce that Government has decided to join the alliance and route its support through NABARD. I propose to allow NABARD to provide Rs.100 crore out of RIDF" (Chidambaram, P., 2005).

The idea of digitally oriented development is as powerful and seductive as the technology upon which it is based. No single technological revolution has changed the lives of current generations in the way that the Internet has. No cultural-technological innovation since Television has had this kind of impact on the world's economy, its politics and its globalizing popular cultures, or even on our cultural conceptions of distance and time. The promise of digital development is that it might have the same reach as the original internet boom of the mid-1990s—only this time, the least privileged communities, those who had missed out on earlier waves of technology, might be able to 'leapfrog' over their more developed competitors. The greatest obstacles to rural development—large distances and inadequate infrastructure—might be obviated by instant access to virtual institutions that provide banking, education, health care, neonatal information, agricultural advice, and so forth.

In rural India, more than half of India's villages lack telephone connectivity, let alone Internet access. The lack of information and communication infrastructure results in people having to waste time and money chasing information and Government Officials. The inaccessibility of information affects the rural poor more than other sectors of the community. Similarly, lack of market information (on commodity prices, various input suppliers, etc.) leads to loss of income and exploitation of rural entrepreneurs by middlemen. Such exploitation and losses further marginalize small and marginal farmers and village artisans (Misra Satyan, 2005). The implications of this scenario on the rural people (with differential impacts on the poor and other vulnerable groups) are three-fold:

- Loss of Income,
- Loss of Time,
- Loss of Opportunity.

Over 50 per cent of the farmers do not have any access to Extension Information in the villages. Only 40.4 per cent farmers have access to some sources of Information (NSSO, 2005). Out of this 22 per cent access the Extension Information through Radio (13%) and Television (9%). Thus the media has been the most important source of Agricultural Extension Information even now. Table 17.1 gives the break-up of "Farmers Accessing Modern Agricultural Technology".

Table 17.1: Percentages of Farmer Households Accessing Modern Agricultural Technology

S. No.	*Source*	*% of hhs*
1.	Participation in Training	0.9
2.	Krishi Vigyan Kendra (KVK)	0.7
3.	Extension Worker	5.7
4.	Television	9.3
5.	Radio	13.0
6.	Newspaper	7.0
7.	Village Fair	2.0
8.	Government Demonstration	2.0
9.	Input Dealer	13.1
10.	Other Progressive Farmers	16.7
11.	Farmers' Study Tour	0.2
12.	Para-technician/private agency/NGO	0.6
13.	Primary Cooperative Society	3.6
14.	Output Buyers/Food Processor	2.3
15.	Credit Agency	1.8
16.	Others	1.7
17.	Any Source (all of the above)	40.4

In this context, Information and Communication Technologies (ICTs) can play a significant role in making information available to the Farming Community at a reasonable cost. ICTs promise to provide innovative solutions to the problems of poverty and inequality by accelerating development and introducing transparency into systems and operations.

In rural India, high transaction costs reduce the efficiency of functioning markets, and preclude some markets all together. Reducing these costs through IT enables rural markets to function, raising productivity and growth. All kinds of digital communication, storage and processing devices are important, and the boundaries between them are blurring. We need to scale up from small experiments, which require organisation building and entrepreneurship (what n-Logue has achieved, for example). Put commercially sustainable information kiosks in 2,50,000 villages (with Government support for infrastructure, perhaps, but keep it out of operations). This will be the biggest step in making India more equitable on information front.

ICT in Agriculture

Agriculture continues to be the occupation and way of life for more than half of Indian population even today. Sustainable prosperity of this class—the farmers, the landless agricultural labourers holds the key for improving the overall Human Resource Development scenario in the country. Indian Agriculture had been on traditional lines till the first waves of Green Revolution in late 60s. The Green Revolution gave a sudden boost to the production and productivity of major cereals in the assured irrigated areas—The Punjab, Haryana and Western U.P. in north, and Godavari, Cauvery deltas in the south. Quick dissemination of Technological Information from the Agricultural Research System to the farmers in the field and reporting of farmers' feedback to the research system is one of the critical inputs in Transfer of Agricultural Technology. The information and communication support during last 55 years has mainly been conventional. The extension personnel of the Department of Agriculture disseminated the technological messages to the farmers manually. This approach has not been able to reach majority of the farmers who are spread across the whole country. This gap remains a challenge for the Extension System even today. To reach over 110 million farmers, spread over 550 districts and over 6,000 blocks is an up hill task. The diversity of agro-ecological situations adds to this challenge further. The success of Green Revolution was mainly achieved due to concerted homogeneous extension approach for the assured irrigated area, coupled with government-supported procurement. Now as we move to address the needs of rainfed eco-systems, the extension strategy becomes more complex. Farmers' needs are much more diversified and the knowledge required to address them is beyond the capacity of the grassroots level extension functionaries.

Today it is possible to find a solution to this situation by using the potential of Information and Communication technologies to meet the location specific information needs of the farmers. The information and communication networks are expanding very fast. The number of Internet connections in India has crossed the 10 million mark and the number of telephone connections is over 100 million. The Internet connectivity has touched almost all the districts in the country and is moving down to the Block and Mandal levels. Pilot projects to connect rural community to the cyber-space are underway at various locations. The initial response of the rural people, particularly women, has been very encouraging. The pilot projects of "Warna Wired Villages" in Kolhapur, Sangli district of Maharashtra, "Gyandoot" intranet project in Dhar district in M.P. and "Information Villages Research Project" of M.S. Swaminathan Foundation, Chennai in Pondicherry, Infothela Mobile Telecentre of IIT Kanpur, Community Radio in Andhra Pradesh, Drishti.com in Haryana, U.P. and Bihar and many more such rural initiatives have successfully demonstrated the acceptability and usage of high-end information and communication connectivity at village level.

Government Initiatives

Government initiatives in promoting use of ICTs in Agriculture can be broadly categorized under following three heads:

I. Central Government: Ministry of Agriculture—Department of Agriculture and Cooperation Initiatives;
II. Central Government: Ministry of Agriculture Department of Agricultural Research and Education Initiatives;
III. Central Government—Ministry of Communication and Information Technology Initiatives; and
IV. State Government Initiatives.

Central Government (Department of Agriculture and Cooperation) Initiatives

Department of Agriculture and Cooperation (DAC), Government of India is charged with the responsibility of giving policy directions and operational guidelines for Indian Agriculture. DAC operates through its various central sector schemes, programmes, technology missions and crop specific development directorates and institutions. DAC has taken a number of important initiatives to promote the use of ICTs to improve the outreach of various programmes, schemes, Institutions and other organs. Some of the major DAC initiatives in this direction are:

Agricultural Marketing Information Network (AGMARKNET)

Agricultural Marketing Information System Network or AGMARKNET is a farmer-centric scheme sponsored by the Directorate of Marketing and Inspection (DMI), Ministry of Agriculture and executed by the Agricultural Informatics Division of NIC for linking all important agricultural produce markets in the country, the State Agricultural Marketing Boards and directorates and DMI for effective information exchange. A step towards Globalisation of Indian Agriculture, the NICNET based AGMARKNET facilitates generation and transmission of prices and arrival information from agricultural produce markets and its web-based dissemination for the use of individual producers, farmers organisations, consumers, traders, exporters, NGOs, communication agencies like Radio, newspapers, TVs, web-sites, food processing industries, Chambers of Commers, policy makers, etc. http://agmarknet.nic.in/

This is a comprehensive searchable database, which tracks prices of agricultural produce in various *mandis* of India on a daily basis. Currently, it covers more than 500 *mandis* of India (one can navigate State-wise or product-wise). It has grown spectacularly over the last two years (it was covering only about 75 markets then) and is expected to cover more than 2,500 *mandis* of India over the next 3-4 years. All the *mandis* of India where Government Officials are stationed will be covered. This is one of the fastest expanding projects of NIC.

Department of Agriculture Network (DACNET)

NICNET based DACNET is an e-governance project executed by NIC to facilitate dissemination and exchange of faster and reliable agricultural information across the Ministry of Agriculture and Directorates and Field Units of Department of Agriculture and Cooperation (DAC). It is aimed at qualitative enhancements of work culture through better transparency, streamlining of existing administrative and technical methods and practices, greater integration and use of the data collected from variety of data sources, moving towards a paperless office environment, providing telecommuting features and better communication through e-mail for better coordination, knowledge sharing and research. DACNET supports Decision Making for planners, integrates Government-to-Government

(G2G) functions, connects Agri-Business Partners (B2B) and Farmers (C2C) and ensures IT Empowerment of the Officials of the Directorate. (*Source*: http://dacnet.nic.in).

ICT Connectivity under ITD (Innovations in Technology Dissemination) Component of NATP

The Government of India has provided ICT infrastructure, connectivity, capacity-building and content generation support to all the blocks, and District Agricultural Technology Management Agencies (ATMAs) in the 28 districts covered under ITD component of NATP (1998-2004). Over 400 blocks of 28 districts in seven participating states have been connected using ICTs.

Kisan Call Centres (KCC)

KCC is another highly innovative programme launched by the Ministry of Agriculture, Government of India. The on-line phone based experts advise programme to Farmers was launched in January 2004. A toll-free telephone number "1551" is accessible to all farmers in the country. The calls are answered by trained Agricultural Graduates in the local language. More complex calls are escalated to SAU Scientists, State Department Officials and other responsible experts, on-line. The KCC operates on all days (including Sundays) during 6.00 a.m. to 10.00 p.m. The calls received during 10.00 p.m. to 6.00 a.m. are recorded on IVR (Interactive Voice Recording) system, and are answered by the experts within next three days. The KCC has recorded over 10 lakh calls and responses during first 18 months of operation (since 21st January 2004). So far, KCC is the most direct (two-way) communication system providing Agricultural Technology, advice and problem-solving mechanism to the Farmers.

Central Government (Department of Agricultural Research and Education) Initiatives

The Department of Agricultural Research and Education is primarily responsible for Research and Education system of Agriculture in the country. The Indian Council of Agricultural Research (ICAR) is the apex agency to conceptualize, plan and implement new initiatives in Agricultural Research and Education System. Agricultural Technology generation system in the country is organized in crop specific research Institutions/ centres/ All India Coordinated Research Directorates spread across the country. All the ICAR institutions in the country are using

ICTs for internal communications and almost all of them are connected with the rest of the world through Internet. The payroll and financial accounting systems of all the institutions are computerized and use of e-mail is very high among the scientific manpower. The research journals of a few institutions are digitized and some of them are subscribing to e-journals as well. The NATP (1998-2004) has provided a very high level of infrastructure, connectivity and capacity-building support to the ICAR Institutions. Over 350 scientists have been trained in ICTs in India and abroad.

The Agricultural Education System

The education system in Agriculture is organized through State Agricultural Universities, one Central University and many affiliated colleges. The ICT usage and teaching is highly variable in terms of infrastructure, faculty support and number of courses offered in the SAUs. Some of the SAUs have state-of-the-art ICT infrastructure with Local Area Network (LAN), WAN and 24-hr Internet support, 2-way Video Conferencing facility (G.B. Pant University of Agricultural Sciences and Technology, Pantnagar; C.C.S. Harayana Agricultural University, Hissar; C.S.A. University of Agricultural Sciences and Technology, Kanpur; Punjab Agricultural University, Ludhiana; M.P. Krishi Vishvavidyalaya, Rahuri; University of Agricultural Sciences, Bangalore; Tamil Nadu Agricultural University, Coimbatore), whereas many of the universities lack basic Computer lab and Internet connectivity facility for Undergraduate and Post Graduate students. Over 50 per cent of Agriculture Graduates passing out from our SAUs and affiliated colleges of general universities do not undergo any course to develop ICT skills. This is a serious handicap and needs urgent attention, as the ICT skills are pre-requisite for most of the jobs, professional requirements immediately.

The ICAR and SAUs Outreach System

The ICAR and all the SAUs are also mandated to disseminate Agricultural Technologies developed and validated by them. This task is performed through the institutional mechanism of Directorates of Extension (DoE) in each university and Zonal Research Stations (ZRSs) and Krishi Vigyan/Gyan Kendras (KVKs). Over 60 per cent of the Districts in our country have a KVK. The communication mechanism of Directorates of Extension, ZRSs and KVKs is thus the most critical link to channel the ICAR and SAU research findings and recommendations to the farmers. This mechanism seems to be pretty weak as less than 5 per cent farmers are reported to have access to KVKs (NSSO, 2005). KVKs are the cutting edge of Agricultural Technology dissemination from the huge research system.

Central Government—Ministry of Communication and Information Technology Initiatives

The National Informatics Centre (NIC) under the Ministry of Communications and Information Technology has been the forerunner in establishing Information Kiosks at village level. The NIC is the key agency to develop and deploy Software systems for e-Governance for all the Government of India Ministries and Departments. NIC has also taken up a few large projects to roll out the ICT connectivity at village level. Two of their primarily rural projects are detailed hereunder:

Wired Villages—WARNA: Agrarian India Re-packages its Schedules, Thanks to a Silent IT Revolution

(*Source*: www.infochange.org and www.mah.nic.in)

An IT revolution has silently swept through 70 villages in Warna, Kolhapur district, Maharashtra which once lived in the shadow of development in the outside world. Warna Wired Village Project changes everything in these villages where livelihoods centre on sugarcane farming and dairying. The basic objective is to use IT as a tool for development, and to bring Government to the village farmer doorstep.

Before this project came to Warna, sugarcane farming and milk procurement were hard work. Now the networking has solved a major problem faced by sugarcane farmers. Once the crop is ready for harvesting, each day's delay in harvesting reduces the sugar content of the cane and as a corollary the price it fetches at the sugar co-operative factory. The one harvester that the co-operative owns would often be monopolised by the rich farmers. Through the network, harvesting dates at each farm and village can be predetermined, and farmers can complain online to the co-operative chief if the harvester does not arrive on time. Each farmer is provided a code number that helps him decide when to take his crop to the factory. His bills are cleared online as well. Co-operative dairies are now selling their products online and milk collection centres have been upgraded to cybercafes! The Warna Group has also opened an office in Europe to market its milk produce. The network provides farmers pricelists of farm produce in the region. So it's easy for them to monitor the rates of agricultural produce in Agricultural Produce Market Committees across the State. They can also access daily weather forecasts, information on cropping patterns, soil conservation and Government schemes. An information package called Disha has also been developed to provide vocational guidance to village children, provide information on career options and educational institutions they can enroll in.

CIC—Community Information Centres
(*Source*: www.cic.nic.in)

IT revolution is based on knowledge and knowledge is the only thing that can be shared without dividing. By sharing, it is not divided on contrary it gets multiplied. The CIC Project has been implemented on pilot basis in 30 blocks of the North Eastern States and Sikkim. The pilot project was inaugurated in August 2000. The main project in North Eastern States was inaugurated in 17th August 2002 with the dedication of 457 CICs to the people of the region.

In view of the successful implementation of the CIC Project in the North Eastern States it was decided to extend the same to other parts of the country. The first State to be taken up was the State of Jammu and Kashmir. This was due to the fact that the barriers for access to ICT, such as difficult geographical terrain and lack of infrastructure, which were found in the North Eastern States, were also an issue for Jammu and Kashmir. It was felt that the basic needs of citizens such as information, education, entertainment and health services could be addressed through the establishment of Community Information Centres in the State. These centres would serve as a platform for e-governance, e-learning and other IT enabled services in the state.

Each CIC has two CIC Operators (CICOs) for managing the centres and providing services to the public. CIC operators have been trained to manage the centres, provide services and impart training on basic computer operations and software packages to visitors. The project is a joint effort by Department of Information Technology (DIT) under Ministry of Communication and Information Technology (MCIT), National Informatics Centre (NIC) and implementing State Governments of the North-Eastern States, J&K and AP.

State Government Initiatives

Agriculture is a State subject under the Constitution of India. The development, extension and marketing infrastructure are the responsibility of State Government. ICT use has been consciously promoted by a few forward looking States during last 5 years. Some of the recent initiatives in this category are:

Jan-Mitra Project (*Rajasthan*)

Jan Mitra brings Government closer to the people in Rajasthan's hinterland.

(*Source*: www.infochange.org, www.iconnect-online.org and www.rajasthan.gov.in)

A UNDP—Rajasthan Government ICT partnership in backward and chronically underserved Jhalawar is providing teeth to the State's commitment to the right to information. The main objective of the Jan-Mitra project is to provide a single-window facility to citizens to access government work, simplify various government procedures through computerisation and use information and communications technology (ICT) to establish direct communications between the administration and the people.

Jan-Mitra is ensuring transparent, accountable and responsive governance and to make the right to information an effective tool in the hands of the rural masses. Most villagers reckon that, for the first time in their lives, the kiosks have brought them closer to their government and *vice versa*. Thanks to Jan Mitra, Registration of births, deaths and land records, grievance redressal, submission of online applications for a clutch of government schemes and services, even getting examination results, all are just a click of a mouse away. And while each service comes with a price tag, the villagers pay up happily, actually counting their savings in terms of time and effort spared.

GAU-SAT-KRU (Gujarat Agriculture University Satellite Krishi Gosthi)

The Directorate of Extension, GAU (now Anand Agricultural University), Anand, Gujarat, is running an innovative one-way video, two-way audio programme since 2002. During the one-hour interactive Video Programme (telecasted by Development Education and Communication Unit, ISRO) the farmers view the programmes at designated Sardar Smriti Kendras (SSKs similar to KVKs in other states). At the end of the programme the scientists from the Studio request the farmers to ask questions/clarifications. The farmer's questions are taken on-line and answered by the panel of scientists on Video. This programme has caught the imagination of the Farmers and more and more farmers are attending the weekly programme at the SSKs. The programme can be viewed from any DRS (Direct Receiving Station) of ISRO (Indian Space Research Organization).

Gyandoot Project (*M.P.*)

Gyandoot was initiated by District Collector of Dhar in Madhya Pradesh in January 2000 as rural e-government project. Computers in 21 major centres in five Blocks of the district were connected through an intranet network. These computers have been established in Gram Panchayats. These are called *Soochanalayas* (Information Centres). From the *Soochanalaya*, user-charge based services are given to the masses and at the same time the information technology related developmental needs of Government departments and Panchayats are met free of cost. This intranet has been named Gyandoot. People outside the district may avail of some of these facilities as well as other important information regarding the district through the corresponding Internet website of WWW.GYANDOOT.NET.

The entire expenditure for the Gyandoot network has been borne by Panchayats. The network has been set up at a total cost of Rs. 25 lakhs. The system is very cost-effective. In the Gyandoot scheme of things, the twenty-one centres where *Soochanalayas* have been established to cater services not only to the twenty-one villages proper but also to surrounding 25-30 villages. *Soochanalayas* have been established in the buildings of such Gram Panchayats which are located either at block headquarters or at prominent haat bazar places or are prominent villages on major roads which are important from the point of the catching bus etc. Thus, each *Soochanalaya* caters services to approximately15 Gram Panchayats and about 25 to 30 surrounding villages. The benefits cover wide-ranging information needs of all villagers, not merely those belonging to co-operatives. Thus, the Gyandoot network benefits over half a million villagers living in 311 Gram Panchayats and over 600 villages.

CARD Computer Aided Administration of Registration Department (*AP*)

CARD brings some of the key implementation issues faced by State and national Governments in their efforts to use IT to improve citizen-government interfaces.

(*Source*: www1.worlbank.org)

Land registration offices throughout Andhra Pradesh now operate computerized counters to help citizens to complete registration requirements within an hour instead of several days, as was necessary under the earlier system. The lack of transparency in property valuation under the old system resulted in a flourishing business of brokers and middlemen leading to corruption. Antiquated procedures such as manual copying and indexing of documents, and storage in paper forms in ill-maintained backrooms have all been replaced.

Registration to document changes in ownership and transactions involving immovable property is governed by the Indian Stamp Act of 1899. Deeds of various kinds are required by law to be written on stamp paper of prescribed value. Certain transactions require a fixed duty. For others the *ad valorem* method is used, whereby the stamp duty is a percentage of the property value or loan that is the subject of the instrument. The *ad valorem* method ensures that inflation will not erode the value of stamp revenues. This method accounts for over 90 per cent of the total revenue from stamp duty. Registration is carried out at the office of the Sub-Registrar of Assurances. In Andhra Pradesh (AP) there are 387 Sub-Registrar offices that register approximately 1.2 million documents per year. The work of the Sub-Registrar is supervised by a hierarchy of District Registrars (28), Deputy Inspectors (6), and the Inspector General (1). The traditional 11-step registration procedure is complex and time consuming, beyond the comprehension of most citizens.

Rural E-Seva—Women Learn and Operate ICT

Helping the rural women as information leaders to bridge the gender divide and providing people of the district various services.
(*Source*: www.idonline.net and www.westgodhavari.org)

The district West Godavari in the State of Andhra Pradesh in India has a total population of over 3.8 million out of which over 70 per cent (2.1 million) lives in rural areas. The women population in this district, which is over 1.8 million, is further marginalized than their male counterparts.

Web enabled rural kiosks termed e-Seva centres at the mandal level, was initiated with twin objectives of helping the rural women as information leaders to bridge the gender divide and providing people of the district various C2C (citizen-to-citizen) and C2G (citizen-to-government) services. These kendras initiated by the district government identifies Self-Help Groups (SHGs) and rural youth who are given training as well as other supports for starting and managing the centres. It not only builds their confidence and empowers them economically but also provide a virtual meeting place and a focal point for synergising and pooling in their efforts. Most of these Self-Help Groups come from the poorest segments of the society and the project helps them achieve economic independence.

Bhoomi Project (Karnataka)
(*Source*: http://revdept-01.kar.nic.in/Bhoomi)

A network for computerization of all land records in the State will streamline operations and eliminate corruption at several levels by use of ICT. Karnataka was formed through the coalescence of the princely Mysore State with Kannada-speaking sections of the Bombay Presidency and princely Hyderabad. For this reason, there are three different systems of maintaining land-records in Karnataka State. The plan document of the Seventh Five Year Plan rightly opined: "Land records form the base for all land reforms and therefore, regular periodic updating of land records is essential in all States".

All 20 million land records of 6.7 million landowners in 176 taluks of Karnataka have been computerised. This system works with the software called "BHOOMI" designed fully in-house by National Informatics Centre, Bangalore. While Government of India largely funds the project; State Government funds some critical components of this project. Implementation of Land record computerization has been difficult in India. Minimizing resistance from staff by harnessing political support was an important contributory factor. Extensive training coupled with a participatory style also helped to diminish resistance. Bhoomi has reduced the discretion of public officials by introducing provisions for recording a mutation request online. Farmers can now access the database and are empowered to follow up. Reports on overdue mutations can point to errant behaviour. These computer-generated documents have statutory authority, and can be used for judicial and financial purpose.

Akshaya Project (Kerala)—District-wide e-literacy Project in India

The first district-wide e-literacy project in India, and one of the largest known Internet Protocol (IP) based wireless networks
(*Source*: www.idonline.net, www.akshaya.net and www.it.kerala.gov.in)

The Akshaya project in the rural Mallapuram district of Kerala, India, is the first district-wide e-literacy project in India, and one of the largest known Internet Protocol (IP) based wireless networks. Through a test network of over 600 village telecentres, known as e-Centres, Akshaya has conducted one of the world's largest computer literacy drives, claiming to reach over 600,000 households, representing more than 3.6 million people, in less than 6 months. The project has created a unique brand of state-funded computer access centres, and simultaneously led to a massive wireless infrastructure, providing a wide range of services and making way to many future opportunities. The project is a state-private partnership, in which the State agency Kerala State IT Mission (KSITM) planned the project, set up the connectivity infrastructure, marketed the idea, developed an e-literacy programme and created an initial suite of applications. The State paid about 85 per cent of the cost of this training as incurred by the entrepreneurs. About 450 Akshaya Centres are working in the district so far.

Kissan Kendras (Kerala)

KISSAN—Karshaka Information Systems Services and Networking.

Karshaka Information Systems Services and Networking is an innovative project by the Department of Agriculture, Government of Kerala. This project for agriculture is a pilot implementation of application of IT through relevant Information Systems and networking to aggregate, share and disseminate information of importance and interest to the farmers, agricultural workers and officials in ways that enhance the total agricultural development and farmers' welfare in the State. The project aims to establish a farmer centric integrated distributed Information System for agriculture, which would focus on catering to the various requirements and services to the farmers across the State.

The goal is to deploy and commission such information systems and networking that helps empowering all the stakeholders in Agriculture through the following five "fundamental rights" of Knowledge Management in the agricultural domain are "Right Information to the Right Person(s) at the Right Time in the Right Place(s) and in the Right Context". The mission of this project is to develop and deploy Information Systems, Network, Processes and Services for Agriculture in ways that lead to (i) amelioration of farmers' distress and enhance their welfare, (ii) increased farm productivity, and (iii) better returns for farm produce and derived products. The core deliverable of the project is the integrated multi-component, multi-modal and geographically distributed Agriculture Information System that is accessible anywhere any-time by all concerned. The project will establish a pilot scale deployment of information systems and networking in the Thiruvananthapuram district and neighbourhood in 10 Panchayat level Krishi Bhavans and relevant farm information agencies, Internet assisted collection of information from authentic sources and dissemination of the same anywhere through different networks.

Private Corporate Sector Initiatives

Private corporate sector has taken major initiatives in using ICTs and establishing Information Kiosks in India. Information Kiosks are integrated in their outreach and community mobilization strategy. The big corporate houses, which have rolled out, well thought out business models having Information Kiosks as the communication instrument include:

ITC-IBD (ITC-International Business Division)—e-Choupals

The International Business Division (IBD) of ITC, one of India's largest exporter of agricultural commodities, conceived e-choupal as an efficient mechanism to integrate the supply chain participants by using information technology. e-Choupal, the web supported initiative of ITC, offers the farmers the information, products and services they need to enhance farm productivity, improve farm price realization and cut transaction costs. Farmers can access latest local and global information on weather, scientific farm practices as well as market prices at the village itself to this web portal—all in local language (Hindi, Telugu, etc.). Choupal has also got the capacity to utilize high quality farm inputs as well as to purchase farm produce from the farmers directly. Giving the literacy and the infrastructure constrant at village level, this model is designed to provide physical service support through a Choupal sanchalak himself is a lead farmer—who acts as the interface between the computer terminal and the farmers. Full contents of this site are therefore made available to the registered sanchalaks only.

The ICT Choupals are currently in operation in 14 States in India though 5200 e-Choupals serving farmers in over 21,000 villages. Recently (in August 2004) ITC also started a Rural Maal (Choupals Sagar at Sehore at Madhya Pradesh). ITC plans to scale up the model to cover major agricultural states in next 5 years, reaching one-lakh villages. It plans to diversify the plans like rice and cotton.

EID-Parry—Parry's Corners

EID Parry (India) Ltd. is a 214 years old company marketing agri-inputs for growers in India. Today, Parry markets over 1.5 million tonnes of fertilizers and helps improve agriculture productivity through products and scientific application techniques. The company also extended several crop protection chemicals for the benefit of the farmers subsequently. EID Parry (India) Ltd. earned the distinction of being the first company to market crop protection chemicals. EID also buys sugarcane and manufactures sugar. EID Parry set up the first sugar factory in India in 1942.

Indiagriline (the web portal of EID Parry) connects farmers and others in rural India to markets and to market price information, knowledge like weather data, agricultural extension services and crop cultivation practises, to social welfare agencies like Primary Health Centres and to their peers. Users have access to buyers, sellers and intermediaries, and relevant experts and infrastructure service providers. The latest technologies have been used to enable them to interact without computer interface interference. The system supports a location-based user interface that considers the given situations of literacy and comfort levels.

Tata Chemicals—Tata Kisan Kendras

Tata Kisan Kendra (TKK) was the initiative taken to help the farmers to improve that quality of life of the farmers. Tata Chemicals Ltd. started TKK with the objective of providing the farmers with infrastructure support, operational supports co-ordination and control of farming activities, and strategic support. Another major objective was to provide the farmer with a package of inputs and services for optimum utilization of balanced primary nutrients; plant protection chemicals; water; seeds; post-harvest services; and to develop a genuine partnership with the farmer. The underlying assumption is that increased and value added services would create a bonding with the farming community and would result in creation of a sustained market for the Tata Group (agricultural inputs and products from the group stable targeted at the rural markets). It is expected that the extension services would result in improved farming, which in turn would increase prosperity and create demand for the products.

TKK today is operating in three states, namely: Uttar Pradesh, Haryana and Punjab. It is active in providing the sophisticated modern technology to the small farmers and making them harness the gain from ICT. The project has generated learning experiences such as customization of the project for the rural users, implementation in a phased manner and community learning.

N-Logue Communications—Chiraag Kendras

N-Logue was established under the aegis of the Telecommunications and Computer Networks (TeNeT for reference please visit: http://www.tenet.res.in) Group of IIT, Madras, which is dedicated to evolving technically superior and cost-effective connectivity solutions for Indian rural population. n-Logue has been set up to provide Voice and Internet services in every under served village and small towns in India. This would naturally exclude the top 100 or so cities and towns in India. The promoters of n-Logue believe that not only do these areas need such services, it can also be provided to them in a way that is both self-sustaining and profitable. The critical success factors for this to happen are appropriate and enabling technologies, dense network penetration and a local franchisee, who will run the enterprise as a business. N-Logue has the technology and a business model that supports such a mission.

N-logue Initial Coverage

- Adayar, Chennai: LSP operates Internet Kiosk,
- Vizhianagaram, AP: 400,000 population town LSP is a business operator in town,
- Thirupati, AP: temple town: LSP sponsored by DHAN foundation, a voluntary organization,
- DHAN foundation to sponsor installations at four other small towns in South India,
- Railways: Guindy, Tambaram in Chennai and Malainagar outside Chennai,
- Cuddalore, Tamil Nadu: total rural set-up, sugar belt; in co-operation with Murugappa Group, a business house with rural clients.

NGO Sector and Other Private Sector Initiatives

Along with the Government and Private corporate sector Non-Government Sector (typically involved in rural development area) has also taken take up some very significant initiatives in using ICTs and establishing Information Kiosks in rural areas. Information Kiosks are generally used by NGOs as an social engineering and empowering strategy. Some of the NGOs have used Information Kiosks exclusively for women empowerment, while others have used ICTs as capacity building and grooming tools to empower rural youth. Some of the NGO initiatives, which have Information

Kiosks as one of the important development instrument at the village level, include:

M.S. Swminathan Research Foundation—Information Villages

The project began in 1998 at the village of Villianur, 20 km from Pondicherry, in Union Territory of Pondicherry, South India. The first efforts of villagers and project staff were to develop mutual understanding. In Pondicherry the organization have certain initial advantages as accessible Government and reasonable telecom infrastructure. The level of poverty is high in rural areas. An earlier programme of the Foundation for community asset building based on biological technologies had been fully operational in this region and the ICT project was expected to complement this programme.

M.S. Swaminathan Research Foundation has implemented this project and have connected twelve villages near Pondicherry in southern India, by a hybrid wired and wireless network consisting of PCs, telephones, VHF duplex radio devices and e-mail connectivity through dial-up telephone lines. This facilitates both voice and data transfer, and have enabled the villagers to get information that they need and can use to improve their lot. The system involves local volunteers to gather information, feed it into an Intranet and provide access through nodes in different villages. Value addition to the raw information, use of the local language (Tamil) and multimedia (to facilitate illiterate users) and participation by local people right from the beginning are the key features of the project. Most of the operators and volunteers providing primary information are women, thus giving them status and influence. All the centres evolved over time to meet the information demands made by the community.

Development Alternatives Group—TARAhaat or Tara Kendras

This project, named "*TARAhaat*" after the all-purpose haat (meaning a village bazar), comprises a commercially viable model for bringing relevant information, products and services via the Internet to the unserved rural market of India.

It's an initiative promoted by Development Alternatives (DA), an NGO. *TARAhaat* is a social enterprise, using Information and Communication Technologies (ICT) to deliver a broad spectrum of services and products designed for the rural and peri-urban citizens of India. TARAhaat delivers their service through a network of franchised community and business centres owned by individual entrepreneurs. The DA Group has a staff of more than 400, including 150 professionals with postgraduate degrees in engineering, sociology, marketing, and management. *TARAhaat* is fortunate to have the entire staff of the DA Group available for the design, implementation and operating management of the portal and its associated services.

TARAhaat was launched as a pilot in late 2000 in Bundelkhand, Madhya Pradesh and Uttar Pradesh. In 2001, *TARAhaat* expanded its operations into Punjab. Today, *TARAhaat* has 41 centres and is in the midst of a major expansion in Punjab, and in the Bundelkhand region.

Drishtee.com—Gram Haat, Gram Daak, Krishi Prasar

Drishtee is a private initiative, known as Drishtee.com Ltd. It partners various local and State Governments in its working model. Investors and donors include the Boston Consulting Group and Microsoft. It is among the pioneers in providing rural services with its presence in almost 5 states of the country with 310 operational kiosks targeting a total population of about 300,000.

Drishtee is a software platform for enabling governance, commerce, education, health and other information services. It facilitates communication and information interchange within a localised intranet between villages and a district centre. Drishtee services are delivered via Information Kiosks that are owned by local villagers. Each kiosk, located at a prominent central location in its district, caters to the needs of the surrounding villages. Typically, the kiosks are financed through a Government-sponsored loan scheme. User fees are charged at the kiosks for the services provided. Kiosk owners are trained to operate the Drishtee system and services. Drishtee has moved with the times and has morphed into a multiple rural services provider. It now offers education services, insurance products and physical products through its franchisee network. The map below (a little dated in terms of figures) provides the coverage of drishtee services in the nation. It has, this year, expanded its reach to several districts of Assam, in the North East.

By the second year of the project, 132 kiosks had already been deployed in States—Haryana, Punjab, Uttar Pradesh, Bihar, Rajasthan, and Orrisa. Improved access to market information, such as crop prices, has increased the bargaining power of farmers *versus* buyers. Rural kiosks encourage citizens to complain about poor public services, increasing the accountability of Government departments. In its first phase, Drishtee was mainly working on connectivity and e-governance and reached coverage of 310 kiosks.

Indian Agri-Business Systems Ltd.—Agri-watch.com

Indian Agri-Business Systems Ltd. (IASL), is India's leading Agri-business Information Services Company. It was promoted in March 2000 by a group of alumni from the Indian Institute of Management, Ahmedabad. The presence of the organization can be felt throughout Nation with its brand name "Agriwatch".

The organization has around 25,000 paid subscribers spread throughout nation benefited with facilities of market rates through the web, magazines, newspapers and SMS services. Further down the line the organization is attached to trading communities in 700 towns and 56 *mandis* across the Nation.

Future Directions: Role of Information Kiosks in Agricultural Extension System

The weak linkages among extension, research, marketing network and farmers limits the effectiveness of research and extension to contribute to agricultural development. The Government of India has identified this problem, and is addressing it through National Programmes. The erstwhile emphasis on usage of vernacular press, radio and television for reaching to the farmers is being augmented with use of state-of-the-art communication technologies such as Internet and satellite communication. Under the new initiative of NATP, adequate attention is being paid to provide ICT connectivity down to the Block level. This connectivity will facilitate two-way communication among all the stakeholders in the Research-Extension-Marketing-Farmers loop. Apart from core Information and Communication Technology (ICT) connectivity, other forms of audio and visual communication like Satellite Communication (SATCOM) are also being promoted on project basis.

The National Agriculture Policy emphasizes the need for broad basing and revitalizing the Agricultural Extension Services to make them innovative, decentralize, farmer driven and farmer accountability. The policy framework for agriculture extension has been formulated to address the changing extension needs of farmers. The new extension regime under the new policy framework recognizes the role of a multi-agency dispensation comprised different strengths. The three arms of the agricultural extension network (as envisaged in policy framework for agriculture extension) are:

(a) Public Extension Services,
(b) Private Extension Services,
(c) Mass Media and Information Technology.

The action plan suggested in the policy framework to facilitate implementation of above agenda focuses on:

- Reducing the size of public extension making it professionally skilled and competent;
- Encouraging private sector participation including cooperatives, corporates, NGOs, para-vets and farmers organizations etc., and
- Augmenting media support to agricultural extension including extensive use of IT/Space Technology, Print and Electronic Media.

Augmenting Media and Information Technology

The use of Information Technology and media is one of the important pillars of Agricultural Extension envisaged in the Policy Framework. As per the policy guidelines, 25 per cent of the funds allocated to State Extension Work Plan have necessarily to be used for augmenting Information Technology and media for information dissemination. Therefore, there are ample opportunities available to the States to cover the IT and communication technology tools to be included in the State Extension Work Plan. The basic philosophy of making these initiatives self-sustaining over a period of time (say, 5 years), will be necessary, while planning any "Action Plan" under this category. This is essential for long-term sustainability of concepts and services at various levels.

Role of ICT as Envisaged in the "New Extension Policy" and in "Reforms in Agricultural Extension"

Information is crucial in agriculture production, in addition to land and capital. Faster transmission of improved farm technologies can revolutionalise agriculture. Matching the speed of technology transfer with that of farmer's acceptance can bring prosperity among farm families. Agriculture is an age-old practice and farmers have vast experience thus their sentiments, perceptions must be respected. ICTs offer a good opportunity to fill this gap. Some of the policy initiatives, which have been indicated in the Policy Framework for Agricultural Extension, are:

- Augmenting Media/IT applications at various levels;
- IT application in agricultural extension and marketing;

- Wider and creative use of Mass Media (AIR, FM, DD, Private Channels and Cable TV Networks, etc.);
- Use of Call Centres for Agricultural Information to Farmers (Kisan Call Centres);
- Farmers' participation in media/IT input;
- FIACs at block and "Information Shops" below of blocks through private sector participation;
- Capacity development for IT usage;
- Application of satellite communication technologies;
- "Information Shops" under the blocks through private initiatives such as FIG Federations, Agri-clinics and Agri-business centres.

Most of the initiatives work with entrepreneurs as partners (at village level). These entrepreneurs are the final links in the chain. The sustainability of all these initiatives is critically linked to the sustainability of these entrepreneurs.

"Harnessing Information Technology for agricultural extension will receive high priority in the new policy agenda. Extensive use of modern information technology will be promoted for communication between researchers, extension workers and their farmer clients to transfer technologies and information more cost effectively. Information technology should be made available, particularly, to those with specific inquiries to guide them in adopting the more knowledge intensive forms of agriculture, which will expand in future."

Information Kiosks in India

The use of ICTs in rural India started in late 90s on pilot basis. A number of projects were taken up mainly in NGO sector to use the Computers as learning and capacity-building tools. One of earliest example in this category is the "Information Village Research Project" of Dr. M.S. Swaminathan Research Foundation Chennai. Soon after a number of projects followed, these included Warna Wired Village Project in Kolhapur and Sangli District of Maharashtra by National Informatics Centre (NIC), Gyandoot Project at District Dhar in Madhya Pradesh and EID-Parry Information Kiosks in Cuddalore District of Tamil Nadu. A brief summary of the Information Kiosks in India is given in Table 17.2.

The Information Kiosks movement in India started in second half of 1990s and within 5 years, there were about 1000 Information Kiosks running as pilots in various States. The success of the pilot projects has been widely appreciated and now there are clearly visible indications of large scale rolling out of Information Kiosks in the country. Few important large-scale upcoming rural connectivity projects include:

Table 17.2: ICT KIOSKS under Different Projects in India

Sl. No.	*Project Name*	*No. of Kiosks*	*State/ Location*	*Remarks*
1.	**e-Seva**	1294	Andhra Pradesh	www.esevaonline.com
2.	**Community Information Centres (CIC)**	687	All North Eastern States and J&K	www.nic.gov.in
3.	**Drishtee**	600	Haryana, Punjab, Bihar and UP	www.drishtee.com
4.	**Srishti**	05	Gujarat	www.sristhi.org
5.	**WARNA Wired villages**	70	Maharashtra	www.mah.nic.in/warna
6.	**ITC e-Choupal**	5200	MP, Rajasthan, AP, Maharashtra and UP	Deccan Chronicle News dt: 31/07/2005.
7.	**ICICI**	2000	Karnataka	www.firextra.com
8.	**TATA Kisan Kendras**	840	UP, Haryana, Punjab	www.tatachemicals.com
09.	**Bhoomi**	282	Karnataka	www.revdept-01.kar.nic.in/bhoomi
10.	**CARD**	214	Andhra Pradesh	www1.worldbank.org
11.	**NDDB (DISK)**	2500	Gujarat	www1.worldbank.org
12.	**n-logue Communications**	2100	Maharashtra, Rajasthan, AP, Tamil Nadu, MP, Karnataka, Gujarat	www.n-logue.com
13.	**Azim Premzi Foundation**	571	Karnataka, AP and Gujarat	www.azimpremjifoundation.org
14.	**Kisan Kerala**	10	Kerala	www.kisankeral.net
15.	**Akshaya Kendra**	2100	Kerala	www.nisg.org
16.	**EID Parry**	30	Tamil Nadu	www.eidparry.com
17.	**e-Mitra/Janamitra**	28	Rajasthan	www.iconnect-online.org

(*Contd.*)

(*Contd.*)

Sl. No.	*Project Name*	*No. of Kiosks*	*State/Location*	*Remarks*
18.	**MSSRF-IVRP**	32	Pondichery	www.mssrf.org
19.	**DHAR-Gyandoot**	34	MP	www.dhar.nic.in
20.	**TARAhaat**	41	MP, UP, Punjab and Haryana	www.tarahaat.com
21.	**DASP**	380	Uttarakhand	http://ua.nic.in
22.	**MKCL**	2500	Maharashtra	www.mkcl.com
23.	**iShakthi (HLL)**	250	Andhra Pradesh	www.hllshakthi.com
24.	**HiWELL (NIIT)**	53	Delhi	www.hole-in-the-wall.com
25.	**MANAGE**	11	Andhra Pradesh	www.manage.gov.in
26.	**Greenstar**	01	Andhra Pradesh	www.e-greenstar.com
27.	**FIACs setup under NATP, ITD Project**	560	AP, Maharashtra, Orissa, HP, Punjab, Bihar and Jharkhand	www.manage.gov.in
28.	**IRMA, Bubaneshwar**	27	Orissa	www.ima-india.org
29.	**SEWA**	25	Gujarat	www.sewa.org
30.	**GRASSO**	150	West Bengal	www.grasso.org
31.	**Kisan Soochna Kendras**	45	Uttarkhand	www.jaikisan.org
32.	**e-cops**	225	Andhra Pradesh	www.apstatepolice.org
33.	**VOICE**	05	Andhra Pradesh	www.ap-it.com
34.	**Telemedicine**	03	Maharashtra	www.doctoranywhere.com
35.	**Tambaram**	05	Tamil Nadu	www.snegham.com
36.	**e-Panchayat**	575	AP, Uttar Pradesh	www.digitalopertunities.org
37.	**ONE**	15	Karnataka	www.i4donline.net
38.	**Ashwini**	32	Andhra Pradesh	www.byerrajufoundation.org
39.	**Mhiti-Mitra (SETU)**	380	Gujarat	www.i4donline.net
40.	**COW**	07	Andhra Pradesh	www.itforchange.net
41.	**Computer-based Training/Employment**	300	Andhra Pradesh	www.itforchange.net

(*Contd.*)

(Contd.)

Sl. No.	*Project Name*	*No. of Kiosks*	*State/Location*	*Remarks*
42.	**Functional Literacy: TATA**	80	Andhra Pradesh	www.itforchange.net
43.	**e-Panchayat**	05	Karnataka	www.itforchange.net
44.	**Cyber Grameen**	01	Andhra Pradesh	www.itforchange.net
45.	**DakNet**	10	Haryana	www.itforchange.net
46.	**Deepalaya's School-Net**	116	Delhi, Haryana	www.itforchange.net
47.	**Delhi Slum Kiosks Project**	10	Delhi	www.itforchange.net
48.	**APARATE**	1000	Karnataka	www.itforchange.net
49.	**Deltagram**	250	India	www.itforchange.net
50.	**i-Community Project: HP Labs India**	20	Andhra Pradesh, Karnataka	www.itforchange.net
51.	**Direct Artisanal Marketing**	20	UP, MP and Delhi	www.itforchange.net
52.	**EDPS2000**	06	Karnataka, Tamil Nadu	www.itforchange.net
53.	**Schoolnet India**	18	Maharashtra (Pune)	www.itforchange.net
54.	**e-Srinkhala**	200	Kerala	www.itforchange.net
55.	**Goa Schools Project**	450	Goa	www.itforchange.net
	Total Number of Rural Information Kiosks	26373		

(a) Interest shown by the Panchayati Raj Ministry for connecting 2,37,000 Panchayats in the country. The Ministry is keen to initiate a knowledge centre in each of its Panchayats (Mission 2007: August 2005).

(b) Department of Information Technology (DIT) has plans to initiate over 1,00,000 Common Service Centres (CSCs) under the National e-Governance Plan (NeGP). In order to implement the CSC programme, the DIT has recently brought a draft framework for rapid prolifiration of CSCs across the country (Mission 2007: August 2005).

(c) The private sector's commitment to complement the knowledge revolution in India by 2007. For example, the ITC is adding 7 new e-Chaupals a day and plans to scale up to 20,000 e-Chaupals by 2010 covering 100,000 villages in 15 states, servicing 25 million farmers (Sivakumar, 2004).

Thus there are clear indications that Information Kiosks movement is getting into the next phase. It is high time to consolidate the learning from all the pilots and make the information empowerment intervention at the village level, more effective.

Summary of Findings

We had undertaken a research study to assess the reach and impact of Information Kiosks on farming community, during June-August, 2005. A summary of our findings is as follows:

The estimated total number of Village Information Kiosks, VIKs (Village Knowledge Centers, VKCs), working in rural areas is around 26373 (4.33% rural internet density), and this number is rising on a very fast pace and likely to touch 50000 by March 2006 and further to over one lakh by March 2007. This forecasting is based on concrete information received from various State Governments on the rural projects already sanctioned and waiting for rolling out across the State. Some of them include—Project e-panchayat in Andhra Pradesh (475), Chhattisgarh, Madhya Pradesh, Uttarakhand and Uttar Pradesh, Kisan Soochna Kendras (3000 KSKs), in Uttarakhand (45 KSKs already operational), and Rajiv Internet Village (Andhra Pradesh)—8000, to be rolled out within this year. On similar lines Government of Maharashtra has decided to establish 10000 information kiosks in rural areas by March 2007. Thus, the rural connectivity is on high growth path and is likely to achieve the target of 16.67 per cent rural Internet Density by 2007.

About 225 farmers are being served by one information kiosks at present. Majority of the kiosks (75%) are serving 70-200 farmers. Thus, the existing 26000 kiosks in the country are expected to be serving 58,54,333 (5.85 million), which is 5.32 per cent of the total farmers of the country. If the present service ratio is considered then the one lakh kiosks (being promoted by Government of India), being put up by 2007 will be able to serve 2.25 crore (22.5 million) farmers, which will account for about 20 per cent of total farmers in the country. A total number of 4,88,889 or around 5,00,000 kiosks are required to serve the entire country for agriculture information

The majority of VKCs (57%) are promoted by private corporate sector and around 43 per cent are established by independent entrepreneurs. The average investment on a VKC is estimated at Rs.65,830. This investment includes amounts invested on computer, related peripherals and Internet connectivity. The concerned promoter generally supplies the hardware and software for the VKCs. The information services through Internet are the major component of the VKC services and the backstopping on the content side is supported by the concerned corporate sector or the concerned project secretariat (for example ITC-IBD, n-logue communications, Gyandoot etc.).

It was found out that majority of the farmers (92%) are aware of VKCs in the respective villages and the remaining farmers have also heard about their existence. An average VKC serves up to 24 villages and up to 225 farmers. The larger villages have more than one VKC as well. Only 31 farmers are visiting a VKC per month, on an average. Thus only 34 per cent farmers are making use of the VKC services. Over 90 per cent of the farmers who use the services of the VKC are coming back again for consulting. The lack of awareness and lack of education are the reasons quoted by other farmers for not using the services. Less than 2 per cent women farmers are visiting the VKCs.

The average age of the farmers utilizing the services of VKCs is in the range of 35-40 years, with 2-3 children and with an average land holding of 22.78 acres. Most of the farmers visiting the VKC are under graduates or diploma holders (62%), around 29 per cent are graduates and a few (9%) post-graduates.

Majority of the queries asked by the farmers are on market prices of agricultural commodities (60%), weather information (13%) cropping practices (8%), soil testing (8%) and the balance on generic subjects like health and education, etc. The average time is spent by a farmer on a single query is between 15-20 minutes. Most of the farmers (81%) are satisfied with the answers/solutions provided by VKCs and majority of them desired to have more services on agricultural information, market rates and online crop advice for the prevailing season.

Most of this information being provided by VKCs is supply-driven and farmers have no say in deciding or requesting the content. Thus, the availability of the information was solely dependent on the project promoters. On questioning, they however, expressed need for more area specific, crop specific, season specific and market specific information.

At present only 24 per cent of the farmers are paying for the information services from VKCs (as the VKCs established by private corporate sector are not charging any fee for information services). On query it was found that over 56 per cent farmers are willing to pay for the services.

Conclusions and Recommendations

The most important suggestion given by farmer was to improve the "Quality, timeliness and reliability of Information". The information available at the Information Kiosks presently is not as per the farmers' needs. The information is not updated regularly by most of the service providers. As the Agricultural Information is not the most saleable service the Kiosk Operators also do not take much interest in it. This suggestion has also been articulated in the report of NSSO 59th round, wherein the farmers suggested to (a) Improvement in quality and reliability of information, (b) Timeliness of information, besides (c) Increase in the frequency of demonstration, (d) Improvement in quality of presentation and improvement of professional competence of Information Provider. "The improvement in quality and reliability of information was the primary concern to most of the farmers. This suggestion was made by more than 20 per cent of the farmer households" (NSSO, 2005). Information Kiosks at village level provide a good mechanism to deliver need-based Agricultural Extension and Marketing Information at a very low cost, and hence it is suggested that good quality Agricultural Information should be made available at the Information Kiosks in the rural areas.

The agricultural sector (both development departments and research and educational organizations) needs to take cognizance of rapidly increasing Internet connectivity and availability of information shops in rural areas. Conscious efforts are required to plan, digitize and host all farmer-centric information on World Wide Web, so as to make the research findings, package of practices and other recommendations and priorities of the Government, available to all the farmers in their local language. This job requires a huge initiative with Central Government intervention to design, develop and host a National Digital Gateway on agriculture, which can serve as a mega portal (a website with many links) to provide need-based information to all the farmers on all aspects of agricultural production, processing, storage, value addition and marketing.

The ICT usage and awareness is increasing at a very fast pace in the rural and semi-urban areas, along with increased availability of the information from private/non-government sources. The functionaries of department of agriculture, horticulture, animal husbandry, etc. and the scientists of Krishi Vignan Kendras (KVKs), are not aware of the dissemination of agricultural information through private channels and hence it is submitted that all the public sector agencies within the districts may be involved in planning and digitization of agricultural information at the district level, so that they can track the agricultural situation in the areas and can also continuously up-date the information required by the farmers.

There is an urgent need to connect all the district and block level offices of agriculture and other line departments, to facilitate regular two-way flow of information between research and development wings of the agriculture.

There is a need of a massive capacity-building programme on use of ICT for the District and Block level officials, in all the States, so that they can meet the challenge of providing latest and demand driven information to the farming community. This recommendation has also been articulated in the Access to Modern Technology for Farming Report of NSSO 2005, it indicates that over 11 per cent of the farming households suggested that there is need for "Improvement of Professional Competence of Information Providers" and "Improvement of Quality of Presentation" by Extension Workers.

REFERENCES

Abdul Kalam, A.P.J. (2005). "Innovation is the Key" Deccan Chronicle, Hyderabad Edition dated 14th March, 2005, The Article was extracted from the President's address a Indian Innovation Awards in New Delhi on March 7, 2005.

Dahlman, C. Utz A. (2005). *India and the Knowledge Economy: Leveraging Strengths and Opportunities*, The World Bank, Washington D.C.

Khairnar, Sunil (2005). *Agricultural Information Management Systems in India*, A Consultancy Report submitted to MANAGE, June 2005.

Misra, Satyan (2005). Information Kiosks in Rural India: Business Provides Franchise Opportunities for Young Entrepreneurs http://main.edc.org/mosaic/Mosaic7/india.asp

Mission 2007 (2005). Mission 2007: Every Village a Knowledge Centre: Policy Recommendations August 2005, National Academy of Agricultural Sciences (NAAS), *Building*, 9-11 August, 2005.

NSSO (2005). *Situation Assessment Survey of Farmers: Access to Modern Technology*, NSS 59th Round (January-December 2003), National Sample Survey Organization, Ministry of Statistics and Programme Implementation, Government of India, June 2005.

Singh, Manmohan (2005). Inaugural Speech of Prime Minister at Launching of "Village Resource Centres (VRC)", a Joint Project of ISRO and MSSRF at Chennai on October 19, 2004.

Sood, A.D. (2003). The KIOSK Networks: Information nodes in the rural landscape http://www.i4donline.net/issue/may03/aditya.htm

Swaminathan, M.S. (2005). "Revamping and Refocusing Research and Fostering New Partnerships for Agricultural Research"—Presidential Address by Prof. M.S. Swaminathan on the 21st Century in 7th Agricultural Science Congress organized by National Academy of Agricultural Sciences on February 16, 2005 at College of Pune.

18

Informatics Blueprint for Integrated Water Resources Planning and Management at Grassroots Level

*A Quintessential Requirement for Adaptation to Climate Change and Sustainable Agricultural Development in India**

MADASWAMY MONI

Introduction

The search for socially desirable, economically viable, and ecologically sound pattern of resource use and ways of life, to promote sustainable development has been going on, ever since its most popular exposition in the well known Brundtland Report entitled "Our Common Future" in 1987. In the field of development, "good governance" and "local ownership" have become important criteria for allocating natural resources, common property resources etc. "Networking of People": Academicians, Researchers, Planners, Decision Makers, Technocrats, Field Workers and NGOs, and "Networking of Knowledge" to understand any combination of related elements that work together to achieve a desired goal, are essential to facilitate "local ownership" and participate in "good governance".

---

* Valedictory Address delivered at the National Conference on "Climate Change and its Impact on Water Resources in India", organized by Department of Geography, School of Earth and Atmospheric Sciences, Madurai Kamaraj University, Madurai, Tamil Nadu (India), 15-17, December 2004. Views expressed in this Paper are personal.

Developing countries are facing a new threat "eco-colonialism" by the developed countries who insist on pre-implementation conditionalities (i.e. relating to environment) on irrigation projects (viz., dam projects), with a possibility of vested interest to dump their surplus foodgrains and other products, in the guise of protecting the environment[1]. The UNDP's Human Development Report–2001 has introduced a Technology Achievement Index (TAI), which is an aggregation of four groups of indicators relating to: (a) the creation of technology (~IPR Divide), (b) diffusion of recent innovations (~ Digital Divide), (c) diffusion of innovations (~ Extension Divide), and (d) diffusion of human skills (~ Education Divide). Both national and international conferences have been discussing and deliberating on issues such as "Sustainable Agricultural Development, Water Resources Development and Earth Care Policies", "Water and Food Security", "Water and Health and Sanitation", "Water and Natural Disaster", "Water Management and Decision-Making", "Water and Ecology", "Water and Women" and finally, "Water and Conflict". According to our noted Scientist, Dr. M.S. Swaminathan[2], bridging the expanding "Nutritional Divide" is fundamental to bridging the "Divides" mentioned above.

Water: A Resource which is ONE and INDIVISIBLE

Water, as a resource, is one and indivisible: rainfall, river waters, surface ponds and lakes and ground water are all part of one system. Water is part of a larger ecological system. Rivers and Underground Aquifers often cut across State boundaries. Floods and droughts affect vast areas of the country, transcending State boundaries. Out of 40 M.ha of the flood prone area in the country, on an average, floods affect an area of around 7.5 M.ha per year.

The basic source of water for both Surface Water (SW) and Ground Water (GW) is the same, namely precipitation. Precipitation, in the form of rainfall, occurs only for a few days in a year and varies from 100 mm in the Western parts of Rajasthan to over 10000 mm at Cherrapunji of Meghalaya. Hence, there is a need to conserve in the soil profile, aquifers, ponds, lakes, reservoirs and rivers, for use during the dry periods. It is estimated that 263 million people live in drought prone area covering 1/3rd of the country's total geographical area of 329 million hectare.

Growth process and the expansion of economic activities inevitably lead to increasing demands for water for diverse purposes: domestic, industrial, agricultural, hydro-power, thermal-power, navigation, recreation, etc. So far, the major consumptive use of water has been for irrigation. Agriculture is the greatest user of water, accounting for about 80 per cent of all consumption. About 70 per cent of irrigation water is wasted in run-off or inefficient irrigation systems[3]. Animal Husbandry Sector and Fisheries Sector also require abundant water. While Water is essential for sustenance of human life, it can as well create problems concerning "Human Health", being a carrier of vectors for diseases such as typhoid, cholera, diarrhoea, malaria, filariasis, shistosomiasis etc. if it is mismanaged (i.e. a warning indeed).

Water has an economic value in all its competing uses and there has been growing demand that water should be recognised as an economic good. A major challenge today is to introduce economies in agricultural, municipal and industrial water use. In addition to this, availability of water is too highly uneven in both space and time. However, Water Scarcity is not a general phenomenon but is a regionally, locally, and seasonally specific problem. This can be substantiated through scientific and socio-economic analytical studies. Therefore, the Water Resource, which is a scarce and precious national and natural resource, should be planned, developed, conserved and managed as such, and on an integrated and environmentally sound basis, keeping in view the socio-economic aspects and needs of sub-national levels.

Water Resources Development Strategies: A Macro-Analysis

Physical Resources

The origin, distribution and utilisation of Water Resources of any country are profoundly influenced by its physiography, climate and soils. Physiographically, India is divided into the following regions: (a) the Northern Mountains, (b) the Great Plains, (c) the Central Highlands, (d) the Peninsular Plateau, (e) the East Coast Belt, (f) the West Coast Belt, and (g) the Islands. The whole country has been further divided into 6 major River Resources Regions, 35 Basins, 112 Catchments, 550 Sub-Catchments, and 3237 (major) Watersheds. Efforts have also been undertaken to further delineate upto micro-watershed level, as it has been done in States like Tamil Nadu.

Development of Water Resources, since our Independence, has been undertaken for specific purposes like irrigation, flood control, hydro-power generation, drinking water supply, industrial and various miscellaneous uses. As a result, a large number of projects comprising dams, barrages, hydro-power structures, canal network etc., have come up all over the country in successive Five Year Plans. The irrigation potential created stands in the vicinity of 93 M.ha by 1999-2000 as against 22.6 M.ha in 1950-51 (pre-plan period).

Since bulk of rainfall occurs during a few days during monsoon, it is necessary to harness the water flow in the river by construction of large storage reservoirs. About 10 Million rainwater harvesting structures are to be constructed within a limited time period, to store the 400 BCM of river flows discharged into the sea without being utilised (Menon, 2001). This would involve acquisition of lakhs of hectares of land, mostly from small landholders, displacing millions of poor farmers. With the failure of monsoon (if occurs), these water bodies get dried up, resulting in crop failures, which will cause large-scale devastation and consequent hardship to people.

Water Use: Agricultural Production

In a span of five decades, food-grain production, which was barely 51 metric tonnes in 1950-51, crossed 208 metric tonnes during 1999-2000. India is not only self-sufficient in meeting the food needs of the vastly increased population but also has built a buffer stock of over 50 metric tonnes of food grains to tide over any shortages which may arise due to bad weather conditions (Mehta and Bandyopadhyay, 2002). India has become the second largest producer of Wheat, Rice and Sorghum, the third largest producer of Sugarcane, and the fourth largest producer of Cotton during 1995-97. In Fruits and Vegetable production, India shares first position with China. In the Animal Husbandry and Dairy Sector too, the progress has been very impressive with milk production increasing from 17 metric tonnes in 1950-51 to over 75 metric tonnes 2001-02.

As a consequence of Water Resources Development Works, apart from the major objectives, India has witnessed developments in inland fish production too. India has now the distinction of being the seventh largest producer of fish in the World and second largest producer of inland fish after China. Amongst the States, the State of West Bengal is the highest producer, followed by the State of Andhra Pradesh and the State of Bihar. It is evident from published statistics of the Government that these three States (Bihar, Andhra Pradesh and West Bengal), if put together, produce about 50 per cent of total inland fish production in the country, while the State of West Bengal alone accounts for about one-third of this production.

The country's population, which is over 1027 Million (2001 AD) at present, is expected to reach a level of around 1390 Million by 2025 AD. Consequently, the food grain production will, however, have to be raised to around 350 metric tonnes by the year 2025 AD. This necessitates "Development of Water Resources" to a substantial order, if the food and fibre needs of our growing population are to be met with. As per the published documents available, a broad assessment of the area that can be ultimately brought under irrigation, both by Surface Water (SW) and Ground Water (GW), has indicated that ultimate irrigation potential of the country would be of the order of 139 M.ha under the categories, viz. Major, Medium and Minor, as follows:

- 58 M.Ha by Major and Medium Irrigation projects,
- 17 M.Ha by Minor Surface Water Schemes, and
- 64 M.Ha by Minor Ground Water Schemes.

It has been perceived that development of Minor Irrigation is quite advantageous, due to small investments, labour intensive, quick maturing, and most of all, a farmer-friendly. Our noted Scientist, Dr. M.S. Swaminathan said, "We need more science, both in public and private sectors, related to agriculture, for falsifying the Neo-Malthusian predictions of widespread food and drinking water insecurity".

Water Use: Access, Equity and Equality Issues

Irrigation constitutes the main component of water use in rural areas. The Land Holdings of India are predominantly of small size. Marginal size holdings (< 1 Hectare) with an average size of 0.4 Hectare constitute 59 per cent, Small holdings of 1 to 2 Hectare size constitute 19 per cent of the total land holdings. Semi-medium holdings of 2 to 4 Hectare account for 13.2 per cent and the rest are medium (7.2%) and large (1.6%). Increasing fragmentation of Land Holdings adversely affects crop production[4].

The question of access, equity, and equality in Water Resources gets directly to the use of irrigation water. Has the expansion of irrigation benefited small farmers? Have the inter-class differences in irrigation widened or narrowed? What is the regional (State) distribution of irrigation by farm size? *Deshpande et al.*[5] point out that the growth rate of all major sources of irrigation was found to be inversely related with the farm size in majority of the States between 1970-71 and 1990-91. Complex issues of equity and social justice in regard to water distribution are required to be addressed.

Per Capita Water Availability: A Situation Analysis

The Hashim Committee Report (1999) reveals that, against the annual precipitation of 4000 Billion Cubic Metres (BCM) occurring over the

Indian landmass, the available run-off is estimated as 1953 BCM. The balance is lost to atmosphere by immediate evaporation and also to the ground as soil moisture. Out of this 1953 BCM, the utilisable flow is only 1086 BCM comprising 690 BCM of surface run-off and 396 BCM of replenishable ground water. While the ground water is being over-exploited, it is possible to harness only about 250 BCM of river flows through major, medium and minor storages, allowing the balance flow of more than 400 BCM (i.e. ~ 60% of the surface run-off) to be wasted to the sea every year. Such an enormous waste of this precious natural resource is going to have a telling effect on the lifestyle of the Indian people.

India faces an increasingly urgent situation: its finite and fragile water resources are stressed and depleting, while demand from various sectors ranging from agriculture to industry and also as a result of population growth, is growing rapidly. The Per Capita availability of fresh water is reducing due to increasing population and is assessed for different years are as under:

Table 18.1: Per Capita Water Availability

Year	*Population (In Million)*	*Per Capita Water Availability (In Cubic Metres)*
1951	361 M	5177
1955	395 M	4732
1991	846 M	2209
2001	1027 M	1820
2025	1394 M (projected)	1341
2050	1640 M (projected)	1140

Source: Publications from the Ministry of Water Resources.

Drinking Water Supply (DWS) in India: Problems at the Grassroots

Rural water supply in India is the largest of its kind in the world and significant progress has been achieved due to sustained efforts by both the Central and State Governments. Adequate drinking water (i.e. 40 litres per capita per day—lpcd) has been made available to about 90 per cent of the habitations, as on 25th November 2002 (Table 18.2). This significant coverage is not without any environmental costs. Heavy dependence on Ground Water (GW) for drinking water supply as well as irrigation coupled with inadequate recharging efforts, ineffective conjunctive use of water resources and the neglect of traditional practices and systems including rainwater harvesting have resulted in the depletion of Ground Water (GW) levels, which have fallen by more than 4 metres during 1981-2000 (i.e. 20 Centimetres per year).

Table 18.2: Coverage of Rural Habitations with DWF

Sl.No.	*Type of Coverage*	*Number of Habitations*	*Percentage of Coverage*
1.	Fully Covered	1,284,203	90.27
2.	Partially Covered	123,359	8.67
3.	Not Covered	15,102	1.06
	Total	**1,422,664**	

The country is facing three major challenges in the water sector: (a) slippage of covered habitations, (b) water quality problems, and (c) sustainability of sources and systems[6].

Despite improved coverage and access to drinking water supply (DWS) in rural India, about 217,211 habitations (as on 1st April 1999) were faced with water quality problems, as categorised below:

Sl.No	*Nature of Quality Problem*	*Number of Affected Habitations*
1.	Excess Fluoride	36,988
2.	Excess Arsenic	3,553
3.	Excess Salinity	32,597
4.	Excess Iron	138,670
5.	Excess Nitrate	4,003
6.	Other Reasons	1,400

As adequate and reliable water quality data is not available, it has been very difficult to identify "hot spots" and optimally target limited resources. It has been reported that it was also not possible to carry out "risk assessment". Hence, interventions have been, in most cases, post-epidemic rather than preventive. Study Reports also reveal that Ground Water (GW) depletion has aggravated water quality problem (like excess fluoride, arsenic and brackishness), which has resulted in diseases like fluorosis and arsenical dermatitis.

Meenakshisundaram (2002) recommends that a survey/census of all drinking water sources requiring rejuvenation must be undertaken so as to facilitate drawing up an Action Plan for augmentation under the "SWAJALDHARA" mode. Remote sensing techniques can be used to identify the location of water bodies (Ponds, Lakes, Tanks etc). The DISNIC—PLAN Project for facilitating "ICT in micro-level planning", sponsored by the Planning Commission during the Tenth Plan, could consider this suggestion.

Need for a Plan of Action to Protect, Allocate and Manage Judiciously

Fresh water, which was once viewed as infinitely renewable natural resource, is no longer available in plenty in India. Developing Countries are still to harness their water potential (Menon, 2001). Water distress is visible in desert and drought prone areas. Apart from the increasing mismatch in the supply-demand scenario, gross mismanagement of water resources, coupled with neglect of traditional water resources environment, have further aggravated the situation. The result is seen in the form of falling water levels, water scarcity, degradation of quality and prevalence of water-borne and water-related diseases in several parts of the country.

In order to ensure the country against water stress and to create a greener India, the previous National Democratic Alliance (NDA) Government has announced its plan during the Year 2003 for "interlinking of various rivers of India". This programme has got the national attention, in view of its manifold objectives: (a) augmentation of irrigated agriculture (35 M.ha, i.e. from 140 M.ha to 175 M.ha), (b) potable water for the rural and urban areas, and industrial water-supply, (c) generating cost-effective hydropower (to the tune of 35,000 to 40,000 MW), (d) inland navigation, (e) ecological up-gradation due to minimum flow guarantee in rivers, (f) sizeable employment generation, (g) flood and drought mitigation, (h) increased tree farming and many other indirect benefits (http://riverlinks.nic.in).

The Common Minimum Programme of the United Progressive Alliance (UPA) Government talks about starting the project with the peninsular rivers[7]. The Common Public desire that the "interlinking of rivers" should become a "social and economic" programme of an equitable distribution of water for domestic and other purposes, with concerns on the issues of environment, ecology and displacement of people etc., rather than becoming a "political programme".

Water is rapidly becoming a scarce resource in India yet continues to be used inefficiently. If India's aspirations for continued economic growth and improved social and environmental conditions are to be met, then fundamental changes in how water is captured, allocated, planned and managed must occur. We need, therefore, a plan of action for "water accounting and budgeting" at national, regional, sub-regional and grassroots level (i.e. village or block level) so as to protect, allocate and manage our available water most judiciously to ensure water securit food security, livelihood security, health security and ecological sec for the People. This calls for an institutional set up for participatory Water Resource Management utilising the facilities available at the School of Earth Sciences, Colleges of Agriculture, Departments of Geography, School of Environment Sciences, and Centres for Water Resources Development Studies of Universities of India.

Water Quality Issues: Is it Unsolvable?

Water Quality (WQ) is a function of chemical, physical, and biological characteristics and it impacts "use-of-water" or "ecosystems-within-the-water". Direct and indirect human activities (i.e. Land-use and agricultural practices) impact WQ problems in Water bodies (viz. rivers, lakes, aquifers etc). Even the "management action" to control point and non-point source pollution and treat wastewaters discharged into the environment, results in WQ problems.

Studies carried out in India reveal that the most important causes of Ground Water Pollution are: (a) "unplanned urban development without adequate attention to sewage and waste disposal", (b) industrialization without provision of proper treatment and disposal wastes and affluent, (c) excessive application of fertilizers for agricultural development, and (d) over-irrigation intrusion due to excessive pumping of fresh water in coastal aquifers. This "over-exploitation" mainly accounts for seawater intrusion. Contamination enters Water bodies through one or more of the following ways:

- *Direct Point Sources*: Transfer of pollutants from municipal industrial liquid waste disposal sites and from municipal and household hazardous waste and refuse disposal sites.
- *Diffuse Agricultural Sources*: Wash off and soil erosion from agricultural lands carrying materials applied during agricultural use, mainly fertilisers, herbicides and pesticides.
- *Diffuse Urban Sources*: Run off from city streets, from horticultural, gardening and commercial activities in the urban environment and from industrial sites and storage areas.

vention and Control of Environmental Pollution (EP) has the
egislation support and review through the Central Pollution
ard (CPCB) and State Pollution Control Boards (SPCBs).
Quality gets monitored on 25 physico-chemical and
ameters, through about 480 stations under Global
onitoring System (GEMS) and Monitoring of Indian

National Aquatic Resources (MINARS) programme. The National Water Quality Monitoring Network Programme covers about 126 rivers (including the tributaries), wells, lakes, creeks, ponds, tanks, drains and canals (http://envfor.nic.in). An "Environmental Information System (ENVIS)" is operational in the Ministry of Environment and Forests. Efforts should be made to undertake "water quality monitoring" at village level and the noticed quality be notified through Viliage Panchayati Raj Institutions. ENVIS should be enhanced to incorporate features of "localism" as well as "localization". The WQ is "quality" in relation to some standards and the different "water use" have different standards.

In Water bodies, future water quality will be very dependent on future human activities, including Water Management Policies, and the direct effect of Climate Change may be very small in relative terms (Hanratty and Stefan, 1998).

Sustainable Development and Management of Water Resources: Water Bodies (Basin), Watershed, and Agro-Ecology Region Planning—A Way Forward for Grassroots Level Prosperity

Utilisable Ground Water (GW) potential of India is limited and is to be utilised judiciously. Development and over exploitation of Ground Water (GW) Resources, in certain parts of the country, have raised the concern and need for judicious and scientific resource management and conservation. Presently, the ground water available at a place is considered as the property of the landowner. In many developed countries, the ownership of the land does not permit the land owner to utilise the GW indiscriminately.

Sustainable Development and Management of Water Resources (SW as well as GW) require a holistic approach (i.e. simultaneous attention to all aspects of Water Use and Management). This requires a thorough understanding of the hydrology of the SW System, GW System, Agro-eco Systems, Soil-Water-Plant-Environment interactions, existing and expected water demands, availability of energy for pumping and social dynamics.

The Agenda-21 of the Rio-Earth Summit (1992), the UN Convention to Combat Desertification (CCD), the UN Convention on Biological Diversity (CBD), the UN Framework Convention on Climate Change (UNFCCC) and its Kyoto Protocol, and the Habitat Agenda adopted by the UN Conference on Human Settlements in 1996, directly or indirectly, suggested integrated planning and management of Water, Land, Minerals, and Biota resources (that land comprises), for sustainable development and use at grassroots level.

The National Water Policy (1987) and the National Water Policy (2002) have addressed to the issues related to develop, conserve, utilise and manage Water Resources in this new millennium. The Hashim Committee Report (1999) dealt with various issues related to "Integrated Water Resources Development Plan" in the country and warrants timely action in respect of: (i) harnessing of the surface flows through major, medium and minor storages, (ii) improving water-use efficiency, and (iii) taking necessary steps for demand and supply. The Nagpur Declaration (2000)[8] on "Natural Resources Planning and Management for Sustainable Development" envisages that for integrated water resources planning and management, both "river basins management" at the macro-level and "watershed management" at the micro-level, should mutually complement each other. Long-term planning needs a clear vision for Land and Water Resources Management at grassroots level.

The All India Soil and Land Use Survey (AISLUS) organisation has delineated the entire country into 3237 major Watersheds (6 major River Resources Regions, 35 Basins, 112 Catchments, 550 Sub-Catchments, and 3237 major Watersheds). Each Watershed contains a complex mixture of soil types, landscapes, climatic regimes, land use characteristics, and agricultural systems. Yet, this complexity is not unmanageable, and therefore, can be sub-divided into more than one Agro-Ecological Region (AER), on the basis of having similar soil types, landscapes, climatic regimes, crop productivity and animal productivity, and hydrologic characteristics.

Cropping Pattern of an area is determined by many factors, which include: type of soil, climate, rainfall characteristics, farmer's requirements of foodgrains for self-consumption as well as market supply, and net rate of financial gain, resulting in various alternative crops that can be grown in that area. Each State of the country is expected to concentrate on agricultural products most suited to its agro-ecological and agro-climatic conditions, as it is not possible to hope to be self-sufficient in all the essential commodities[9].

AER Planning is useful in conjunction with Watershed Planning for developing a list of priority needs in monitoring, research, education, and implementation activities for various basins (water bodies) in the country. Using the AER characteristics along with existing Water Quality Monitoring Information, it is possible to determine:

- Where the steepest and most erosive landscapes are located,

- Where the water quality problems are most severe,
- Where land use patterns are most conducive to non-point source pollution production, and
- Where the research knowledge-base concerning sources of pollution and methods of managing pollution is weakest.

Sustainability in Agricultural Development depends upon the judicious use of natural resources, viz. soil, water, livestock, plants, fisheries, forests, climate and rainfall, and topography. This requires that (a) both SW and GW supplies are maintained at the desired level, and (b) the quality of Land and Water Resources does not deteriorate with time.

For each major Watershed, it is suggested to engage all stakeholders within the Watershed, in discussions and coordinated planning efforts that cross State boundaries. In this connection, the following recommendations of the Core Group-V Report (2000)[10] on "Agricultural Resources Information System (AgRIS)", of the Standing Committee on Soils and Agriculture (Ministry of Agriculture) are noteworthy:

- "A Decision Support System (DSS) on water allocation in an irrigation system to remove the existing disparities in the availability between the head-reach and tail-end farms and between large and small farms, to achieve "equity and social justice";
- "AER Planning is useful in conjunction with Watershed Planning for developing a list of priority needs in monitoring, research, education, and implementation activities for various basins (water bodies) in the country";
- "A DSS on Water Bodies (Basin) using Watershed and AER Planning Concepts is to be taken up";
- "National Livestock Policy (NLP) studies classifies the country into four livestock farming system, viz. Cattle Farming System (CFS), Buffalo Farming System (BFS), Sheep Farming System (SFS), and Goat Farming System (GFS). To observe the possible linkage between livestock systems and cropping systems, demographic and agro-climatic features, DSS on Cattle Farming System (CFS), DSS on Buffalo Farming System (BFS), DSS on Sheep Farming System (SFS), and DSS on Goat Farming System (GFS) may be developed".

Effective and Economical Management of Water Resources: Need for Frontiers of Knowledge Through Intensified Research Efforts

The National Water Policy (2002) envisaged that for effective and economical management of our Water Resources, the frontiers of knowledge need to be pushed forward in several directions by intensifying research efforts in various areas, including the following:

- hydrometeorology;
- snow and lake hydrology;
- surface and ground water hydrology;
- river morphology and hydraulics;
- assessment of water resources;
- water harvesting and ground water recharge;
- water quality;
- water conservation;
- evaporation and seepage losses;
- recycling and re-use;
- better water management practices and improvements in operational technology;
- crops and cropping systems;
- soils and material research;
- concrete, fibre reinforced concrete, new methodologies in tunnelling technologies;
- instrumentation, advanced numerical analysis in structures and back analysis;
- seismology and seismic design of structures;
- the safety and longevity of water-related structures;
- economical designs for water resource projects;
- risk analysis and disaster management;
- use of remote sensing techniques in development and management;
- use of static ground water resource as a crisis management measure;
- sedimentation of reservoirs;
- use of sea water resources;
- prevention of salinity ingress;
- prevention of water logging and soil salinity;
- reclamation of water logged and saline lands;
- environmental impact;
- regional equity.

With a view to devising a strategy for the future, it was learnt that the Central Ministry of Water Resources has commissioned a series of six reports on Water Resources Management (WRM) in 1990s, in collaboration with the Ministry of Urban Affairs, the Ministry of Rural Affairs, and the World Bank:

- Water Resource Management: Initiating and Sustaining Water Sector Reforms;
- Inter-Sectoral Water Allocation, Planning and Management;
- Groundwater Regulation and Management—An Important Source of Drinking Water and Food Security;
- Irrigation and Drainage in Indian Agriculture;
- Rural Water Supply and Sanitation; and
- Urban Water Supply and Sanitation.

The main feature which revolved around these reports included "strengthening data, technological and information systems". What will be the role of Departments of Geography, which are spread throughout the country and educating more than 12,000 students every year? Will they involve themselves in developing "Knowledge Information System and Management" on Watersheds, as a part of "ICT Diffusion Developmental Action Research"?[11]. Geo-Informatics on DAMs (which is numbering about 4000) safety and relief measures at village level is a must and can be developed through the Departments of Geography (200 in numbers) located throughout the country.

Integrated Watershed Development Approach (IWDA): A Holistic Approach for Sustainable Development of Rainfed Agriculture

Agricultural sector is the mainstay of the Indian rural economy around which, socio-economic privileges and deprivations revolve. Any change in its structure is likely to have a corresponding impact on the existing pattern of social equality at the grass-roots. The Indian Agriculture contributes about 25 per cent of GDP, provides employment to about 65 per cent of the labour force, and accounts for about 20 per cent of total exports.

Critical Situation of Agriculture

Agricultural development faces resource constraint for "irrigation and drainage" infrastructure, which continues to attract high public investment. It is quite relevant to note "the critical situation of Indian Agriculture", as appraised by the Planning Commission, during its Ninth Plan Mid-term appraisal, which was captured as follows:

- Low public investment in irrigation and poor maintenance of rural infrastructure (specially canals and roads);
- Decline in investments in rural electrification, and in its availability. This has vastly affected production in eastern India, where huge Groundwater potential remains untapped;
- Rising level of subsidies for power, water, fertilisers, and food, which cut the public sector investments in Agriculture and also induce inefficient use of scarce resources (i.e. water)?
- Aggravation of environmental problems leading to loss of soil fertility and groundwater, which in turn reduces "return on capital", which leads the farmers demanding further subsidies to maintain the same level of production.

Most parts of Eastern Region (ER)[12] lags rest of India on some social indicators like: Poverty ratio (highest in Bihar), Infant fertility (highest in Orissa), Overall literacy (lowest in Bihar) and Female literacy (lowest in Orissa and Bihar), and Poverty incidence varies within ER. In areas where investments in agricultural infrastructure have lagged, such as Bihar, Orissa, Assam and Madhya Pradesh, rates of growth in agricultural productivity and poverty reduction also lagged[13]. Given the strong linkage between output growth (particularly agricultural growth) and reduction in poverty, accelerated agricultural growth is critical, to reduce poverty in ER States. The existing food security has been achieved through "increase in irrigated agriculture" and "introduction of HYV of crops". Even then, the Rainfed Agriculture is crucial to the country's economy and food security because 67 per cent of the net sown area is rainfed and about 44 per cent of the total food production is contributed by the rainfed agriculture.

Farming Without Irrigation: A Hard Way of Life

Rainfed agriculture is generally characterized by (a) low levels of productivity and low-input usage, (b) decrease of the agricultural intensification, and (c) instability in production and the farmers' income. Small and Marginal Farmers (SMFs) in rainfed areas, continue cultivating their land in order to retain their ownership (or user) rights. Rainfed Agriculture is hence complex, diverse and risk-prone. The rainfed areas have not benefitted from the "green revolution effects" of irrigated agriculture. Of this 67 per cent, about 30 per cent area is under dryland agriculture, wherein the annual rainfall is up to 400 mm only. One-sixth area of the country is drought-prone.

Rainfall is erratic and poorly distributed in Rainfed Areas, and soils are often severely eroded, infertile, and deficient in organic matter. It is true that "the lesser the rain in an area, the greater the trouble for the farmers and villagers there". A large number of farmers are Small and Marginal Farmers (SMFs) and depend on Rainfed Agriculture for their livelihood. Farming without irrigation is a hard way of life (http://lnweb18.worldbank.org/sar/sa.nsf). At the same time, they have an abundance of biomass and cattle-waste. It is noteworthy that they have never used cattle-waste in the cultivation process. Adopting composting and discontinuing chemical based agricultural practices, has gained widespread visibility (http://www.geocities.com/ ncdc_vns/org-frm.htm).

Watershed: A Geo-Hydrological Unit for Development of Rainfed Farming

The agricultural investment in India has traditionally concentrated on the development of irrigable lands[14]. Agricultural policies have systematically discouraged small-scale rainfed farming[15] and made large-scale mono-cropping plantations economically attractive to urban investor. Until the early 1980s, relatively little economically promising new technology was available to farmers in rainfed areas. Watershed[16] is a geo-hydrological unit of area for the development of Rainfed farming. Watershed development has become a trusted tool for the overall development of the village and people living within a watershed area.

In the words of noted economist, Professor C.H. Hanumantha Rao, "watershed development has been conceived basically as a strategy for protecting the livelihoods of the people inhabiting the fragile ecosystems experiencing soil erosion and moisture stress". Watershed projects, launched during 1984 in India, have contributed significantly to the evolution of rainfed technologies and approaches for (a) increasing agricultural production in Rainfed Areas through improved land and crop management; (b) On-site moisture conservation; and (c) Production of fodder and fuel in non-cultivated areas. The basic components of the Watershed Approach consist of:

- Community Development (Human Resource Development),
- Soil and Land Management,
- Water Management,

- Afforestation,
- Pasture/Fodder Development,
- Livestock Management,
- Rural Energy Management, and
- Farm and Non-Farm Value Addition Activities;

Watershed Management (WSM): A Well Accepted Single Window and Sound Strategy for Efficient Management of Land, Water and Vegetation Resources

If the Water Resource is managed properly, then not only drinking water problem can be mitigated but also the economic condition of villagers in general, and farmers in particular, will also be changed. The Bahidwadi Village in the District of Ahmednagar Tehsil has become tanker-free and the villagers are harvesting both Khariff and Rabi crops as water table has gone up from 8 m to 5 m, after the completion of a Watershed[17]. Watershed Management (WSM) has been found to a well accepted single window and sound strategy for efficient management of land, water and vegetation resources through community participation for sustainable productivity, resource conservation, ground water recharge, drought moderation, employment generation and social equity. Study reports reveal that benefits derived from Watershed Methodology, could be categorized as follows:

(a) Increase of crop yield by 25-40 per cent in dry land farming;
(b) Reduction of soil loss due to erosion by 30 per cent;
(c) Large Scale Vegetation Cover of barren hill slopes;
(d) Large tracts of marginal lands brought under dry land horticulture;
(e) Development of Agro-Horti and Agro-Forestry systems;
(f) Harvesting of Water Resources through *nala* bunds, farm ponds, gully embankments;
(g) Regeneration of grass lands for more fodder and grass; and
(h) Considerable increase of income of farmers.

This is due the fact that the development is based on type of soil, depth of soil, vegetative cover, harvestable rain water in that area, watering that area, water budgeting, and treatment given to soils from the ridge to the valley.

The Government of India has accorded highest priority to the holistic and integrated development of rainfed areas (http://agricoop.nic.in/annrep/fnchap3.htm) for:

- Meeting the projected foodgrains requirement,
- Bridging the regional disparity in terms of production and productivity between the irrigated and Rainfed Areas,
- Restoring ecological balance,
- Generating employment opportunities in rural areas.

The Central Government launched the National Watershed Development Project for Rainfed Areas (NWDPRA) during the 8th Plan (covering 25 States and 2 UTs) and continued during the 9th Plan (covering 28 States and 2 UTs) for increasing agricultural productivity and production through sustainable use of natural resources. Also, the Watershed Development Projects in Shifting Cultivation Areas (WDPSCA) was launched in N.E.R. States during the 8th Plan (1994-95) and continued during the 9th Plan also. The Union Budget (1999-2000) suggested establishing a "Watershed Development Fund"[18] at NABARD for integrated watershed development in 100 priority districts of 14 States through participatory approach. Many externally aided projects on integrated watershed development were taken up in States, as given below:

- Integrated Watershed Development Project—Phase II (The World Bank/IDA Credit)—in Shivalik Hills in the States of Haryana, Punjab, Himachal Pradesh, Jammu & Kashmir, and Uttrakhand,
- Karnataka Watershed Development Project (The World Bank/IDA Credit),
- Assam Rural Infrastructure and Agriculture Support Project (ARIASP) (The World Bank/IDA Credit),
- Diversified Agriculture Support Project (DASP) in Uttar Pradesh and Uttrakhand,
- German Assisted Project (Kreditanstant for Widderaufbau) in Maharashtra,
- DANIDA Aided Projects (Comprehensive Watershed Development Project) in Tamil Nadu, Karnataka, Orissa, and Madhya Pradesh,
- DFID (U.K.) Aided project in Karnataka (Karnataka Watershed Development Project—KAWAD),

- Indo-German Bilateral Project on Watershed Management in the States of Bihar, Gujarat, Madhya Pradesh, Orissa, Maharashtra, Andhra Pradesh, Uttar Pradesh, Himachal Pradesh, Tamil Nadu, and Rajasthan.

There has to be a shift in the approach and strategy of the Watershed programme from the conventional soil conservation approach of SAFE DISPOSAL OF RUN-OFF to RAINWATER CONSERVATION AND HARVESTING based on indigenous systems and practices, suitably supported by appropriate technology[19]. About 18 million hectares of Rainfed area[20], which is suffering from low yields of crop production and soil erosion due to inadequate vegetative cover, are suitable for agro-forestry[21].

Development of the vast rainfed areas (90 million hectares) would require over Rs. 37,000 crores, and scientific treatment for soil and water conservation for 12 million hectares of arable and 3 million hectares of non-arable land would require about Rs.7,500 crores.[22] The perspective plan that the Tenth Five Year Plan Working Group had proposed is to cover 88.5 million hectares of Rainfed and degraded lands in the next four Five Year Plans (i.e. upto 2022 A.D.). What about the investment? Will it come from the Government or from the Private? The investment question, no doubt, is a pertinent issue and some loud thinking is called for in this regard.

Diffusion of Information and Locally Adapted Technologies: An Essential Ingredient for Local Empowerment

Experiences in Sub-Saharan Africa (stretching from Senegal in the west to Sudan, Somalia and Ethiopia in the east) have thrown some helpful guidance for planning in Rainfed Areas. They provide promising approaches to natural resources management involving close contact between the community concerned and the outsider agency involved, over a period of several years. This time element is important because it takes a long time to develop a series of appropriate activities, which can address priorities of the local people in rainfed areas, use their conventional wisdom, and introduce adaptations which are within their grasp. Alternative way is to build upon successful practices elsewhere, which would certainly permit a far wider diffusion of information and locally adapted technologies.

We have some success stories, like the transformation brought out in Jhabua district of Madhya Pradesh, e.g. availability of drinking water, low levels of crop failure, availability of fodder, increase in milk yields, and more importantly employment and food security[23]. In this context, it must be noted that in the Indian approaches followed so far in watershed development programmes, the projects are mostly confined to soil and water conservation measures and plantation activities. The rejuvenation of the Rainfed Areas is largely perceived as a "land-based" programme with the implicit assumption that groundwater recharge as a result of soil and water conservation, will automatically result in better livelihoods of the local people. The reality, on the other hand, is that investments are made without any prioritization of the poor and totally insensitive to the reality of the caste and class structures of the villages. The work of a dedicated NGO in the "Pushkar Gap Project"[24] is a revealing study.

All these clearly show that "a better life for disadvantaged rural people may be more feasibly sought through a different sort of change; through quiet personal revolutions in the perceptions, values and choices of professionals concerned with research, technology and action for rural development"[25]. Such changes, however, require perseverance and painstaking efforts, taken resolutely in a consistent direction. For bringing about departures in Rainfed areas, new kinds of research and extension systems are needed, not only technical support but also to make major changes in terms of the definition of their objectives. Thus the systems devised for extension need to move away from providing "experts" to tell the local people what they should be doing, but to move towards the development of staff who can act as "Catalyst, Convenor and Colleague"[26].

Despite 50 years of development planning, Rainfed agriculture continues to be the largest and the most important segment of crop production in India. Both the infrastructural facilities and access-to-markets capabilities influence developments in the Rainfed agriculture (e.g. oilseeds development in semiarid tropics) in India. The need for a second green revolution, by making grey areas green, was thus felt (i.e. making Rainfed areas more productive)[27]. There has been a paradigm shift: from a "commodity" centred "Green Revolution" towards an "integrated natural resources management" centred "Evergreen Revolution" in India[28]. While physical access to food was the main consideration during the 1980s and 1990s, economic and ecological access to food security would be the major challenge in the near future. Fusion of technologies is expected to benefit "Rainfed farming" facilitating "bubble-up" economics, i.e. small farmer based rural economy to build up national economic development, in stark contrast to the failure of "trickle down" economics[29].

The emergence of the new multilevel paradigm in favour of the decentralization of power provides an important opportunity to relocate the rights and responsibilities for resource management in the hands of local people. However, much depends on the mode of functioning, and how much resource will be devolved to the local people. Further, the mere devolution of control over resources to local communities will not be enough to ensure a more sustainable development in our dry lands. Ability to steer through such development with devotion and dedication is fundamental. It is here that the role of highly motivated and dedicated NGOs becomes relevant.

Jodha's (2001)[30] numerous writings on "Fragile Environments" have pointed to the neglect, by the mainstream policy-makers, of people's adaptation strategies to high-risk-low-productivity environments through folk technologies and various formal and informal institutional arrangements. To revive these traditional systems, it has been emphasized that the incorporation of these elements is essential, viz. community stake, local control and functional knowledge of the natural resources base. Broadly, the process of globalisation would create circumstances, which are beyond the control of communities in fragile Rainfed areas. Since, however, it is not possible to wish away the process of globalisation; the solution would be to influence and modify these processes by adapting to the changes led by globalisation. The specific focus of the steps could be on minimising economic losses, preventing exclusion possibilities, ensuring local participation in the resource harnessing decision, creating compensatory mechanisms for environmental services offered by fragile areas and their people to the rest of society and the economy etc. (Jodha, 2001).

This "Blue Revolution" is a step towards water conservation and management, smallholder irrigation, drainage and waste water reuse, to achieve sustainable public health and food security, and ushering "2nd Green Revolution", "Gray Green Revolution", or "Evergreen Revolution" in India. To achieve quick and spectacular results, it is necessary to establish models of appropriate transfer of technology at District, Block, Panchayat and Farm levels spread over the entire country.

Indigenous Technical Knowledge (ITK): Stage Set for Exploitation for Sustainable Use and Management of Natural Resources in Agriculture

Indigenous Technical Knowledge (ITK) refers to the unique, traditional, local knowledge existing within and developed around the specific conditions of women and men indigenous to particular geographic area (Gremier & Louise, 1998). Mehta and Bandopadhyay (2002)[31] list out that the documentation activities have been undertaken in the following areas:

- Agro-Animal Based Yarns, Natural Dyes and Weaves,
- Crops and Cropping Systems,
- Ethno Botany and Agro Biodiversity,
- Farm Implements,
- Fishery,
- Food Product Development,
- Grain and Seed Storage,
- Horticulture,
- Low Cost Housing Material,
- Pests and Disease Management,
- Post Harvest Technology,
- Rainwater Management,
- Soil and Water Management,
- Soil Fertility Management,
- Tillage and Interculture,
- Veterinary Sciences and Animal Husbandry.

The pressure on land has increased to an extent that the relationship between the living beings and the soil has become critical. Our soil resources face deterioration aggravated due to:

- Deforestation,
- Accelerated Soil Erosion,
- Deterioration of Soil Physical Environment,
- Increasing waterlogging and salinity in Canal Irrigated Areas,
- Declining water table due to over-exploitation of Ground Water,
- Poor Management of Rain Water,
- Lower efficiency in Water use and fertiliser (and agro-chemicals), and
- Rapid Industrialisation Coupled with Pollution and Environmental Degradation.

Mehta and Bandopadhyay's (2002) study clearly shows that the emerging importance and potential of ITK could usher in WIN-WIN situation for natural resources conservation and management for sustainable agricultural development in India. Ram Kumar (1998)[32] also asserts that the value of ITK has to recognise and blend with scientific information, if the livestock sector is to be improved and sustained. The importance of traditional wisdom and indigenous knowledge and practices to sustainability is being brought centre-stage through pooling of traditional knowledge, distilling and evaluating it in the context of modern scientific and technological environment, and harnessing it for sustainable agricultural growth.

A key element of the FARM Programme[33] is the recognition of communities indigenous knowledge, which when complemented by specialist "formal knowledge" promote participatory learning and research for achieving sustainable use and management of natural resources in agriculture and attainment of household food security through innovative approaches, in Rainfed areas.

Climate Change: What does it Mean to the Common Public in General? The Farmer in Particular?

Climate Change (CC) means changes in temperatures, precipitation, water stress, floods, cyclones and droughts. Current Climatic Models (CMs) are predicting global warming of about 1.4 to 5.8 degrees centigrade during 21st Century, which will create many challenges to agriculture in terms of economic growth, poverty eradication, land degradation, access to water and food security. Available Research Studies provide directional evidences that Climate Change (CC) would influence the bio-physical vulnerability of Indian farmers.

The atmospheric concentrations of key anthropogenic greenhouse gases [Carbon Dioxide (CO_2), Methane (CH_4), Nitrous Oxide (N_2O), and Tropospheric Ozone (O_3)] reached their highest levels in 1990s, primarily due to the combustion of fossil of fuels, agriculture and Land use changes[34]. Due to the long span of greenhouse gases in the atmosphere, and regardless of efforts to reduce their emissions, Climate Change (CC) would occur for at least in the middle of the next century. Adaptation is therefore a necessity. Planning adaptation must begin with an understanding of vulnerable populations and regions. Vulnerability[35] and Adaptation[36] are central to International Policy on Climate Change (IPCC), in both United Nations Framework Convention to Climate Change (UNFCCC) and the Kyoto Protocol.

Hazard +	:	potential threat to humans and their welfare
Vulnerability =	:	exposure and susceptibility to losses
Risk Disaster	: :	probability of hazard occurrence realization of a risk

Recent experiences, worldwide, show that any change in the critical variables (temperature, radiation, and precipitation) can induce vulnerability of food production in a major way. The IPCC reports show that impacts of Climate Change (CC) in the future would be borne disproportionately by the Poor. Agro-management options have to be identified for sustaining yields under climatic variations. Effective policies and measures are required to safeguard and improve food security and rural livelihoods, as the vulnerable are the landless and the rural poor.

Adaptation to Climate Change for Agricultural Productivity

The issues of Climate Change (CC) and its potential impacts on agriculture have been a major research topic in recent times. Potential Climate Change (CC) impact on Agriculture will be on Crop Yield variability and Irrigation Demands. Most yield analysis in India showed higher vulnerability to Rainfed crops. Crop Yield Variability is one of the often-quoted indicators of agricultural impact, due to climatic change. Most yield analysis in India showed higher vulnerability to Rainfed crops. Sinha and Swaminathan (1991)[37] estimated that a 2°C increase in mean air temperature could decrease rice yield by about 0.75 ton/hectare in the High Yield Areas and by about 0.06 ton/hectare in the Low Yield Coastal Regions. Rao and Sinha (1994)[38] in their crop-simulation study have estimated that under a 2 × CO_2 climate change scenario, the wheat yields could decrease by 28 per cent—68 per cent without considering the CO_2 fertilization effects. Kumar and Parikh (1998)[39] have shown that even with adaptation by farmers of their cropping patterns and inputs, in response to Climate Change (CC), the loss in farm-level net-revenue is estimated to range between 9 per cent and 25 per cent for a temperature rise of 2°C to 3.5°C.

There have been a lot of "deliberations" that Climatic Changes (CCs) could affect food supply and access, through their direct and indirect effects on crops, soils, livestock, fisheries and pests. Long-run investments in land, machinery, irrigation, seeds, fertilisers, etc., are strongly influenced by climate and climate variability[40]. Climate Change (CC) may further exacerbate the regional differences in the Eastern Region, because these regions with limited irrigation infrastructure are also the areas where agriculture is most vulnerable. Regional climatic data about precipitation, temperature, and extreme events are required to anticipate and plan Climate Change (CC) Adaptation Strategies.

According to Ravindranath (2002)[41], Agriculture sector contributes to greenhouse warming (20% of radiative forcing) but at the same time, provides opportunity for low cost mitigation options by undertaking CO_2 sequestration and Methane and Nitrous Oxide emission reduction. Ravindranath (2002) suggests adaptive measures, which should receive highest priority, as follows:

- Increase irrigated areas based on sustainable water resources;
- Adopt soil and water conservation measures and prevent land degradation;
- Increase water use efficiency of both Dryland and Irrigated agriculture;
- Develop drought-resistant and pest-resistant crop varieties;
- Develop slat-resistant crop varieties for coastal zones;
- Promote mixed-cropping practices;
- Develop crop varieties adapted to projected Climate Change;
- Improve weather information system.

Use of Renewable Energy Sources: A Premier Step to Check Growth of CO_2 Concentration Level

In India, the agricultural production is emerging as an important consumer of commercial energy. Solar energy has wide prospectus for utilisation in Agriculture, as the sun's energy is available when the Agriculture operations are done. Solar energy is a natural gift and is of prime importance in the field of all energy. Sun is recognised as the father of all sources of energy. Renewable Energy can be used to replace various activities in rural areas and in crop production. In agriculture, the maximum energy is used in "land operations" and "irrigation". The Rural Energy Technology (RET) can substitute the conventional energy sources to a great extent as given below:

Options	Operations in Rural Areas based on RET,
Solar Energy	Water Pumping for irrigation, Post-Harvest Operations, and Domestic Cooking,
Biogas	Fuel for Domestic Use,
Biomass	Domestic Cooking.

The restricted supply of electric power and the frequent shortage of diesel oil pose the most disturbing and immediate problem, to an individual farmer. Due to its perennial availability, the solar energy will remain a favoured option for sustainable development of agriculture. Serious efforts are required to save as much of our installed electrical power of about 100,000 MW, through use of renewable energy sources (RESs), which will check the growth of the CO_2 concentration level. The cost of power generation from RESs is comparable if the environment cost is also accounted for. The extensive use of RESs to complement the existing energy sources is an option to sustainable energy flow in rural areas[42].

Agricultural Ecology Network for Capacity-Building at Grassroots Level: A Pre-requisite for Adaptation to Climatic Change

With Climate Change (CC) and Globalization occurring simultaneously, the Indian agriculture is confronted by the impacts of both the processes. What are lacking in the available literature are the potential interactions between the effects of Climate Change and the On-going Economic Changes. There is an urgent need to integrate the Socio-Economic Components into Biophysical model, which will require the base line data on the following:

- *Climatic Scenarios*: Spatial-temporal pattern of temperatures and precipitation;
- *Socio-economic*: Population (density, growth, urban, coastal, Below Poverty Line); Income (GDP, AgGDP, Growth); Land (Ownership, Use, Net Cultivated Area, Gross Cultivated Area), Labour Force; Asset;
- *Resource*: Land, Livestock and Capital assets, Energy use, Irrigation;
- *Bio-diversity*: Diversification of Cropping Pattern.

Change Over Base Line

- Demographic variables (Population density, Sex ratio, etc.),
- Land Use and Land Degradation,

- *Pollution*: Magnitude, Health Hazards,
- *Economic*: Poverty, Income, Sectoral Composition,
- *Social*: Literacy, disaster deaths.

Agricultural Production on a sustainable nature depends on the judicious use of natural resources like soil resources, water resources, animal resources, crop/plant genetic resources, in an acceptable technology management under the prevailing Socio-economic infrastructure at grassroots in India. Climate adaptation and sustainability goals (resilient sustainable development) can be jointly advanced by changes in policies that lessen pressure on resources, improve management of environmental risks, and enhance adaptive capacity. Negative impacts of Climate Changes (CCs) can be limited by:

- Changes in crops and crop varieties,
- Improved water management and irrigation systems,
- Adapted planning schedules and tillage practices, and
- More scientific land use planning.

Indian Agriculture has provided food, feed, nutrition, employment and environmental securities to the ever increasing population of human being and livestock. Increased climate variability and long-term changes in Climate will hamper Farmer's adaptive efforts[43]. Grassroot level research should be directed towards environmental issues such as Climate Change (CC), Loss of Biodiversity, Soil Degradation and Water-Level-Pollution versus farm-level technologies for improving Crop, Fisheries, Forestry and Livestock production.

A Climate Change Network Project (Agro-Climatic Zone-wise) has been launched by the Indian Council of Agricultural Research (ICAR), during the 10th Plan for comprehensive understanding of impacts, adaptation, and mitigation strategies for various crops and cropping systems. There have been suggestions from various corners that, to promote networking, training, capacity-building and documentation/ dissemination of best practices, technologies and success stories, it was essential to envisage a National Centre for Agricultural Ecology (NCAE) at New Delhi. Lester Brown (2002)[44] suggests an economy for the Earth—"Eco-economy"—that respects the principles of ecology which can be integrated into eco-systems, in a way that will stablise the relationship between the economy and the earth, enabling economic progress to continue.

Integrated Water Resources Planning and Management at Grassroots Level for Adaptation to Climate Change and Sustainable Agricultural Development

Need for an Informatics Blueprint

The Dublin Statement (1992) on Water and Sustainable Development and Agenda 21 of the UN Conference (1992) in Rio de Janeiro stressed on the need of/for intensive efforts to appropriately develop and efficiently manage the world's scarce fresh water resource, based on a participatory approach involving Users, Planners and Policy Makers at all levels. The Global Water Partnership (1996) recognised the advantage of ICT developed and made available globally.

Water Resources Development and Management is an Information System. An Information System (IS) is defined as components that work together to provide desired information, in the proper format, at an appropriate time. System, which can have numerous subsystems and sub-subsystems, is the term used to describe any combination of related elements that work together to achieve a desired goal. The National Water Policy (2002) which has envisaged a well developed information system, for water related data in its entirety, at the National/State level, is a prime requisite for resource planning.

Exchange of Information between Hydrological Science and Water Managers

In the Water Sector, it is important to distinguish between the needs of those who wish to estimate the potential magnitude of Climate Change impacts on hydrology and water resources—to meet IPCC concerns, for example—and the more pragmatic needs of Water Managers who need to consider how best to adapt to Climate Change. The two sets of requirements are linked, but there are some important differences in emphasis[45]. Such research questions need developments in the following areas:

- Creation of credible climate change scenarios.
- Characterization of natural climatic and hydrological variability.
- Improved hydrological models (Particularly important is development and application of process-based models of hydrological processes that include realistic representations of processes that generate stream-flow and recharge and determine water quality).

- Characterization of uncertainty. How important are the different sources of uncertainty—in emissions, global climate response, and regional climate change—for estimated effects of climate change?
- Impacts on real-world water systems (Characteristics of the Water Management System are a very important buffer between hydrological effect and impact on users and the environment).
- Effects of adaptation. Most impact studies have ignored adaptation by water managers, and in opposition, it often is asserted that water managers will be able to adapt. However, how will managers make adaptation decisions in practice on the basis of incomplete information, and what would be the effects of inefficient adaptation on the impacts of climate change?

Efficient adaptation to Climate Change in the Water Sector requires effort in five main areas:

- *Data for Monitoring*. Adaptive water management requires reliable data on which to make decisions, calibrate models, and develop projections for the future. These data should cover not just hydrological characteristics but also indicators of Water Use.
- *Understanding Patterns of Variability*. An understanding of patterns of variability—in particular, the stability of a "baseline" climate—is important for medium-term water management. It is increasingly recognized that even in the absence of climate change, the recent past may not be a reliable guide to the hydrological resource base of the near future.
- *Analytical Tools*. Effective water management requires numerous tools to assess options and the future. These tools include scenario analysis and risk analysis, which are used in some parts of water management but currently are by no means widespread.
- *Decision Tools*. Scenario and risk analysis provides information on possible futures and their consequences. They must be supplemented with tools such as Bayesian and other decision-making tools to make decisions on the basis of the information provided. Again, techniques for decision-making under uncertainty are not widely used in water management at present, and some of the approaches being used are not very sophisticated.
- Management techniques research into specific aspects of many demand-side approaches in particular, as well as into

opportunities for Seasonal Flow Forecasting and Innovative Water Supply and Treatment Technologies (such as desalination).

The above efforts are needed to improve water management even in the absence of Climate Change (CC), and there is an overwhelming need to improve the exchange of information between Hydrological Science and Water Managers. Water Managers have long been accustomed to dealing with change, although until recently this has been primarily change resulting from changes in demand and altered legislative or statutory requirements. Climate Change does not in itself stimulate development of new adaptive strategies, but it encourages a more adaptive, incremental, risk-based approach to water management. More precisely, it provides further encouragement for a trend that already is gathering pace.

Decision Support System (DSS) on Interaction Effects

The interaction effects of CO_2, rainfall and temperature can be best studied through the use of Crop Growth Simulation Models (i.e. InfoCrop of Indian Agricultural Research Institute, New Delhi). The InfoCrop Model deals with crops such as: chickpea, cotton, groundnut, maize, mustard, pearl millet, pigeon pea, potato, rice, sorghum, soyabean, sugarcane and wheat, and has taken into consideration the "growth and yield loss" due to insects, weeds and diseases.

This IARI DSS Model facilitates to explore the window of opportunities for land use planning and food security of a region by integrating natural resources inventory, bio-physical simulation of a agricultural production potential, and socio-economic goals and constraints. This model uses GIS technology, RS technology, Database technology, Crop and Livestock Model, and also an Optimization model. Efforts are now being made to upscale for the national level as well as to down scale to "village and farm level"[46].

E-Governance and Sustainable Development: A Paradigm Shift for Grassroots Level Development

Emergence of Information Technology on the national agenda and the announcement of IT policies by various State Governments have recognised the "Convergence of Core Technologies and E-Governance" as the tool for sustainable development and globalisation of economy. Models of Digital Governance (e-Governance) are continuously evolving and improvising to harness the potential offered by the Information and

Communication Technologies (ICTs) and deal with new realities in the area of governance. There are no rigid and finite models of Digital Governance. In fact, several developing countries are putting into practice innovative e-Governance models that may be technologically simple but are changing the way information is distributed in the society.

Based on primary experimentation and secondary research, a few generic models (http://www.digital governance.org), which have emerged and are being practiced, have been identified, on the basis of emergence of knowledge societies and knowledge networkers, role of information in governance process, and link between ICT and governance, are as follows:

Broadcasting/Wider-Dissemination Model: National and Local Governments in developing countries need to aggressively adopt this model if they want to enhance participation of citizens in the governance processes.

Critical Flow Model: Different organizations can use it differently depending on the aspect of governance they want to address. By focusing on the critical aspect of information and locating its likeable users, the model corrects information failure, raising awareness about the bad governance practices, and acts as a hindrance to bad governance practices.

Comparative Analysis Model: Developing countries could very effectively use this comparative model as ICT opens their access to the global and local knowledge products at a relatively low-cost. The model however becomes ineffective in absence of a strong civil society interest and public memory that is essential to force decision-makers to improve existing governance practices.

E-Advocacy/Lobbying and Pressure Group Model: The model enhances the scope of participation of individuals and communities in debates, which affect them and help them, build a global alliance.

Interactive-Service Model: The potential of ICT for the governance is fully leveraged in this model and can lead to greater objectivity and transparency in decision-making processes. This Model facilitates establishing decentralised forms of governance (G2C2G or G4C4G).

All these models share in common, the inherent characteristics of the new technologies, which are: (a) enabling equal access to information to anyone who is a linked to the digital network and (b) de-concentration of information across the entire digital network. Geometry of Information Flows is a detailed human-centric picture of information flows in a society. It focuses on "Who are the People" getting connected and are benefiting when there is an increase in information access and flow.

Studying and influencing "Geometry of Information Flows" is important because the aim is to use ICT for development purposes—those that bring real, significant changes in the lives of disadvantaged communities rather than simple embedding of ICT in the society. Somewhat detailed steps, as available from published materials, for e-Governance projects are given below:

- Who are our Target groups that we want to reach out to, through ICT for Development projects?
- What are the key information needs of the disadvantaged community?
- What are the existing channels by which information reaches to the disadvantaged community?
- What is the weakest link in the chain of information flows: from source to the disadvantaged community?

Guiding Principles of Designing ICT for Development Projects are:

- Focus on the Disadvantaged Communities, who otherwise will be excluded,
- Provide that information or service which otherwise will not be provided,
- Focus on utilizing and where possible building upon what is existing rather than thrusting a new intervention,
- Create an outcome which in absence of ICT, will not be produced efficiently or timely,
- Understand the difference between direct benefits and trickle-down benefits for the disadvantaged community.

E-government is more about putting together the IT infrastructure to make the average citizen's interface with the Government easier, while e-governance has more to do with the laws and regulation of the Internet. E-government differs from the basic tenets of ICT (computerization and connectivity) by adding an element of interaction with citizens or the private sector, enhancing transparency, and providing "voice" for those outside the Government. Initially it was referred to as G2C (i.e. Government to Citizen) and now is defined as G4C (i.e. Government for Citizen). The famous voice of Abraham Lincoln runs like this: "Government of the People, Government for the People, and Government by the People". This is the appropriate definition for "e-Governance"/ "e-Government". The ongoing World Bank aided "National Hydrology Information Network", thus, needs evaluation, in view of the emerging "convergence and e-Governance" paradigm in the Country.

Digital development in Rural Areas: A Journey started in 1985 with the establishment of NICNET in districts of India.

In India, "district" is the basic administrative unit at the sub-state level and also consistent with the decentralized planning process prevailing at the grassroots. With the establishment of NICNET in districts numbering about 520 in 1985-87, National Informatics Centre (NIC) has launched its "district information system (DISNIC)"[47] in about 28 sectors, viz. agriculture, animal husbandry, education, health, industries, rural development, micro-level planning, etc., as an informatics tool for development planning and responsive administration.

Indian village is a cognizable unit located in a specific agro-ecological and sociological environment. Its potentials and constraints for development are well known. As of today, the development planning is a highly compartmentalized activity managed by atomized Government departments handling agriculture, rural enterprises, forests, fisheries, water, health, education, culture, technology and livelihoods—almost in isolation of each other through different projects that rarely converge.

Village problems are inter-related and the resources (natural and human) are integrated. People are both the "end and means" of development and also bound by a common space, history, culture and know-how. Indian villages are still complex, intertwined and multi-faceted. The database project to facilitate micro-level planning (DISNIC-PLAN) was one among the 27 sectors in 1990s. The DISNIC-PLAN new initiative, during 2004–07, will support building up databases and decision support systems, and communication systems to facilitate: sustainability of resources, poverty alleviation, empowerment of women, full employment, production systems planning, infrastructure planning and habitat planning. Production potentials of village through "circular-flows" and "chain-effect" should be understood.

The objective of the project was to come out with an "Informatics Blueprint" that covers villages (i.e. Informatics for Development Programme at the grassroots level in India). "GEOGRAPHY is the study of manifestation of the constant tussle between the MAN and the EARTH", and has two bases: (i) the scientific study of the earth as a system: the Land, Air and Water; and (ii) the humanistic study of man's way of life controlled/determined by the Land. Water and Climate, as components of physical environment, decides the economic activity and the way of life including socio-cultural development[48].

Digital Initiatives in Indian Agricultural Sector

According to the National IT Task Force (1999) recommendation (No.79), "the Government shall take all necessary steps to boost IT for Agriculture and Integrated Rural Development". The Ministry of Agriculture and National Informatics Centre (NIC) emphasized informatics for agricultural development in the National Conference on "Informatics for Sustainable Agricultural Development (ISDA-95), way back in 1995. India is expected to become a "Knowledge Society" by 2008 and by which time, any farmer in a remote village can demand and get information, viz. landuse planning for cropping strategy for farmers fields based on integrated information on soil, water, weather, fertiliser and pest management models; how and where to get proper seeds or good quality nursery plants; prevailing prices of farm equipments, agricultural produce, products and series of such set of information, which can lead to high productivity and optimum cost benefit to the farmers.

As a step towards "reaching" technology to the small holders (Resource-Poor-Farmers) of the Country, the Central Ministry of Agriculture have taken initiatives to build up an Informatics Network—AGRISNET: A NICNET-based Agricultural Informatics and Communication—during the Tenth Plan. This ICT Network envisages to facilitate an integrated approach of "Internet Technology" and "Sustainable Agricultural, Rural and Backward Area Development" with its farm and non-farm linkages. During the Ninth Plan (i.e. during 1997-2002), the AGRISNET was suggested to have nodes upto "block level", as "block" is the planning unit for agricultural development. To usher in "Agricultural Governance"[49] in the country, it is essential to make the AGRISNET as the "rural infrastructure" reaching upto 6.5 lakh villages. This digital initiative is based on the recommendations of ISDA-95 Conference (Informatics for Sustainable Agricultural Development), which includes among the others, the following Informatics Networks:

- AGRISNET—an Infrastructure network upto block level agricultural offices facilitating agricultural extension services and agri-business activities to usher in rural prosperity,

- AGMARKNET—with a road map to network 7,000 Agricultural produce wholesale markets and 32,000 rural markets,
- ARISNET—Agricultural Research Information System Network,
- SeedNET—Seed Informatics Network,
- CoopNet—to network 93,000 Agricultural Primary Credit Societies (PACS) and Agricultural Cooperative Marketing Societies to usher in ICT enabled services and rural transformation,
- HORTNET—Horticultural Informatics Network,
- FERTNET—Fertilisers (Chemical, Bio and Organic Manure) Informatics Network facilitating "Integrating Nutrient Management" at farm level,
- VISTARNET—Agricultural Extension Information System Network,
- PPJN—Plant Protection Informatics Network,
- APHNET—Animal Production and Health Informatics Network networking about 42,000 Animal Primary Health Centres,
- FISHNET—Fisheries Informatics Network,
- LISNET—Land Information System Network linking all institutions involved in land and water management for agricultural productivity and production systems, which has now evolved as "Agricultural Resources Information System" project during the Tenth Plan being implemented through NIC,
- AFPINET—Agricultural and Food Processing Industries Informatics Network,
- ARINET—Agricultural and Rural Industries Information System Network to strengthen Small and Micro-Enterprises (SMEs),
- NDMNET—Natural Disaster Management Knowledge Network in India,
- WeatherNET—Weather Resource System Information Network of India.

India is a land of diversity with different types of terrain, various agro-climatic conditions, different levels of socio-economic conditions, and varied levels of regional development. Informatics Networks, besides Computer Networks, are increasingly considered as development tools for achieving:

- Reaching the Unreached: Public Services,
- From Digital Divide to Digital Opportunities for sustainable Development and Economic Growth,

- Fostering agricultural growth, poverty reduction and sustainable resources use,
- Sustainable Development and Earth Care Policies—Water, Energy, Education, Health, Agriculture and Rural Development, Bio-diversity,
- A Cluster of Villages—Sustainable Societies in Viable Rural Space.

Internet-based communication, transactions, feedback, and follow-ups, have thrown up enormous opportunities and challenges for: learning and counter-learning, moves and counter moves, mobilisation and counter-mobilisation, and opinion making and breaking. "Networking of People" and "Networking of Information" through localization are essential. Distributed databases (WWW information resources, Multimedia databases, Mobile databases, Genome Databases and Digital Libraries) are the converging points of networking the database technologies to bring "Power to the People" for planning and bringing in responsive administration (e-Governance) in developing countries.

Emergence of Information and Communication Technologies (ICTs) on the national agenda and the announcement of ICT policies by various State Governments, since 1995, have recognised the "Convergence of Core Technologies and E-Governance" as the tool for sustainable development and globalisation of economy. The extent of the use and application of scientific, technical and social information to advance development determines the progress of a nation. There are no rigid and finite models of Digital Governance. In fact, several developing countries are putting into practice: innovative e-Governance models that may technologically simple but are changing the way information is distributed in the society (Broadcasting/Wider-Dissemination Model, Critical Flow Model, Comparative Analysis Model, E-Advocacy/Lobbying and Pressure Group Model, and Interactive-Service Model).

ICT accelerates globalisation and make access to knowledge and information much easier for the people to attain "rural prosperity" in respect of: Creation of Technologies (—IPR Divide), Diffusion of recent Innovation (—Digital Divide), Diffusion of Old Innovation (—Extension Divide), and Diffusion of human skills (—Educational Skills). One of the major problems of using ICT for rural prosperity is language barrier, i.e. "localization". Over the years, the Government of India, its institutions and collaborators have striven towards enabling the Internet to support "multiple" Indian scripts and languages. A "national policy on localisation" is anvil.

Agricultural Resources Information System—A Needed Informatics Blueprint

The scientific management of Agricultural Resources becomes a very important part of modern agricultural production in the 21st Century and is a new direction in the domain of agricultural engineering. Agriculture is highly dynamic in nature, because of the changing phenomenon of agricultural crops, which is further complicated by the interaction of crops with environment. A wide range of agricultural practices could be effected including potential environmental benefits that could come from minimizing adverse impacts, by reducing external inputs and greater use efficiency. Precision farming is one of the most scientific and modern approaches to sustainable agriculture that is gaining momentum in the new millennium. This approach serves the dual purpose of enhancing productivity and reducing ecological degradation, as it deals with the question of where and when. Farmers, Land and Natural Resources (supported by the Land) have intrinsic and dynamic relationship. Resources Application and Agronomic Practices are to match with soil attributes and crop requirements, as they vary across a site"[50]. A stocktaking and diagnostic survey is needed early in the planning process to provide information about the wide range of factors, among the others, influencing agricultural performance:

- Agro-climatic data
- Agro-economic data
- Agro-forestry Resources
- Animal Resources
- Capital Resources
- Climate Resources
- Crops and Cropping Systems
- Environment Data
- Fisheries Resources
- Forestry Resources
- Institutional Resources
- Land Owners Data
- Plant Resources
- Socio-economic and Infrastructure data
- Soil Resources
- Spices Resources
- Water Resource

Mollet (1984)[51] provides details on the contents of Agricultural Resources Information System and recommends data and information on basic resources such as (i) soil resources, (ii) water resources, (iii) climate resources, and other data sets (collated from Remote Sensing as well as conventional means) such as (iv) basic data on crops, (v) animal husbandry and fisheries, (vi) genetic (plant, animal and fisheries) materials, (vii) land ownership, (viii) socio-economic data, (viii) infrastructure for agricultural development. Zhu Zesheng and Sun Ling (1996)[52] has proposed a seven-layered architectural model has been proposed agricultural resources management: application layer, management layer, decision models layer, decision data layer, production data layer, weather data layer, and environmental data layer. This layered approach is essential to build up required Agricultural Research Information System (AgRIS) in districts, in order to develop decision support systems (DSSs), as listed below, appropriate advisories to the Small and Marginal Farmers (SMFs):

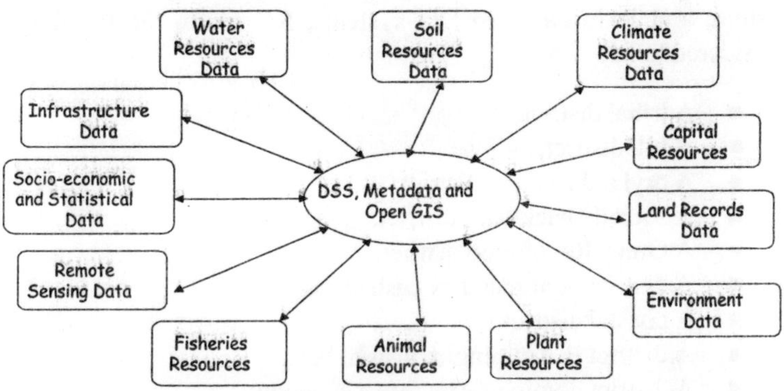

Fig. 18.1: Components of Agricultural Resources Information Systems

- Crop Suitability based on Factor Endowment
- Land Suitability Assessment;
- Land Productivity Assessment;
- Population Supporting Capacity;
- Land Evaluation and Land Use Planning;
- Land Degradation Risk Assessment;
- Quantification of Land Resources Constraints;
- Land Management;
- Agro-ecological Characterization for Research and Planning;
- Agricultural Technology Transfer;

- Agricultural Inputs Recommendations;
- Farming Systems Analysis and Development;
- Environmental Impact Assessment;
- Monitoring of Land Resources Development;
- Livestock (cattle, buffalo, goat, and sheep) Farming Systems;
- Water Allocation in an Irrigation System;
- Fodder Resources Development;
- Water Bodies (Basin) Planning Systems using Watershed and Agro-Eco Region Planning Concepts.

The Report of the Core Group-V (AgRIS Report, 2000) of the Standing Committee on Agriculture and Soils suggested strengthening "Agricultural Resources Information System" in all districts (regions) of the country, "irrespective of past or future growth regions" (Fig. 18.1). The Report also recommended pilot projects on the basis of Agricultural Production Systems (viz. Arid Agro-Eco-system, Coastal Agro-Ecosystem, Hill and Mountain Agro-Eco-system, Irrigated Agro-Eco-system, and Rainfed Agro-Eco-system), for which the typologies considered to be:

- A tribal district,
- A hill district,
- A dryland farming district,
- A socially backward district,
- A Green Revolution district,
- A district dominated by cash crops,
- A coastal district,
- A district in a mining/industrial belt,
- A district dominated by forest economy,
- A dairy farming district,
- A district dominated by one or two urban centres,
- A district in arid-zone,
- A district which is flood prone but having vast wasteland that could be used to generate forest cover.

The Pilot project demonstration in districts of above mentioned typologies, in each state, will facilitate development of Decision Support Systems (DSSs) on "production practices and systems" which need to be adapted to respond to new market demands and export opportunities, poverty alleviation or growing labour shortages, depending on the setting (NATP, 1998). Development of Agricultural Resources Information

Systems, using Geomatics Technology in pilot districts with public funding, has been sanctioned as a scheme during the Tenth Plan (2004-07).

Development of Agricultural Resources Information System (AgRIS) is the way forward for "Integrated Water Resources Planning and Management at Grassroots level" in days to come. This is a quintessential requirement for adopting "More Crops per Drop" as well as for adapting to Climate Changes in order to facilitate sustainable agricultural development in India. The Government Digital Initiatives: DISNIC Programme as well as AgRIS Programme at grassroots level will provide the necessary "Informatics Blueprint" for adaptation to climatic changes for sustainable agricultural development in India.

Summing up

The paper makes a powerful plea for an Informatics Blueprint for Integrated Water Resources Planning and Management at grassroots level in relation to climate change and sustainable agricultural development. In this context, a comprehensive analysis has been attempted, which starts with a broad-brush picture of existing water resource development strategies, and policies, with special reference to water in Agricultural production, its access and equality issues, problems of drinking water supply and its quality issues at the grassroots level. Against this background, the paper pleads for a *plan of action* to protect, allocate and manage water resources judiciously with a water accounting and budgeting system at various spatial levels to ensure water security, food security, livelihood security, health security and ecological security for the people. A key recommendation in this context is the setting up of an institution, bringing together the School of Earth Sciences, Colleges of Agriculture, Departments of Geography, School of Environmental Sciences, and Centres of Water Resources Development as well as University Departments specializing in this field.

The focus in the latter half of the paper is on sustainable development and management of water resources in the light of the Agenda 21 scenario and ICT Diffusion Developmental Action Research. This will be pioneered by the Geography Departments in the country (turning out 15,000 qualified geographers every year) working with Geo-hydrological units for development of Rainfed farming, with a sound strategy for watershed management of Land, Water and Vegetation resources. Some success stories and studies of Rainfed farming, facilitating "bubble-up" economies, i.e. small farmer-based rural economy and their methodologies have been cited.

The last part of the paper focusses on climate change, its biophysical vulnerability on Indian farmers and adaptation strategies for meeting the change, including use of renewable energy resources. A few generic models with link between ICT and governance, e.g. Broadcasting/Wider-Dissemination Model, Critical Flow Model, Comparative Analysis Model, E-advocacy/Lobbying and Pressure Group Model, Interactive Service Model etc. The paper concludes with the need for wider diffusion of the latest digital development strategy for Rural areas, ushered in by the NIC, through its new initiative of the DISNIC Plan during 2004-07, which will support building up data bases and decision support communication systems, to facilitate sustainability of resources, poverty alleviation, improvement of women, full employment, production systems planning, infrastructure planning and habitat planning. Thus the DISNIC Plan project will come out with an *Informatics Blueprint* providing Informatics for Developmental Programme at the grassroots level. With Water, the *elixir of life*, and now a *scarce commodity*, as the starting point and the central pillar of development strategy, the paper covers a wide spectrum of issues and problems relating to Integrated Water Resources Planning and Management, providing a Blueprint for accelerating broad-based rural growth, leveraged through the new Information and Communication Technologies to achieve sustainable development in the truest sense of the phrase. The Government Digital Initiatives: DISNIC Programme as well as AgRIS Programme at grassroots level will provide the necessary "Informatics Blueprint" for adaptation to climatic changes for sustainable agricultural development in India.

St. Thiruvalluvar, the Tamil Poet and a Philosopher of 1 B.C. Century, gives an account of importance and relation among "Climate", "Water Resources", "Water Management", "Society", and "Economy" in the Chapter 2 (Kural Venpa 11-20) and the "Treatise on Agriculture" in the Chapter 104 (Kural Venpa 1031—1040), in a very succinct manner. This Chapter talks about the "Van Sirappu: Glory of Rain" and decisively portrays that:

"Neerindru Amaiyaadhu Ulahu": "Water resource is very basic for the Earth"

"Vinhindru Poippin: If the Water from Sky fails"

"Puyalennum Vaari Valamkundrik kaal: if the Sky becomes dry"

"Visumbin Thuli Veezhin Allal: but for the drops of the Sky"

"Nedunkadalum Thanneermai Kundrum: the Ocean will shrink"

These treatises on the "glory of rain" reiterates that water is a prime natural resource, a basic human need and a precious national asset.

NOTES

1. Menon, M.S. (2001). "Report of the World Commission on Dams: A Framework for Underdevelopment?", published in *The Hindu*, August 14, 2001.
2. Swaminathan, M.S. (2002). "Food Security and Community Grain Banks", *Yojana*, January 2002.
3. World Bank (2003). Concept Paper circulated in the "*Multi-Stakeholder Regional Consultation for International Assessment on Role of Agricultural Science and Technology in Reducing Hunger, Improving Rural Livelihoods and Stimulating Environmentally Sustainable Economic Growth*", 12-13 May, 2003, New Delhi.
4. Michael, A.M. (2002). "Blue Revolution: A Step Towards Water Conservation and Management, Small Holder Irrigation, Drainage and Waste Water Usage, to achieve sustainable public health engineering and food security, and ushering in 2nd Green Revolution in India", presented as a Theme paper in the 2nd International Conference on "Sustainable Agriculture, Water Resources Development and Earth Care Policies" organised by Bhoovigyan Vikas Foundation (An Earth Care Foundation), New Delhi on 18-20 December, 2002.
5. Deshpande, R.S., Venkatachalam, L. and Narayanamurthy, A. (2002). "Policies for Access, Equity and Equality: Physical Resources and Human Resources", presented as a Theme paper in the 2nd International Conference on "Sustainable Agriculture, Water Resources Development and Earth Care Policies" organised by Bhoovigyan Vikas Foundation (An Earth Care Foundation), New Delhi on 18-20 December, 2002.
6. Meenakshisundaram, S.S. (2002). "Drinking Water Supply in India: A Strategic Policy Analysis", presented as a Theme Paper in the 2nd International Conference on Sustainable Agriculture, Water Resources Development and Earth Care Policies", 18-20 December, 2004, orgnaised by Bhoovigyan Vikas Foundation, New Delhi.
7. Gargi Parsai (2004). "Interlinking of Rivers: How, When and Where" published in *The Hindu* dated Tuesday, October 19, 2004.
8. National Association of Geographers India (NAGI)'s Annual Congress at Nagpur in January 2000.
9. Abdul Kalam, A.P.J. and Rajan, Y.S. (1998). *India—2020: A Vision for The New Millennium—Food, Agriculture and Processing* (Chapter-IV), Viking Penguin Books India (Pvt.) Ltd., New Delhi, India.
10. Report of the Core Group—Von Agricultural Resources Information System, constituted under the Standing Committee on Agriculture and Soils of the PC-NNRMS, Department of Agriculture and Cooperation and Department of Space, Government of India, 2000.
11. Moni, M. and Sundaram, K.V. (2003). "*Rainfed Agriculture: A Step towards Blue Revolution to Usher in Sustainable Agriculture and Rural*

Development", Presented at the XXV Indian Geography Congress, Held on 27-29 December, 2003, Ghaziabad (Uttar Pradesh), India.

12. The Eastern Region (ER) of India comprises of 130 districts covering States of Orissa, Bihar, Chhattisgarh, Jharkhand, Eastern UP, West Bengal and Andhra Pradesh.
13. Roy, B.C. and Suresh Pal (2001). Incremental Capital-Output Ratio in Indian Agriculture, *Agricultural Economics Research Review*, 14(1): 34-46.
14. Such investment has focused on extensive and costly Irrigation Systems and Green Revolution (GR) technologies involving high-yielding crop varieties, inorganic fertilizers, and pesticides.
15. Gopal, K.S. and Sashi Kumar (1995). *ILEIA Newsletter,* Vol. 11, No. 4, 1995) and http://www. oneworld/ ileia
16. Watershed refers to a "contiguous area draining into a single water body or a water course" or "it is a topographical area having a common drainage". This means that the rainwater falling on an area coming within a ridgeline can be harvested and will flow out of this area through single point. Some refer it as a catchment area or river basin.
17. Shankar Chaterjee (2002). "*Management of Water Resources Changed Economic Conditions of Farmers: A Study in Maharashtra*", presented at the 2nd International Conference on Sustainable Agriculture, Water Resources Development and Earth Care Policies, organised by Bhoovigyan Vikas Foundation (An Earth Care Foundation), 18-20 December, 2004, New Delhi.
18. The Total Corpus of the WDF is Rs. 200 crores which includes Rs. 100 crores from NABARD and Rs.100 crores from the Department of Agriculture and Cooperation, Ministry of Agriculture, Government of India.
19. Planning Commission (2001). The Working Group Report on Rainfed Farming and Natural Resources Management, Report No. 15/2001, Government of India, New Delhi.
20. Rainfed farming means a high-cost cultivation in India.
21. National Forest Policy (Draft) 2001, Planning Commission, Government of India: Agroforestry helps improving food accessibility, employment generation through diversification in agriculture, conservation and development of natural resources and overall development of the region.
22. Planning Commission: The Ninth Five-Year Plan Document (1997-2002: Volume II), Government of India.
23. Anil Agarwal, Sunita Narain and Indira Khurana (2001). "*Making Water Everybody's Business: Practice and Policy of Water Harvesting*", Centre for Science and Environment, New Delhi, 2001.
24. Pushkar Gap Project: A people led model with "holistic" and "bottom-up" approach for combating deserticification and poverty alleviation—by the Consortium of Indian Scientists for Sustainable Development, New Delhi.
25. Robert Chambers (1984). "Beyond the Green Revolution: A Selective Essay"—in *Understanding Green Revolutions—Agrarian Change and*

Development Planning in South Asia. Essays in honour of B.H. Farmer edited by Tim P. Bayliss Smith and Sudhir Wanmali, Cambridge University Press, London, 1984.

26. Chambers, R., Pacey, A. and Thrupp. L.A. (1989). *Farmer First: Farmer Innovation and Agricultural Research*, Intermediate Technology Publications, London, 1989.
27. Paroda, R.S. (2000). "Farm Research: Policy Support Essential" published in *Survey of Indian Agriculture 2000*, The Hindu Publications.
28. M.S. Swaminathan (2002). " Who will feed India in 2020" in the Dr. Salim Ali Memorial Lecture.
29. Madaswamy, Moni (2003). "*Digital Opportunities—A Positive Force for Agricultural Growth, Poverty Reduction and Sustainable Resource Use in India*", Theme Paper presented at the National Conference on "Land Care Movement for Food, Water and Livelihood Security" organised by Soil Conservation Society of India, at Chennai on 21-23, January 2003; and also see *Geospatial Today*, Vol. 1 and Issue 6.
30. Jodha, N.S. (2001). *Life on the Edge—Sustaining Agriculture and Community Resources in Fragile Environments*, Oxford University Press, 2001.
31. Mehta, S.L. and Bandyopadgyay, A. (2002). "Sustainable Agriculture: Farming Systems and Natural Resources", presented as a Theme paper in the 2nd International Conference on "Sustainable Agriculture, Water Resources Development & Earth Care Policies" organised by Bhoovigyan Vikas Foundation (An Earth Care Foundation), New Delhi on 18-20 December 2002.
32. Ram Kumar, S. (1998). "Realising the Reality: the First Step in Sustainable Agricultural Development", *News Letter of the Agricultural Research and Extension Network*, 37: 22-24.
33. The Farmer-centred Agricultural Resource Management (FARM) Programme is an initiative of eight Asian countries, viz. China, India, Indonesia, Nepal, Philippines, Sri Lanka, Thailand and Vietnam, and supported by UNDP and implemented by FAO, during 1990s.
34. Inter-governmental Panel on Climate Change (2001). "Climate Change 2001: Synthesis Report. A Contribution of Working Groups I, II, and III to the Third Assessment Report of the Inter-governmental Panel on Climate Change", Cambridge University Press, Cambridge, United Kingdom.
35. Vulnerability The extent to which climate change may damage or harm a system; it depends not only on a system's sensitivity, but also on its ability to adapt to new climatic conditions. Sensitivity—is the degree to which a system will respond to a change in climatic conditions (e.g. the extent of change in ecosystem composition, structure and functioning, including net primary productivity, resulting from a given change in temperature or precipitation).

36. Adaptability—The degree to which adjustments are possible in practices, or structures of system to projected or actual changes of climate; adaptation can be spontaneous or planned, and can be carried out in response to or in anticipation of changes in conditions.
37. Sinha, S.K. and Swaminathan, M.S. (1991). "Deforestation, Climate Change and Sustainable Nutrition Security: A Case Study of India", *Climatic Change,* 19, 1991.
38. Rao, D.G. and Sinha, S.K. (1994). "Impact of Climate Change on simulated Wheat production in India", in Rosenzweig C., Iglesis, A. (Eds.): *Implications of Climate Change for International Agriculture: Crop Modelling Study*", Washington D.C., United States Environment Protection Agency.
39. Kumar, K.S. and Parikh, J. (1996). "Climate Change Impacts on Indian Agriculture: the Ricardian Approach" in Dinar *et al.* (Eds.): *Measuring the Impacts of Climate Change on Indian Agriculture*, World Bank Technical Paper No. 402, Washington DC, World Bank.
40. Mac Callaway (2002). "A Framework for conducting benefit cost assessment of adaptation options in the Agricultural sector", UNEP Collaborating Centre on Energy and Environment (Denmark) in South Asia Expert Workshop on "Adaptation to Climate Change for Agricultural Productivity" organised by Government of India, UNEP and CGIAR, 1-3 May 2002, New Delhi.
41. Ravindranath, N.H. (2002). "Benefits of Linking Adaptation and Mitigation in Agriculture and Forestry", South Asia Expert Workshop on "Adaptation to Climate Change for Agricultural Productivity" organised by Government of India, UNEP and CGIAR, 1-3 May 2002, New Delhi.
42. Mathur, A.N. (2002). "Sustainable Agricultural System: Reduced Reliance on Non-Renewable Energy Sources and Substitution of Renewable Energy Sources", presented as a Theme paper in the 2nd International Conference on "Sustainable Agriculture, Water Resources Development and Earth Care Policies" organised by Bhoovigyan Vikas Foundation (An Earth Care Foundation), New Delhi on 18-20 December 2002.
43. World Bank (2003). Concept Paper circulated in the "Multi-Stakeholder Regional Consultation for International Assessment on Role of Agricultural Science and Technology in Reducing Hunger, Improving Rural Livelihoods and Stimulating Environmentally Sustainable Economic Growth", 12-13 May 2003, New Delhi.
44. Lester, R. Brown (2002). *Eco-Economy: Building for the Earth,* Orient Longman Private Limited, India, 2002.
45. IPCC (2001). Third Assessment Report of the Inter-Governmental Panel on Climate Change (Working Group II): Impacts, Adaptation and Vulnerability, UNEP & WMO, 2001.
46. Samra, J.S. (2002). "Impacts, Adaptation and Mitigation of Climate Change in Indian Agriculture", presented as a Theme paper in the 2nd International Conference on "Sustainable Agriculture, Water Resources Development

and Earth Care Policies" organised by Bhoovigyan Vikas Foundation (An Earth Care Foundation), New Delhi on 18-20 December, 2002.

47. DISNIC—District Information System Programme of NIC, National Informatics Centre, Government of India, 1987.
48. Professor Gopalakrishnan, Head of the School of Earth and Atmospheric Sciences, Madurai Kamaraj University in his Address to the National Workshop on DISNIC-PLAN held on 27-28 August 2004 at Madurai.
49. Agricultural Governance means e-Governance in Agricultural Sector.
50. McBratney and Pringle (1997). Spatial variability in soil-implication for precision agriculture. In J.V. Stafford (Ed.) *Precision Agriculture* 1947. Bioss Scientific Publ. Ltd., Oxford, United Kingdom. Vol. I: 3-31.
51. Mollet, J.A. (1984). *Planning for Agricultural Development*, Croom Helm (London and Canberra), St.Martin's Press, New York, 1984.
52. Zhu Zesheng and Sun Ling (1996). *GIS Implementation of Management System of Agricultural Resources* in http://www.esri.com/library/userconf/proc96/TO350/ PAP341/P341.htm.

Index